S0-BHW-576

his book comes with access to more content online.

Quiz yourself and study
with flashcards!

Register your book or ebook at
www.dummies.com/go/getaccess

Select your product, and then follow the prompts
to validate your purchase.

You'll receive an email with your PIN and instructions.

CISSP®

7th Edition

by Lawrence C. Miller, CISSP and Peter H. Gregory, CISSP

A Wiley Brand

CISSP® For Dummies®, 7th Edition

Published by: **John Wiley & Sons, Inc.,** 111 River Street, Hoboken, NJ 07030-5774, www.wiley.com

Copyright © 2022 by John Wiley & Sons, Inc., Hoboken, New Jersey

Published simultaneously in Canada

For general information on our other products and services, please contact our Customer Care Department within the U.S. at 877-762-2974, outside the U.S. at 317-572-3993, or fax 317-572-4002. For technical support, please visit https://hub.wiley.com/community/support/dummies.

Wiley publishes in a variety of print and electronic formats and by print-on-demand. Some material included with standard print versions of this book may not be included in e-books or in print-on-demand. If this book refers to media such as a CD or DVD that is not included in the version you purchased, you may download this material at http://booksupport.wiley.com. For more information about Wiley products, visit www.wiley.com.

Library of Congress Control Number: 2022930207

ISBN 978-1-119-80682-0 (pbk); ISBN 978-1-119-80689-9 (ebk); ISBN 978-1-119-80690-5 (ebk)

SKY10032986_020222

Contents at a Glance

Table of Contents

PART 2: CERTIFICATION DOMAINS 43

Introduction

Since 1994, security practitioners around the world have been pursuing a well-known and highly regarded professional credential: the Certified Information Systems Security Professional (CISSP) certification. And since 2001, *CISSP For Dummies* has been helping security practitioners enhance their security knowledge and earn the coveted CISSP certification.

Today, there are approximately 140,000 CISSPs worldwide. Ironically, some skeptics might argue that the CISSP certification is becoming less relevant because so many people have earned it. But the CISSP certification isn't less relevant because more people are attaining it; more people are attaining it because it's more relevant now than ever. Information security is far more important than at any time in the past, with extremely large-scale data security breaches and highly sophisticated cyberattacks becoming all too frequent occurrences in our modern era.

Many excellent and reputable information security training and education programs are available. In addition to technical and industry certifications, many fully accredited postsecondary degree, certificate, and apprenticeship programs are available for information security practitioners. And there certainly are plenty of self-taught, highly skilled people working in the information security field who have a strong understanding of core security concepts, techniques, and technologies. But inevitably, there are also far too many charlatans who are all too willing to overstate their security qualifications, preying on the obliviousness of business and other leaders to pursue a fulfilling career in the information security field (or for other, more-dubious purposes).

The CISSP certification is widely regarded as *the* professional standard for information security professionals. It enables security professionals to distinguish themselves from others by validating *both* their knowledge and experience. Likewise, it enables businesses and other organizations to identify qualified information security professionals and verify the knowledge and experience of candidates for critical information security roles in their organizations. Thus, the CISSP certification is more relevant and important than ever before.

About This Book

Some people say that a CISSP candidate requires a breadth of knowledge many miles across but only a few inches deep. To embellish on this statement, we believe that a CISSP candidate is more like the Great Wall of China, with a knowledge base extending over 3,500 miles — with maybe a few holes here and there, stronger in some areas than others, but nonetheless one of the Seven Wonders of the Modern World.

The problem with lots of CISSP preparation materials is defining how high (or deep) the Great Wall is. Some material overwhelms and intimidates CISSP candidates, leading them to believe that the wall is as high as it is long. Other study materials are perilously brief and shallow, giving the unsuspecting candidate a false sense of confidence while attempting to step over the Great Wall, careful not to stub a toe. To help you avoid either misstep, CISSP For Dummies answers the question, "What level of knowledge must a CISSP candidate possess to succeed on the CISSP exam?"

Our goal in this book is simple: to help you prepare for and pass the CISSP examination so that you can join the ranks of respected certified security professionals who dutifully serve and protect organizations and industries around the world. Although we've stuffed it chock-full of good information, we don't expect that this book will be a weighty desktop reference on the shelf of every security professional — although we certainly wouldn't object.

Also, we don't intend for this book to be an all-purpose, be-all-and-end-all, one-stop shop that has all the answers to life's great mysteries. Given the broad base of knowledge required for the CISSP certification, we strongly recommend that you use multiple resources to prepare for the exam and study as much relevant information as your time and resources allow. *CISSP For Dummies*, 7th Edition, provides the framework and the blueprint for your study effort and sufficient information to help you pass the exam, but by itself, it won't make you an information security expert. That takes knowledge, skills, and experience!

Finally, as a security professional, earning your CISSP certification is only the beginning. Business and technology, which have associated risks and vulnerabilities, require us, as security professionals, to press forward constantly, consuming vast volumes of knowledge and information in a constant tug-of-war against the bad guys. Earning your CISSP is an outstanding achievement and an essential hallmark in a lifetime of continuous learning.

Foolish Assumptions

It's been said that most assumptions have outlived their uselessness, but we assume a few things nonetheless! Mainly, we assume the following:

>> You have at least five years of professional experience in two or more of the eight domains covered on the CISSP exam (corresponding to chapters 3 through 10 of this book). Actually, this is more than an assumption; it's a requirement for CISSP certification. Even if you don't have the minimum experience, however, some experience waivers are available for certain certifications and college education (we cover the specifics in Chapter 1), and you can still take the CISSP exam and apply for certification after you meet the experience requirement.

>> You have general IT experience, perhaps even many years of experience. Passing the CISSP exam requires considerable knowledge of information security and underlying IT technologies and fundamentals such as networks, operating systems, and programming.

>> You have access to the Internet. Throughout this book, we provide lots of URLs for websites about technologies, standards, laws, tools, security associations, and other certifications that you'll find helpful as you prepare for the CISSP exam.

>> You are a "white hat" security professional. By this, we mean that you act lawfully and will have no problem abiding by the (ISC)² Code of Ethics (which is a requirement for CISSP certification).

Icons Used in This Book

Throughout this book, you occasionally see icons in the left margin that call attention to important information that's particularly worth noting. You won't see smiley faces winking at you or any other cute little emoticons, but you'll definitely want to take note! Here's what to look for and what to expect.

CROSS REFERENCE

This icon identifies the CISSP Common Body of Knowledge (CBK) objective that is covered in each section.

REMEMBER

This icon identifies general information and core concepts that are well worth committing to your nonvolatile memory, your gray matter, or your noggin — along with anniversaries, birthdays, and other important stuff. You should certainly understand and review this information before taking your CISSP exam.

TIP

Tips are never expected but always appreciated, and we sure hope that you'll appreciate these tips! This icon flags helpful suggestions and tidbits of useful information that may save you some time and headaches.

WARNING

This icon marks the stuff your mother warned you about. Well, okay, probably not, but you should take heed nonetheless. These helpful alerts point out confusing or difficult-to-understand terms and concepts.

TECHNICAL STUFF

You won't find a map of the human genome or the secret to cold fusion in this book (or maybe you will), but if you're an insufferable insomniac, take note. This icon explains the jargon beneath the jargon and is the stuff that legends — or at least nerds — are made of. So if you're seeking to attain the seventh level of NERD-vana, keep an eye out for these icons!

Beyond the Book

In addition to what you're reading right now, this book comes with a free, access-anywhere Cheat Sheet that includes tips to help you prepare for the CISSP exam and your date with destiny (your exam day). To get this Cheat Sheet, simply go to www.dummies.com and type *CISSP For Dummies Cheat Sheet* in the Search box.

You also get access to hundreds of practice CISSP exam questions, as well as dozens of flash cards. Use the exam questions to identify specific topics and domains that you may need to spend a little more time studying and to become familiar with the types of questions you'll encounter on the CISSP exam (including multiple-choice, drag-and-drop, and hotspot). To gain access to the online practice, all you have to do is register. Just follow these simple steps:

1. **Register your book or e-book at Dummies.com to get your personal identification number (PIN).**

   Go to www.dummies.com/go/getaccess.

2. **Choose your product from the drop-down list on that page.**

3. **Follow the prompts to validate your product.**

4. **Check your email for a confirmation message that includes your PIN and instructions for logging in.**

   If you don't receive this email within two hours, please check your spam folder before contacting us through our support website at http://support.wiley.com or by phone at +1 (877) 762-2974.

Now you're ready to go! You can come back to the practice material as often as you want. Simply log in with the username and password you created during your initial login; you don't need to enter the access code a second time.

Your registration is good for one year from the day you activate your PIN.

Where to Go from Here

If you don't know where you're going, any chapter will get you there, but Chapter 1 may be a good place to start. If you see a particular topic that piques your interest, however, feel free to jump ahead to that chapter. Each chapter is individually wrapped (but not packaged for individual sale) and written to stand on its own, so feel free to start reading anywhere and skip around! Read this book in any order that suits you (though we don't recommend upside down or backward).

1

Getting Started with CISSP Certification

Chapter **1**

(ISC)² and the CISSP Certification

I n this chapter, you get to know the (ISC)² and learn about the CISSP certification, including professional requirements, how to study for the exam, how to get registered, what to expect during the exam, and (of course) what to expect after you pass the CISSP exam!

About (ISC)² and the CISSP Certification

The International Information System Security Certification Consortium (ISC)² (https://www.isc2.org) was established in 1989 as a not-for-profit, tax-exempt corporation chartered for the explicit purpose of developing a standardized security curriculum and administering an information security certification process for security professionals worldwide. In 1994, the Certified Information Systems Security Professional (CISSP) credential was launched.

The CISSP was the first information security credential accredited by the American National Standards Institute (ANSI) to the ISO/IEC 17024 standard. This international standard helps ensure that personnel certification processes define specific competencies and identify required knowledge, skills, and personal attributes. It also requires examinations to be independently administered and designed to properly test a candidate's competence for the certification. This process helps a certification gain industry acceptance and credibility as more than just a marketing tool for certain vendor-specific certifications (a widespread criticism that has diminished the popularity of many vendor certifications over the years).

TECHNICAL STUFF

The International Organization for Standardization (ISO) and International Electrotechnical Commission (IEC) are two organizations that work together to prepare and publish international standards for businesses, governments, and societies worldwide.

The CISSP certification is based on a Common Body of Knowledge (CBK) identified by the (ISC)² and defined through eight distinct domains:

- » Security and Risk Management
- » Asset Security
- » Security Architecture and Engineering
- » Communication and Network Security
- » Identity and Access Management (IAM)
- » Security Assessment and Testing
- » Security Operations
- » Software Development Security

You Must Be This Tall to Ride This Ride (And Other Requirements)

The CISSP candidate must have a minimum of the equivalent of five cumulative years of professional (paid), *full-time*, direct work experience in two or more of the domains listed in the preceding section. Full-time experience is accrued monthly and requires full-time employment for a minimum of 35 hours per week and 4 weeks per month to get credit for 1 month of full-time work experience. Part-time experience can also be credited if you are employed fewer than 35 hours per week but at least 20 hours per week; 1,040 hours of part-time experience would be the equivalent of 6 months of full-time experience. Credit for work experience can also be earned for paid or unpaid internships. You'll need documentation from the organization confirming your experience or from the registrar if you're interning at a school.

The work experience requirement is a hands-on one; you can't satisfy the requirement just by having "information security" listed as one of your job responsibilities. You need to have *specific* knowledge of information security and to perform work that requires you to apply that knowledge regularly. Some examples of full-time information security roles that might satisfy the work experience requirement include (but aren't limited to)

>> Security analyst

>> Security architect

>> Security auditor

>> Security consultant

>> Security engineer

>> Security manager

Examples of information technology roles for which you can gain partial credit for security work experience include (but aren't limited to)

>> Systems administrator

>> Network administrator

>> Database administrator

>> Software developer

For any of these preceding job titles, your particular work experience might result in your spending some of your time (say, 25 percent) doing security-related tasks. This is legitimate for security work experience. Five years as a systems administrator, for example, spending a quarter of your time doing security-related tasks, earns you 1.25 years of security experience.

Furthermore, you can get a waiver for a maximum of one year of the five-year professional experience requirement if you have one of the following:

>> A four-year college degree (or regional equivalent)

>> An advanced degree in information security from one of the National Centers of Academic Excellence in Cyber Defense (CAE-CD)

>> A credential that appears on the (ISC)^2-approved list, which includes more than 45 technical and professional certifications, such as various SANS GIAC certifications, Cisco and Microsoft certifications, and CompTIA Security+ (For the complete list, go to https://www.isc2.org/Certifications/CISSP/Prerequisite-Pathway.)

See Chapter 2 to learn more about relevant certifications on the (ISC)^2-approved list for an experience waiver.

TIP

In the U.S., CAE-CD programs are jointly sponsored by the National Security Agency and the Department of Homeland Security. For more information, go to www.nsa.gov/resources/educators/centers-academic-excellence/cyber-defense.

If you don't have the minimum required experience to become a CISSP, you can still take the CISSP certification exam and become an associate of (ISC)^2. Then you'll have six years to meet the minimum experience requirement and become a fully certified CISSP.

Preparing for the Exam

Many resources are available to help the CISSP candidate prepare for the exam. Self-study is a major part of any study plan. Work experience is also critical to success, and you can incorporate it into your study plan. For those who learn best in a classroom or online training environment, (ISC)^2 offers CISSP training seminars.

We recommend that you commit to an intense 60-day study plan leading up to the CISSP exam. How intense? That depends on your personal experience and learning ability, but plan on a minimum of 2 hours a day for 60 days. If you're a slow learner or reader, or perhaps find yourself weak in many areas, plan on four to six hours a day — and more on the weekends. But stick to the 60-day plan. If you need 360 hours of study, you may be tempted to spread this study over a 6-month period for 2 hours a day. Consider, however, that committing to six months of intense study is much harder (on you, as well as your family and friends) than two months. In the end, you'll likely find yourself studying only as much as you would have in a 60-day period anyway.

Studying on your own

Self-study might include books and study references, a study group, and practice exams.

Begin by downloading *The Ultimate Guide to the CISSP* from the (ISC)² website at `https://www.isc2.org/Certifications/CISSP`. This guide provides a good overview of the CISSP certification and the exam, as well as links to several helpful CISSP study resources.

Next, read this (ISC)²-approved book, and review the online practice at `www.dummies.com`. (See the introduction for more information.) *CISSP For Dummies* is written to provide a thorough and essential review of all the topics covered on the CISSP exam. Then read any additional study resources to further your knowledge and reinforce your understanding of the exam topics. You can find several excellent study resources in the official *CISSP Certification Exam Outline*. Finally, rinse and repeat: Do another quick read of *CISSP For Dummies* as a final review before you take the actual CISSP exam.

WARNING

Don't rely on *CISSP For Dummies* (as awesome and comprehensive as it is!) or any other book — no matter how thick it is — as your *sole* resource to prepare for the CISSP exam.

Joining a study group can help you stay focused and provide a wealth of information from other security professionals' broad perspectives and experiences. It's also an excellent networking opportunity (the talking-to-real-people type of network, not the TCP/IP type of network)! Study groups or forums can be hosted online or at a local venue. Find a group that you're comfortable with and flexible enough to accommodate your schedule and study needs. Or create your own study group!

Finally, answer *lots* of practice exam questions. Many resources are available for CISSP practice exam questions. Some practice questions are too hard, others are too easy, and some are just plain irrelevant. Don't despair! The repetition of practice questions helps reinforce important information that you need to know to successfully answer questions on the CISSP exam. For this reason, we recommend taking as many practice exams as possible. Start with the online practice at `www.dummies.com` (see the introduction for more information).

WARNING

No practice exams exactly duplicate the CISSP exam. And forget about brain dumps. Using or contributing to brain dumps is unethical and is a violation of the (ISC)² nondisclosure agreement, which could result in your losing your CISSP certification permanently.

Getting hands-on experience

Getting hands-on experience may be easier said than done, but keep your eyes and ears open for learning opportunities while you prepare for the CISSP exam.

If you're weak in networking or applications development, for example, talk to the networking group or developers in your company. They may be able to show you a few things that can help you make sense of the volumes of information that you're trying to digest.

TIP

Your company or organization should have a security policy that's readily available to its employees. Get a copy, and review its contents. Are critical elements missing? Do any supporting guidelines, standards, and procedures exist? If your company doesn't have a security policy, perhaps now is a good time for you to educate management about issues of due care and due diligence as they relate to information security. Review your company's plans for business continuity and disaster recovery, for example. Those plans don't exist? Perhaps you can lead this initiative to help both yourself and your company.

Getting official (ISC)² CISSP training

Classroom-based CISSP training is available as a five-day, eight-hours-a-day seminar led by (ISC)²-Authorized Instructors at (ISC)² facilities and (ISC)² Official Training Providers worldwide. Private onsite training is also available, led by (ISC)²-Authorized Instructors and taught in your office space or a local venue. This option is convenient and cost-effective if your company sponsors your CISSP certification and has 10 or more employees taking the CISSP exam. If you generally learn better in a classroom environment or find that you have knowledge or experience in only two or three of the domains, you might seriously consider classroom-based training or private onsite training.

If it's not convenient or practical for you to travel to a seminar, online training seminars provide the benefits of learning from an (ISC)²-Authorized Instructor at your computer. Online training seminars include real-time, instructor-led seminars offered on a variety of schedules, with weekday, weekend, and evening options to meet your needs, as well as access to recorded course sessions for 60 days. Self-paced training is another convenient online option that provides virtual lessons taught by authorized instructors with modular training and interactive study materials. Self-paced online training can be accessed from any web-enabled device for 120 days and is available any time and as often as you need.

You can find information, schedules, and registration forms for official (ISC)² training at https://www.isc2.org/Certifications/CISSP.

The American Council on Education's College Credit Recommendation Service has evaluated and recommended three college credit hours for completing an Official (ISC)² CISSP Training Seminar. Check with your college or university to find out whether these credits can be applied to your degree requirements.

Attending other training courses or study groups

Other reputable organizations offer high-quality training in both classroom and self-study formats. Before signing up and spending your money, we suggest you talk to someone who has completed the course and can tell you about its quality. Usually, the quality of a classroom course depends on the instructor; for this reason, try to find out from others whether the proposed instructor is as helpful as they are reported to be.

Many cities have self-study groups, usually run by CISSP volunteers. You may find a study group where you live, or if you know some CISSPs in your area, you might ask them to help you organize a self-study group.

Always confirm the quality of a study course or training seminar before committing your money and time.

Taking practice exams

Taking practice exams is a great way to get familiar with the types of questions and topics you'll need to be familiar with for the CISSP exam. Be sure to take advantage of the online practice exam questions that are included with this book. (See the introduction for more information.) Although the practice exams don't simulate the adaptive testing experience, you can simulate a worst-case scenario by configuring the test engine to administer 150 questions (the maximum number you might see on the CISSP exam) with a time limit of 3 hours (the maximum amount of time you'll have to complete the CISSP exam). Learn more about computer-adaptive testing for the CISSP exam in the "About the CISSP Examination" section later in this chapter and on the (ISC)² website at https://isc2.org/Certifications/CISSP/CISSP-CAT.

To study for the CISSP exam successfully, you need to know your most effective learning styles. Boot camps are best for some people, for example, whereas others learn better over longer periods. Furthermore, some people get more value from group discussions, whereas reading alone works better for others. Know thyself, and use what works best for you.

Are you ready for the exam?

Are you ready for the big day? We can't answer this question for you. You must decide, based on your learning factors, study habits, and professional experience, when you're ready for the exam. Unfortunately, there is no magic formula for determining your chances of success or failure on the CISSP examination.

In general, we recommend a minimum of two months of focused study. Read this book, and continue taking the practice exam on the Dummies.com website until you consistently score 80 percent or better in all areas. *CISSP For Dummies* covers *all* the information you need to know to pass the CISSP examination. Read this book (and reread it) until you're comfortable with the information presented and can successfully recall and apply it in each of the eight domains. Continue by reviewing other study materials (particularly in your weak areas), actively participating in an online or local study group, and taking as many practice exams from as many sources as possible.

Then, when you feel like you're ready for the big day, find a romantic spot, take a knee, and — wait, wrong big day! Find a secure Wi-Fi hotspot (or other Internet connection), take a seat, and register for the exam!

Registering for the Exam

The CISSP exam is administered via computer-adaptive testing at local Pearson VUE testing centers worldwide. To register for the exam, go to the (ISC)² website (https://www.isc2.org/Register-For-Exam) and click the Register link, or go directly to the Pearson VUE website (www.pearsonvue.com/isc2).

On the Pearson VUE website, you first need to create an account for yourself; then you can register for the CISSP exam, schedule your test, and pay your testing fee. You can also locate a nearby test center, take a Pearson VUE testing tutorial, practice taking the exam (which you should definitely do if you've never taken a computer-based test, and then download and read the (ISC)² nondisclosure agreement (NDA).

TIP

Download and read the (ISC)² NDA when you register for the exam. Sure, the text is legalese, but it isn't unusual for CISSPs to be called upon to read contracts, license agreements, and other "boring legalese" as part of their information security responsibilities, so get used to reading it (and also get used to not signing legal documents without actually reading them)! You're given five minutes to read and accept the agreement at the start of your exam, but why not read the NDA in advance so that you can avoid the pressure and distraction on exam day and simply accept the agreement? If you don't accept the NDA in the allotted five minutes, your exam will end, and you'll forfeit your exam fees!

When you register, you're required to quantify your relevant work experience, answer a few questions regarding any criminal history and other potentially disqualifying background information, and agree to abide by the (ISC)² Code of Ethics.

The current exam fee in the United States is $749. You can cancel or reschedule your exam by contacting Pearson VUE by telephone at least 24 hours in advance of your scheduled exam or online at least 48 hours in advance. The fee to reschedule is $50. The fee to cancel your exam appointment is $100.

WARNING

If you fail to show up for your exam or you're more than 15 minutes late for your exam appointment, you'll forfeit your entire exam fee!

TIP

Great news! If you're a U.S. military veteran and are eligible for Montgomery GI Bill or Post-9/11 GI Bill benefits, the Veterans Administration will reimburse you for the full cost of the exam, whether you pass or fail. In some cases, (ISC)² Official Training Providers also accept the GI Bill for in-person certification training.

About the CISSP Examination

The CISSP examination itself is a grueling 3-hour, 100- to 150-question marathon. To put that into perspective, in three hours, you could run an actual (mini) marathon, watch *Gone with the Wind, Titanic,* or one of the *Lord of the Rings* movies, or cook a 14 pound turkey. Each of these feats, respectively, closely approximates the physical, mental (not intellectual), and emotional toll of the CISSP examination.

The CISSP exam is an adaptive exam, which means that the test changes based on how you're doing. The exam starts out relatively easy and gets progressively harder as you answer questions correctly. That's right; The better you do on the exam, the harder it gets. But that's not a bad thing! Think of it as being like skipping a grade in school because you're smarter than the average bear. The CISSP exam assumes that if you can answer harder questions about a given topic, logically, you can answer easier questions about that same topic, so why waste your time?

You'll have to answer a minimum of 100 questions. After you've answered the minimum number of questions, the testing engine will either conclude the exam (if it determines with 95 percent confidence that you're statistically likely to pass or fail the exam) or continue asking up to a maximum of 150 questions until it reaches 95 percent confidence in either result. If you answer all 150 questions, the testing engine will determine whether you passed or failed based on your answers. If you run out of time (exceed the 3-hour time limit) but have answered the minimum number of questions (100), the testing engine will determine whether you passed or failed based on your answers to the questions you completed.

The CISSP exam contains 25 pre-test items. They are included for research purposes only. (Taking the test is kind of like being a test dummy — for dummies.) The exam doesn't identify which questions are real and which are trial questions, however, so you'll have to answer all questions truthfully and honestly and to the best of your ability!

There are three types of questions on the CISSP exam:

>> **Multiple choice:** Select the *best* answer from four choices, as in this example:

Which of the following is the FTP control channel?

A: TCP port 21

B: UDP port 21

C: TCP port 25

D: IP port 21

The FTP control channel is port 21, but is it TCP, UDP, or IP?

>> **Drag and drop:** Drag and drop the correct answer (or answers) from a list of possible answers on the left side of the screen to a box on the right side of the screen. Here's an example:

Which of the following are message authentication algorithms? Drag and drop the correct answers from left to right.

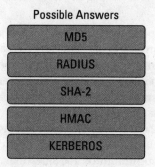
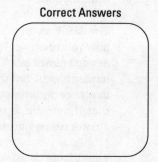

© *John Wiley & Sons, Inc.*

MD5, SHA-2, and HMAC are all correct. You must drag and drop all three answers to the box on the right for the answer to be correct.

>> **Hotspot:** Select the object in a diagram that best answers the question, as in this example:

Which of the following diagrams depicts a relational database model?

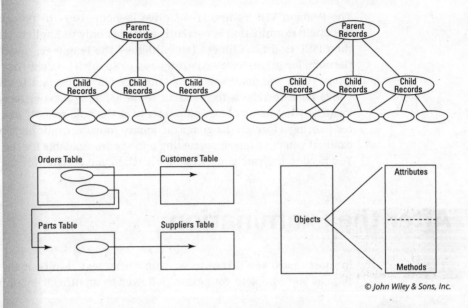

© John Wiley & Sons, Inc.

Click one of the four panels to select your answer choice.

As described by (ISC)², you need a scaled score of 700 (out of 1000) or better to pass the examination. All three question types are weighted equally, but not all questions are weighted equally. Harder questions are weighted more heavily than easier questions, so there's no way to know how many correct answers are required for a passing score. But wait — it gets even better! On the adaptive exam, you no longer get a score when you complete the CISSP exam; you'll get either a pass or fail result. Think of this situation as being like watching a basketball game with no scoreboard or a boxing match with no indication of who's winning until the referee raises the victor's arm.

All questions on the CISSP exam require you to select the best answer (or answers) from the choices presented. The correct answer isn't always a straightforward, clear choice. (ISC)² goes to great pains to ensure that you really, *really* know the material.

TIP

A common, effective test-taking strategy for multiple-choice questions is to read each question carefully and eliminate any obviously wrong choices. The CISSP examination is no exception.

WARNING

Wrong choices aren't necessarily obvious on the CISSP examination. You may find a few obviously wrong choices, but they stand out only to someone who has studied thoroughly for the exam.

The Pearson VUE computer-adaptive, 3-hour, 100- to 150-question version of the CISSP examination is currently available only in English. If you prefer to take the CISSP exam in Chinese (simplified — the language, not the exam), French, German, Japanese, Korean, Portuguese, or Spanish because that's your native language (or if you don't speak the language but *really* want to challenge yourself), you'll have to take a form-based, 6-hour, 250-question version of the CISSP exam — what many of us would refer to as the "old school" exam. You're permitted to bring a foreign-language dictionary (nonelectronic *and* nontechnical) to the exam, if you need one. Also, testing options are available for the visually impaired. You need to indicate your preferences when you register for the exam.

After the Examination

In most cases, you'll receive your unofficial test results at the testing center as soon as you complete your exam, followed by an official email from (ISC)².

WARNING

In some rare instances, your unofficial results may not be available immediately. (ISC)² analyzes score data during each testing cycle; if there aren't enough test results early in the testing cycle, your results could be delayed up to eight weeks.

If, for some reason, you don't pass the CISSP examination — say that you read only this chapter of *CISSP For Dummies,* for example — you'll have to wait 30 days to try again. If that happens, we strongly recommend that you read the rest of this book during those 30 days! If you fail a second time, you'll have to wait 90 days to try again. If *that* happens, we most strongly recommend and highly urge you to read the rest of this book — perhaps a few times — during those 90 days! Finally, if you fail on your third attempt, you'll have to wait 180 days. You'll have no more excuses; you'll definitely need to read, reread, memorize, comprehend, recite, ingest, and regurgitate this book several times!

After earning your CISSP certification, you must remain an (ISC)² member in good standing and renew your certification every three years. You can renew the CISSP certification by accumulating 120 Continuing Professional Education (CPE) credits or by retaking the CISSP examination. You must earn a minimum of 40 CPE credits during each year of your 3-year recertification cycle. You earn CPE credits for various activities, including taking educational courses or attending seminars and security conferences, belonging to association chapters and attending meetings, viewing vendor presentations, completing university or college courses, providing security training, publishing security articles or books, serving on relevant industry boards, taking part in self-study, and doing related volunteer work. You must document your annual CPE activities on the secure (ISC)² website to receive proper credit. You're also required to pay a $125 (U.S.) annual maintenance fee to (ISC)². Maintenance fees are billed in arrears for the preceding year, and you can pay them in the secure members' area of the (ISC)² website.

WARNING

Be sure to be truthful on your CPE reporting, and retain evidence of your training. (ISC)² audits some CPE submissions.

TIP

As soon as you receive your certification, register on the (ISC)² website, and provide your contact information. (ISC)² reminds you of your annual maintenance fee, board of directors elections, annual meetings, training opportunities, and events, but *only* if you maintain your contact info — particularly your email address.

» Discovering the joy of giving back

» Working with others in your local security community

» Getting the word out about CISSP certification

» Bringing about change in your organization

» Advancing your career with other certifications

» Finding a mentor and being a mentor

» Achieving security excellence

Chapter **2**

Putting Your Certification to Good Use

Although this book is devoted to helping you earn your CISSP certification, we thought it would be a good idea to include a few things you might consider doing after you've earned your CISSP. If you're still exploring the CISSP certification, the information in this chapter will help you better understand many of the benefits of being a CISSP, including your role in helping others.

So what do you do after you earn your CISSP? You can do plenty of things to enhance your professional career and the global community. Here are just a few ideas!

Networking with Other Security Professionals

Unless you work for a large organization, there probably aren't many other information security (infosec) professionals in your organization. You may be the only one! Yes, it can feel lonely at times, so we suggest that you find ways to make connections with infosec professionals in your area and beyond. Many of the activities described in this chapter provide networking opportunities. If you haven't been much of a social butterfly before, and your professional network is somewhat limited, get ready to take your career to a whole new level as you meet like-minded security professionals and potentially build lifelong friendships.

THE POWER OF ONLINE BUSINESS NETWORKING

We promise that we have no affiliations with LinkedIn when we say it, but hear this: LinkedIn is one of the best business networking tools to come along since the telephone and the business card. LinkedIn can help you expand your networking horizons and help you make contacts with other business professionals in your company, your profession, your region, and far beyond.

Chances are that you aren't new to LinkedIn, so we'll skip the basics here. People in the infosec business are a bit particular, however, and that's what we want to discuss. Infosec professionals tend to be skeptical. After all, we're paid to be paranoid, as we sometimes say, because the bad guys (and gals) *are* out to get us. This skepticism relates to LinkedIn in this way: Most of us are wary of making connections with people we don't know. So as you begin to network with other infosec professionals on LinkedIn, tread lightly, and proceed slowly. It's best to start making connections with people you actually know and people you've actually met. If you make connection requests with infosec people you haven't met, there's a pretty good chance that they'll ignore you or decline the request. They're not being rude; they're just aware of the fact that many scammers out there will build fake connections in the hope of earning your trust and pulling some kind of ruse later.

Similarly, if you've been one of those open networkers in the past, don't be surprised if others are a bit reluctant to connect with you, even those you've met. As you transition into an infosec career, you'll find that the rules are a bit different.

Bottom line: LinkedIn can be fantastic for networking and learning, but do know that infosec professionals march to the beat of a different drummer.

It's not *what* you know, but *who* you know. (Well, *what* you know matters too!)

If you're just getting started in your infosec career (regardless of your age or other career experience), you'll likely meet other infosec professionals that have at some point in their careers been in your shoes, who will be happy to help you find answers and solutions to some of those elusive questions and challenges that may be perplexing you. You may find that you're initially doing more taking than giving, but make sure that you're at least showing your appreciation and gratitude for their help — and remember to give back later in your career when someone new to infosec asks to pick *your* brain for some helpful insight.

As you venture out in search of other infosec professionals, put your smile on, and bring plenty of business cards. (Print your own if your employer doesn't provide any.) You're sure to make new friends and experience growth in the security business that may delight you.

Being an Active (ISC)² Member

Being an active (ISC)² member is easy! Besides volunteering (see the following section), you can participate in several other activities, including the following:

>> **Attend the (ISC)² Congress.** For years, (ISC)² rode the coattails of ASIS (formerly the American Society for Industrial Security; we blame Kentucky Fried Chicken for becoming KFC and starting the trend of businesses and organizations dropping the original meaning behind their acronyms!) and occupied a corner of the ASIS annual conference. But in 2016, (ISC)² decided that it was time to strike out on its own and run its own conference. In 2017, one of your authors (first name starts with *P*) attended and spoke at the very first stand-alone (ISC)² Congress and found it to be a first-class affair every bit as good as those other great national and global conferences. Find out about the next (ISC)² Congress at https://congress.isc2.org.

>> **Vote in (ISC)² elections.** Every year, one-third of the (ISC)² board of directors is elected to serve three-year terms. As a CISSP in good standing, you've earned the right to vote in the (ISC)² elections. Exercise that right! The best part is becoming familiar with other CISSPs who run for board positions so you can select those who will best advance the (ISC)² mission. You can read the candidates' biographies and understand the agendas they'll pursue if elected. With your vote, you're doing your part to ensure that the future of (ISC)² rests in good hands with directors who can provide capable leadership and vision.

>> **Attend (ISC)² events.** (ISC)² conducts several in-person and virtual events each year, from networking receptions to conferences and educational events. (ISC)² often holds gatherings at larger industry conferences such as RSA and BlackHat. Check the (ISC)² website regularly to find out more about virtual events and live events in your area.

>> **Join an (ISC)² chapter.** (ISC)² has more than 150 chapters in more than 50 countries. You can find out more at `www.isc2.org/chapters`. You have many great opportunities to get involved in local chapters, including chapter leadership, chapter activities, and community outreach projects. Chapter events are also great opportunities to meet other infosec professionals.

>> **Partake in free training.** (ISC)² offers lab-style courses, immersive courses, and express training at the Professional Development Institute that can help expand your horizons. Find out more at `www.isc2.org/Development`.

>> **Enjoy exclusive resources and discounts.** (ISC)² membership has many perks in the form of discounts and access to exclusive content and services. Find out more at `www.isc2.org/Member-Resources/Exclusive-Benefits`.

>> **Wear your digital badge proudly.** You can set up your digital badges and use them on LinkedIn, business cards, blogs, and elsewhere. Best of all – they're free. Learn more at `https://credly.com`.

TIP

It's important for (ISC)² to have your correct contact information. As soon as you become a CISSP (or even before), make sure that your profile is accurate and complete so that you'll receive announcements about activities.

Considering (ISC)² Volunteer Opportunities

(ISC)² is much more than a certifying organization: It's also a *cause*, and you might even say it's a *movement*. It's security professionals' *raison d'être*, the reason we exist — professionally, anyway. As one of us, consider throwing your weight into the cause.

Volunteers have made (ISC)² what it is today, and they make valuable contributions toward your certification. You can't stand on the sidelines and watch others do the work. Use your talents to help those who'll come after you. You can help in many ways. For information about volunteering, see the (ISC)² Volunteering website (`www.isc2.org/Membership/Volunteer-Grow`).

TIP

Most sanctioned (ISC)² volunteer activities are eligible for CPE credits. Check with (ISC)² for details.

Writing certification exam questions

The state of technology, laws, standards, and practices within the CISSP Common Body of Knowledge (CBK) is continually changing and advancing. To be effective and relevant, CISSP exams need to have fresh new exam questions that reflect how security is done today. Therefore, people working in the industry — such as you — need to write new questions. If you're interested in being a question writer, visit the (ISC)² website to apply.

Speaking at events

(ISC)² now holds more security-related events worldwide than it has at any other time in its history. More often than not, (ISC)² speakers are local volunteers — experts in their professions who want to share with others what they know. If you have an area of expertise or a unique perspective on CISSP-related issues, consider educating others via a speaking engagement. For more information, visit the (ISC)² website at www.isc2.org/Membership/Volunteer-Grow, and find the speaking opportunities that interest you.

TIP

If you speak at an (ISC)² Congress, your conference fees are waived. You need to pay only for transportation, lodging, and meals.

Helping at (ISC)² conferences

(ISC)² puts on a fantastic annual conference called the (ISC)² Congress. This conference is an excellent opportunity to learn new topics and meet other infosec professionals. But the conference doesn't run itself; it's powered by volunteers! Go to the (ISC)² Congress website at https://congress.isc2.org to find information about volunteering.

Reading and contributing to (ISC)² publications

(ISC)² publishes quarterly online magazines called *InfoSecurity Professional INSIGHTS* and *Cloud Security INSIGHTS* that are associated with *InfoSecurity Professional* magazine. You can find out more at www.isc2.org/InfoSecurity-Professional/InfoSecurity-Professional-Insights.

The *(ISC)² Blog* is a free online publication for all (ISC)² members. Find the blog, as well as information about writing articles, at https://blog.isc2.org.

Supporting the (ISC)² Center for Cyber Safety and Education

The Center for Cyber Safety and Education, formerly the (ISC)² Foundation, is a not-for-profit charity formed by (ISC)² in 2011. The center is a conduit through which security professionals can reach society and empower students, teachers, and the general public to secure their online lives through cybersecurity education and awareness programs in the community. The center was formed to meet those needs and expand altruistic programs, such as Safe and Secure Online, the Information Security Scholarship Program, and industry research (the center's three core programs). Find out more at www.iamcybersafe.org.

Participating in bug-bounty programs

As an (ISC)² member, you can earn CPE credits and contribute to a safer world by participating in Bugcrowd's bug-bounty programs. You even have a chance to be honored in the organization's hall of fame. Find out more at www.bugcrowd.com/customers/isc-2.

Participating in (ISC)² focus groups

(ISC)² has developed focus groups and quality-assurance testing opportunities. (ISC)² is developing new services, and it needs to receive early feedback during the requirements and design phases of its projects. Participating in these groups and tests can influence future (ISC)² services that will aid current and future certification holders. (ISC)² doesn't have a web page dedicated to this topic; you'll be notified of opportunities by email.

Joining the (ISC)² community

(ISC)² has developed a new interactive community that's full of discussion groups. With more than 16,000 members in the first year, the community is well designed and easy to use. The community has more resources than we can list here! You can sign up and join discussions at https://community.isc2.org.

Getting involved with a CISSP study group

Many communities have CISSP study groups that consist of volunteer mentors and instructors who help those who want to earn the certification.

If your community doesn't have a CISSP study group, consider starting one. Many communities have them already, and the organizers can give you advice on starting your own. You can find out more from nearby (ISC)² chapters and other local security groups, or visit `https://community.isc2.org/t5/Study-Groups/ct-p/CertificationStudyGroups` to find a group near you.

Helping others learn more about data security

In no way are we being vain or arrogant when we say that we (the writers of this book and you, the readers) know more about data security and safe Internet use than perhaps 99 percent of the general population, for two main reasons:

>> Security is our profession.

>> Security is not always easy to do.

A legion of volunteer opportunities is available to help others keep their computers (and mobile computing devices) secure and use the Internet safely. Here is a concise list of places where you can help:

>> Service clubs

>> Senior centers

WHY VOLUNTEER?

Why should you consider volunteering for (ISC)² or for any other professional organization? Here are two main reasons:

- Volunteerism of any kind is about *giving back* to a larger community. Consider the volunteers who helped you earn your CISSP certification. There are many of them, but they aren't always visible.

- Volunteering looks good on your résumé. We consider this fact to be a byproduct of volunteering, not the primary reason for it.

Volunteering for (ISC)² or any other cause should be a reflection of your character, not simply an activity to embellish your résumé. Although your intention in volunteering may be to help others, volunteering will also change you — for the better.

Consider it a good idea to periodically check the (ISC)² website for other ways you can help.

>> Schools (be sure to read about Safe and Secure Online earlier in this chapter)

>> Alumni associations and groups

>> Your place of employment

Using a little imagination, you can undoubtedly come up with additional opportunities. The world is hungry for the information you possess!

Becoming an Active Member of Your Local Security Chapter

In addition to (ISC)², many security organizations worldwide have local chapters, perhaps in or near your community. Here's a short list of some organizations that you may be interested in:

>> **International Systems Security Association (ISSA):** www.issa.org

>> **ISACA:** www.isaca.org

>> **Society for Information Management (SIM):** www.simnet.org

>> **InfraGard:** www.infragard.net

>> **Open Web Application Security Project (OWASP):** https://owasp.org

>> **ASIS International:** www.asisonline.org

>> **High Technology Crime Investigation Association (HTCIA):** https://htcia.org

>> **Risk and Insurance Management Society (RIMS):** www.rims.org

>> **Society of Information Risk Analysts (SIRA):** www.societyinforisk.org

>> **The Institute of Internal Auditors (IIA):** www.theiia.org

>> **International Association of Privacy Professionals (IAPP):** https://iapp.org

>> **Disaster Recovery Institute International (DRII):** https://drii.org

>> **Computer Technology Investigators Network (CTIN):** www.ctin.org

Local security groups provide excellent opportunities to find peers in other organizations and discover more about your profession. Many people find that the contacts they make as part of their involvement with local security organizations can be especially valuable when they're looking for new career opportunities.

You certainly can find many more security organizations with local chapters beyond the ones we include in the preceding list. Ask your colleagues and others about security organizations and clubs in your community.

TIP

Many communities have local information security groups and clubs that are not affiliated with national or global organizations. Through word of mouth, you might find one of these groups located near you.

Spreading the Good Word about CISSP Certification

As popular as the CISSP certification is, some people still don't know about it, and many who may have heard of it don't understand what it's all about. Tell people about your CISSP certification, and explain the certification process to your peers. Here are some facts that you can share with anyone and everyone you meet:

>> The CISSP certification started in 1994.

>> CISSP is *the* top-tier information security professional certification.

>> More than 142,000 security professionals in more than 170 countries have the CISSP certification.

>> CISSP was the first credential accredited by the ANSI (American National Standards Institute) to ISO (International Organization for Standardization) Standard 17024.

>> The average CISSP salary is $131,030 (U.S.).

>> The organization that manages the CISSP certification has other certifications for professionals who specialize in various fields of information security. The organization also promotes information security awareness through education programs and events.

Promote the fact that you're certified. How can you promote it? After earning your CISSP, you can simply put the letters *CISSP* after your name on your business cards, stationery, email signature, résumé, blog, and website. While you're at it, put the CISSP logo or your digital badge on there, too (and be sure to abide by any established terms of use).

TIP

Many other certifications available from (ISC)² are described later in this chapter.

Leading by example

Like it or not, security professionals, particularly those with the CISSP certification, are role models for those around them. From a security perspective, whatever we do — along with how we do it — is viewed as the standard for correct behavior.

REMEMBER

Being mindful of this fact, we need to conduct ourselves as though someone is looking — even if no one is — at everything we do.

Using Your CISSP Certification to Be an Agent of Change

As a certified security professional, you're an agent of change in your organization: The state of threats and regulations is ever-changing, and you must respond by ensuring that your employer's environment and policies continue to defend your employer's assets against harm. Here are some of the essential principles for being a successful change agent:

>> Identify and promote only essential changes.

>> Promote only those changes that have a chance to succeed.

>> Anticipate sources of resistance.

>> Distinguish resistance from well-founded criticism.

>> Involve all affected parties the right way.

>> Don't promise what you can't deliver.

>> Use sponsors, partners, and collaborators as co-agents of change.

>> Change metrics and rewards to support the changing world.

>> Provide training.

>> Celebrate all successes.

REMEMBER

Your job as a security professional doesn't involve preaching; instead, you need to recognize opportunities for improvement and reduced risks to the business. Work within your organization's structure to bring about change in the right way. That's the best way to reduce security risks.

Earning Other Certifications

In business and technology, no one's career stays in one place. You're continuously growing and changing, and ever-changing technology also influences organizations and your role within them.

You shouldn't consider your quest for certifications to be finished when you earn your CISSP — even if it *is* the highest-level information security certification out there! Security is a journey, and your CISSP certification isn't the goal, but a (major) milestone along the way. CISSP should be part of your *security lifestyle*.

Other (ISC)² certifications

(ISC)² has several other certifications, including some that you may aspire to earn after (or instead of) receiving your CISSP. These certifications are

>> **Associate of (ISC)²:** If you can pass the CISSP or SSCP certification exams but don't yet possess the required professional experience, you can become an Associate of (ISC)². Read about this option on the (ISC)² website.

>> **CCSP (Certified Cloud Security Professional):** This certification on cloud controls and security practices was co-developed by (ISC)² and the Cloud Security Alliance.

>> **SSCP (Systems Security Certified Practitioner):** This certification is for hands-on security techs and analysts. SSCP has had a reputation for being a "junior" CISSP certification, but don't be fooled — it's anything but that. SSCP is highly technical, more so than CISSP. For some people, SSCP may be a stepping stone to CISSP, but for others, it's a great destination all its own.

>> **CSSLP (Certified Secure Software Lifecycle Professional):** Designed for software development professionals, the CSSLP recognizes software development in which security is part of the software requirements, design, and testing so that the finished product has security designed and built in, rather than added afterward.

>> **HCISPP (HealthCare Information Security and Privacy Practitioner):** Designed for information security in the healthcare industry, the HCISPP recognizes knowledge and experience related to healthcare data protection regulations and the protection of patient data.

>> **CAP (Certification and Accreditation Professional):** Jointly developed by the U.S. Department of State's Office of Information Assurance and (ISC)², the CAP credential reflects the skills required to assess risk and establish security requirements for complex systems and environments.

CISSP concentrations

(ISC)² has developed follow-on certifications (think *accessories*) that accompany your CISSP. (ISC)² calls these certifications *concentrations* because they represent the three areas you may choose to specialize in:

>> **ISSAP (Information Systems Security Architecture Professional):** Suited for technical systems security architects

>> **ISSEP (Information Systems Security Engineering Professional):** Demonstrates competence for security engineers

>> **ISSMP (Information Systems Security Management Professional):** About security management (of course!)

All the concentrations require that you first be a CISSP in good standing, and each has a separate exam. Read about these concentrations and their exams on the (ISC)² website at www.isc2.org/Certifications/CISSP-Concentrations.

Non-(ISC)² certifications

Organizations other than (ISC)² have security-related certifications, one or more of which may be right for you. None of these certifications competes directly with CISSP, but some of them overlap with CISSP somewhat.

Nontechnical/nonvendor certifications

Many other certifications are not tied to specific hardware or software vendors. Some of the best include

>> **CISA (Certified Information Systems Auditor):** Consider this certification if you work as an internal auditor or your organization is subject to one or more security regulations, such as Sarbanes-Oxley, HIPAA, GLBA, or PCI. ISACA manages this certification. Find out more about CISA at www.isaca.org/cisa.

>> **CISM (Certified Information Security Manager):** Similar to (ISC)²'s Information Systems Security Management Professional (ISSMP) certification (which we talk about in the section "CISSP concentrations" earlier in this chapter), you may want the CISM certification if you're in security management. Like CISA, ISACA manages this certification. Read more about it at www.isaca.org/cism.

>> **CRISC (Certified in Risk and Information Systems Control):** This certification concentrates on organization risk management, controls, and information security. Find out more at www.isaca.org/crisc.

>> **CGEIT (Certified in the Governance of Enterprise IT):** Look into this certification if you want to demonstrate your skills and knowledge in the areas of IT management and governance. Effective security in an IT organization depends on *governance,* which involves the management and control of resources to meet long-term objectives. You can find out more about CGEIT at www.isaca.org/cgeit.

>> **CDPSE (Certified Data Privacy Solutions Engineer):** This relatively new certification from ISACA is all about technical skills within the growing privacy profession. For more information, visit www.isaca.org/cdpse.

>> **CPP (Certified Protection Professional):** Primarily a security management certification, CPP is managed by ASIS International. The CPP certification (www.asisonline.org/certification) designates people who have demonstrated competence in all areas constituting security management.

>> **PSP (Physical Security Professional):** ASIS International also offers this certification, which caters to professionals whose primary responsibility focuses on threat surveys and the design of integrated security systems. Read more at www.asisonline.org/certification.

>> **CIPP (Certified Information Privacy Professional):** The International Association of Privacy Professionals (IAPP) has this and other country-specific privacy certifications for security professionals with knowledge and experience in personal data protection. Find out more at https://iapp.org/certify/cipp (login required).

>> **CIPP/US (Certified Information Privacy Professional/U.S.):** Privacy in the United States is growing fast, and IAPP has developed a U.S. version of the CIPP. Read more at https://iapp.org/certify/cippus.

>> **CIPP/C (Certified Information Privacy Professional/Canada):** Privacy in Canada is growing in importance, so much that IAPP has a Canadian version of CIPP. Find out more at https://iapp.org/certify/cippc.

>> **CIPP/E (Certified Information Privacy Professional/Europe):** Privacy in Europe is so important in our industry that the IAPP has developed a version of the CIPP especially for European privacy matters. See more at https://iapp.org/certify/cippe.

>> **CIPP/A (Certified Information Privacy Professional/Asia):** IAPP has an Asia version of the CIPP certification that focuses on privacy laws and practices in Asian countries. Find out more at https://iapp.org/certify/cippa.

- » **CIPM (Certified Information Privacy Manager):** This certification is designed for privacy program leaders in organizations; it focuses on building a privacy team and privacy operations. Find out more at `https://iapp.org/certify/cipm`.

- » **CCISO (Certified Chief Information Security Officer):** This certification demonstrates the skills and knowledge required for the typical CISO position. Read more at `https://ciso.eccouncil.org`.

- » **CBCP (Certified Business Continuity Planner):** A business continuity planning certification offered by the Disaster Recovery Institute. You can find out more at `https://drii.org/certification/cbcp`.

- » **DRCE (Disaster Recovery Certified Expert):** This certification recognizes knowledge and experience in disaster recovery planning. For more information about DRCE and related certifications, visit `www.bcm-institute.org/certification`.

- » **PMP (Project Management Professional):** A good project manager — someone you can trust with organizing resources and schedules — is a wonderful thing, especially on large projects. The Project Management Institute (`www.pmi.org`) offers this certification.

- » **PCI QSA (Payment Card Industry Qualified Security Assessor):** The Payment Card Industry Security Standards Council developed the QSA certification for professionals who audit organizations that store, transmit, or process credit card data. This certification is for PCI auditors. Find out more at `www.pcisecuritystandards.org`.

- » **PCI ISA (Payment Card Industry Internal Security Assessor):** This certification, also from the Payment Card Industry Security Standards Council, is for security professionals within organizations that store, transmit, or process cardholder data. Find out more at `www.pcisecuritystandards.org`.

- » **GIAC (Global Information Assurance Certification):** The GIAC family of certifications includes categories in Audit, Management, Operations, and Security Administration. GIAC non-vendor-specific certifications complementing CISSP are GIAC Certified Forensics Analyst (GCFA) and GIAC Certified Incident Handler (GCIH). Find more information at `www.giac.org/certifications`. Several vendor-related GIAC certifications are mentioned in the next section.

Technical/vendor certifications

We won't even pretend to list all the technical and vendor certifications here, but these are some of the best-known vendor-related security certifications:

» **AWS Certified Security – Specialty:** AWS offers numerous certifications in architecture, data analytics, and (of course) security. Find out more at `https://aws.amazon.com/certification/certified-security-specialty`.

» **CCIE (Cisco Certified Internetworking Expert) Security:** Cisco offers several product-related certifications for specific products, including ASA firewalls and intrusion prevention systems. Find out more at `www.cisco.com/certifications`.

» **Check Point Security Administration certifications:** You can earn certifications related to Check Point's firewall and other security products. Visit `www.checkpoint.com/certification`.

» **CEH (Certified Ethical Hacker):** We know, we know — an "ethical hacker" is a contradiction in terms to some people, but it provides real business value for others. Read about it carefully before signing up. This certification is offered by the International Council of E-Commerce Consultants (EC-Council). You can find out more at `https://cert.eccouncil.org`.

» **ENSA (Network Security Administrator):** Also from EC Council, this certification recognizes the defensive view as opposed to the offensive view of CEH. You can read more at `https://cert.eccouncil.org`.

» **LPT (Licensed Penetration Tester):** Another EC Council certification takes penetration testing to a higher level than CEH. Learn more at `https://cert.eccouncil.org`.

» **CHFI (Certified Hacking Forensics Investigator):** Also from EC Council, this certification recognizes the skills and knowledge of a forensic expert who can detect computer crime and gather forensic evidence. Find out more here: `https://cert.eccouncil.org`.

» **CSFA (CyberSecurity Forensic Analyst):** This certification demonstrates the knowledge and skills required for conducting computer forensic examinations. Part of the certification exam is an actual forensics assignment in the lab. Check out `www.cybersecurityforensicanalyst.com/` for more information.

» **CompTIA Security+:** A security competency certification for PC techs and the like. We consider this certification an entry-level certification that may not be for you. Still, you may advise your aspiring colleagues who want to get into information security that this certification is an excellent place to start. You can find out more at `www.comptia.org/certifications/security`.

» **OSCP (Offensive Security Certified Professional):** Offered by Offensive Security, OSCP is considered one of the top penetration testing certifications available. Many people consider CEH the entry-level pen testing cert and OSCP the top dog. Find out more at `www.offensive-security.com`.

You can find many other security certifications. Use your favorite search engine and search for phrases such as "security certification" to find information.

Choosing the right certifications

Regularly, technology and security professionals ask us which certifications they should earn next. Our answer is almost always the same: Your decision depends on where you are now and where you want your career to go. There is no single "right" certification for everyone; determining which certification you should seek is a very individual thing.

When considering other certifications, ask yourself the following questions:

>> **Where am I in my career right now?** Are you more focused on technology, policy, operations, development, or management?

>> **Where do I want my career to go in the future?** If (for example) you're stuck in operations, but you want to be focusing on policy, let that goal be your guide.

>> **What qualifications for certifications do I possess right now?** Some people tackle certifications based on the skills they already possess, and they use those newly earned certifications to climb the career ladder.

>> **What do I need to do in my career to earn more qualifications?** You need to consider what certifications you may be qualified to earn right now and what experience you must develop to earn future certifications.

If you're honest with yourself, answering these questions should help you discern what certifications are right for you. We recommend that you take time every few years to do some long-term career planning; most people will find that the answers to the questions we've listed here will change.

You might even find that some of the certifications you have no longer reflect your career direction. If so, permit yourself to let those certifications lapse. There's no sense hanging on to old certifications that no longer exhibit (or help you attain) your career objectives. Each of us has done this at least once, and we may again someday.

REMEMBER

Most non\technical certifications require you to prove that you *already* possess the required job experience to earn them. People make this common mistake: They want to earn a certification to land a particular kind of job. But that's not the purpose of a certification. Instead, a certification is evidence that you already possess both knowledge and experience.

Finding a mentor, being a mentor

If you're somewhat new to infosec (and even if you're not!), and you find yourself asking many questions about your career, perhaps you would benefit from a mentor. A *mentor* is someone who has lived your professional lifestyle and been on the security journey for many years.

We suggest you shop around for a mentor and decide on one after talking with a few prospects. Mentors often have different approaches, from casual discussions to more structured learning.

If you're not sure where to find a mentor, start with one or more of your area's local security organizations or activities. You may have to find a long-distance mentor if you live outside a major city, but the experience can still be rewarding!

As you transition in your career from a security beginner to a security expert, consider being a mentor yourself. You'll find that although you'll be helping another aspiring security professional get their career started, you'll also learn quite a bit about security and yourself along the way.

TIP

Being mentored is not just for beginners. Even accomplished leaders have mentors who help them on their professional journeys.

Building your professional brand

You are defined by more than just your job title and your certifications. As you take your career further into information security expertise (and perhaps leadership), you'll want to establish your brand above and beyond the job you are in today. Infosec professionals tend to stay in their positions for three to four years — a small fraction of a career. Instead of remaking your brand each time you change employers, elevate your brand to set it apart from your employers. Here are some of the ways you can spread your wings:

>> **Create a LinkedIn profile.** LinkedIn has become the de-facto platform for building your brand. If you haven't done a lot with LinkedIn, we suggest that you pick up a copy of *LinkedIn For Dummies,* 4th Edition, by Joel Elad (John Wiley & Sons, Inc.) and go all in.

>> **Join (ISC)² and other communities.** You might find your niche through the (ISC)² communities discussed earlier in this chapter, where you can help and be helped.

>> **Use other social media.** If you are serious about building your brand, you might also consider creating a professional Twitter and/or Instagram account.

>> **Start a blog.** Your opinions and insights matter, and a blog is a great way to express yourself through articles and other information about yourself and your contributions to the profession.

>> **Print personal business cards.** If you are a business-card type of person, consider getting your own business cards. Go plain or go fancy. Peter prefers the minimalist approach, as you can see in Figure 2-1.

FIGURE 2-1:
Make your own personal business cards.

Photo courtesy of authors

REMEMBER

Building your brand is about contributing to the profession, not seeing what you can find for the taking.

Personal Branding For Dummies, 2nd Edition, by Susan Chritton (Wiley), is a great way to learn more about your brand and how you can use it to help others and get ahead.

Pursuing Security Excellence

We think that the best way to succeed in a security career is to pursue excellence every day, whether you're already in your dream security job or just starting.

The pursuit of excellence may sound like a lofty or vague term, but you can make a difference every day by doing the following:

>> **Do your best job daily.** No matter what you do for a living, be the very best at it.

>> **Maintain a positive outlook.** Happiness and job satisfaction are due in large part to your attitude. Having a good attitude helps make each day better and allows you to do a better job. Because optimism is contagious, your positive outlook will encourage your co-workers, and pretty soon, everyone will be whistling, humming, or doing whatever else they do when they like their jobs. Have *an attitude of gratitude*.

>> **Continually improve yourself.** Take the time to read about security practices, advances, developments, and changes in the industry. Try to figure out how innovation in the industry can help you and your organization reduce risk even more, with less effort.

>> **Understand your value.** Take the time to understand how your work adds value to the organization; try to develop more ways to add value and reduce risk.

>> **Understand the big picture of security in your organization.** Whether or not you're responsible for some aspect of security, take the time to understand your organization's principles to increase security and reduce risk. Use the security and risk management principles in Chapter 3, and see how those principles can help improve security even more. Think about your role in advancing the cause of asset and information protection in your organization.

>> **Understand information security on a global scale.** Take the time to understand big-picture trends globally: what nation-states pose the greatest threats, developments in security and privacy laws, workforce trends, and changing attitudes about information security. This information will help you stay current in this rapidly evolving industry.

If you make the pursuit of excellence a habit, you can change for the better over time. You end up with an improved security career, and your organization gets better security and reduced risk.

2

Certification Domains

Understand security and risk management concepts and principles.

Make your knowledge of asset security one of your assets.

Design and implement secure software, systems, and facilities.

Master communication and network security fundamentals.

Recognize identity and access management techniques.

Conduct security assessments, scans, testing, and audits.

Apply security operations concepts and controls.

Ensure secure software development throughout the development life cycle.

Management domain

» **Understanding compliance, ethics, governance, security policies and procedures, business continuity planning, and risk management**

» **Implementing security education, training, and awareness**

Chapter **3**

Security and Risk Management

The Security and Risk Management domain addresses many fundamental security concepts and principles, as well as compliance, ethics, governance, security policies and procedures, business continuity planning, risk management, and security education, training, and awareness. This domain represents 15 percent of the CISSP certification exam.

Understand, Adhere to, and Promote Professional Ethics

Ethics (or moral values) help describe what you should do in a given situation based on a set of principles or values. Ethical behavior is important in maintaining credibility as an information security professional and is a requirement for maintaining your CISSP certification. An organization often defines its core values (along with its mission statement) to help ensure that its employees understand what is acceptable and expected as they work to achieve the organization's mission, goals, and objectives.

CROSS REFERENCE

This section covers Objective 1.1 of the Security and Risk Management domain in the CISSP Exam Outline (May 1, 2021).

Ethics are not easily discerned, and a fine line often hovers between ethical and unethical activity. Unethical activity doesn't necessarily equate to illegal activity, and what may be acceptable in some organizations, cultures, or societies may be unacceptable or even illegal in others.

Ethical standards can be based on a common or national interest, individual rights, laws, tradition, culture, or religion. One helpful distinction between laws and ethics is that laws define what we *must* do, and ethics define what we *should* do.

Many common fallacies abound about the proper use of computers, the Internet, and information, which contribute to this gray area:

>> **The Computer Game Fallacy:** Any system or network that's not properly protected is fair game.

>> **The Law-Abiding-Citizen Fallacy:** If no physical theft is involved, an activity really isn't stealing.

>> **The Shatterproof Fallacy:** Any damage done will have a limited effect.

>> **The Candy-from-a-Baby Fallacy:** It's so easy, it can't be wrong.

>> **The Hacker's Fallacy:** Computers provide a valuable means of learning that will in turn benefit society.

REMEMBER

The problem here lies in the distinction between *hackers* and *crackers.* Although both may have a genuine desire to learn, crackers learn at the expense of others.

>> **The Free-Information Fallacy:** Any and all information should be free and thus can be obtained through any means.

Almost every recognized group of professionals defines a code of conduct or standards of ethical behavior by which its members must abide. For the CISSP, that code is the (ISC)² Code of Ethics. The CISSP candidate must be familiar with the (ISC)² Code of Ethics and Request for Comments (RFC) 1087, "Ethics and the Internet," for professional guidance on ethics (and information that you need to know for the exam).

(ISC)² Code of Professional Ethics

As a requirement for (ISC)² certification, all CISSP candidates must subscribe to and fully support all portions of the (ISC)² Code of Ethics. Intentionally or knowingly violating any provision of the (ISC)² Code of Ethics may subject you to examination by a peer-review panel and revocation of your hard-earned CISSP certification.

The (ISC)² Code of Ethics consists of a preamble and four canons. The canons are listed in order of precedence; thus, any conflicts should be resolved in the following order:

Preamble:

>> The safety and welfare of society and the common good, duty to our principals, and to each other, requires that we adhere, and be seen to adhere, to the highest ethical standards of behavior.

>> Therefore, strict adherence to this Code is a condition of certification.

Canons:

>> Protect society, the common good, necessary public trust and confidence, and the infrastructure.

>> Act honorably, honestly, justly, responsibly, and legally.

>> Provide diligent and competent service to principals.

>> Advance and protect the profession.

TIP

The best way to comply with the (ISC)² Code of Ethics is to never take part in any activity that imparts even the *appearance* of an ethics violation. Making questionable moves puts your certification at risk, and it may also convey to others that such activity is acceptable. Remember to lead by example!

Organizational code of ethics

Just about every organization has a code of ethics, or a statement of values, that it requires its employees or members to follow in their daily conduct. As a CISSP-certified information security professional, you are expected to be a leader in your organization, which means that you exemplify your organization's ethics (or values) and set a positive example for others to follow.

In addition to your organization's code of ethics, two other computer security ethics standards you should be familiar with for the CISSP exam and adhere to are the Internet Activities Board's "Ethics and the Internet" (RFC 1087) and the "Ten Commandments of Computer Ethics" created by the Computer Ethics Institute.

Internet Architecture Board: Ethics and the Internet (RFC 1087)

Published by the Internet Architecture Board (IAB) (https://www.iab.org) in January 1989, RFC 1087 characterizes as unethical and unacceptable any activity that purposely

>> Seeks to gain unauthorized access to the resources of the Internet

>> Disrupts the intended use of the Internet

>> Wastes resources (people, capacity, computer) through such actions

>> Destroys the integrity of computer-based information

>> Compromises the privacy of users

Other important tenets of RFC 1087 include

>> Access to and use of the Internet is a privilege and should be treated as such by all users of [the] system.

>> Many of the Internet resources are provided by the U.S. government. Abuse of the system thus becomes a Federal matter above and beyond simple professional ethics.

>> Negligence in the conduct of Internet-wide experiments is both irresponsible and unacceptable.

>> In the final analysis, the health and well-being of the Internet is the responsibility of its users who must, uniformly, guard against abuses which disrupt the system and threaten its long-term viability.

Ten Commandments of Computer Ethics

The Ten Commandments of Computer Ethics were originally created by the Computer Ethics Institute in 1992. The commandments are still used by many organizations, including (ISC)², to develop their own code of ethics.

1. Thou shalt not use a computer to harm other people.

2. Thou shalt not interfere with other people's computer work.

3. Thou shalt not snoop around in other people's computer files.

4. Thou shalt not use a computer to steal.

5. Thou shalt not use a computer to bear false witness.

6. Thou shalt not copy or use proprietary software for which you have not paid.

7. Thou shalt not use other people's computer resources without authorization or proper compensation.

8. Thou shalt not appropriate other people's intellectual output.

9. Thou shalt think about the social consequences of the program you are writing or the system you are designing.

10. Thou shalt always use a computer in ways that ensure consideration and respect for your fellow humans.

Understand and Apply Security Concepts

Confidentiality, integrity, availability — collectively referred to as the CIA triad — as well as nonrepudiation and authenticity, are foundational security concepts (all discussed in the following sections) that the CISSP candidate must understand. The CIA triad forms the basis of information security (see Figure 3-1). The triad is composed of three fundamental information security concepts:

>> Confidentiality

>> Integrity

>> Availability

FIGURE 3-1: The CIA triad.

© John Wiley & Sons, Inc.

As with any triangular shape, all three sides depend on one another (think of a three-sided pyramid or a three-legged stool) to form a stable structure. If one piece falls apart, the whole thing falls apart.

This section covers Objective 1.2 of the Security and Risk Management domain in the CISSP Exam Outline (May 1, 2021).

Confidentiality

Confidentiality limits access to information to subjects (users and machines) that require it. Confidentiality is usually accomplished by several means, including

» **Access and authorization:** Ranging from physical access to facilities containing computers to user account access, role-based access controls, and attribute-based access controls, the objective is to make sure that only people who have proper business authorization are permitted to access information. This topic is covered in Chapter 7.

» **Vulnerability management:** This aspect includes everything from system hardening to patch management and the elimination of vulnerabilities from applications. What we're trying to avoid is any possibility that someone can attack the system and get to the data.

» **Thorough system design:** The overall design of the system excludes unauthorized subjects from access to protected data.

» **Sound data management practices:** The organization has established processes that define the use of the information it manages or controls.

These characteristics work together to ensure that secrets remain secrets.

Privacy is a concept that is closely related to confidentiality and is most often associated with personal data. The objective of privacy is the confidentiality and proper handling of personal data. Various U.S. and international laws exist to protect the privacy (confidentiality and proper use) of personal data.

Personal data most commonly refers to personally identifiable information (PII) or personal health information (PHI). PII includes names, addresses, Social Security numbers, contact information (in some cases), and financial or medical data. PHI consists of many of the same data elements as PII, but also includes an individual patient's medical records and health-care payment history. Personal data, in more comprehensive legal definitions (particularly in Europe), may also include race, marital status, sexual orientation or lifestyle, religious preference, political affiliations, and any number of other unique personal characteristics that may be collected or stored.

TIP

The U.S. Health Insurance Portability and Accountability Act (HIPAA), discussed later in this chapter, defines PHI as protected health information. In its more general context, PHI refers to personal health information.

Integrity

Integrity safeguards the accuracy and completeness of information and processing methods throughout the information life cycle. It ensures that

>> Unauthorized users or processes don't make modifications to data.

>> Authorized users or processes don't make unauthorized modifications to data.

>> Data is internally and externally consistent, meaning that a given input produces an expected output.

Key characteristics of data integrity include completeness, timeliness, accuracy, and validity. Some of the measures taken to ensure data integrity are

>> **Authorization:** Determines whether data has proper authorization to enter a system. The integrity of a data record includes whether it should even be in the system.

>> **Input control:** Verifies that the new data entering the system is in the proper format and in the proper range.

>> **Access control:** Controls who (and what) is permitted to change the data and when the data can be changed.

>> **Output control:** Verifies that the data leaving the system is in the proper format and complete.

Availability

Availability ensures that authorized users have reliable and timely access to information, and to associated systems and assets, when and where needed. Availability is easily one of the most-overlooked aspects of information security. In addition to denial-of-service attacks, threats to availability include single points of failure, inadequate capacity (such as storage, bandwidth, and processing) planning, equipment malfunctions, and business interruptions or disasters. The characteristics of a system that determine its availability include

>> **Resilient hardware design:** Features may include redundant power supplies, network adapters, processors, and other components. These features help ensure that a system will keep running even if some of its internal components fail.

>> **Resilient software:** The operating system and other software components need to be designed and configured to be as reliable as possible, incorporating techniques such as multithreading, multiprocessing, and multiprogramming.

>> **Resilient architecture:** We're talking big picture here. In addition to resilient hardware design, we suggest that other components have redundancy, including routers, firewalls, switches, telecommunications circuits, and whatever other items may otherwise be single points of failure.

>> **Sound configuration management, change management, and preventive maintenance processes:** Availability includes not only the components of the system itself, but also good system management practices. After all, availability means avoiding unscheduled downtime, which is often a consequence of sloppy configuration management and change management practices or of neglected preventive maintenance.

>> **Established business continuity and disaster recovery plans:** Organizations need to ensure that natural and human-made disasters do not negatively affect the availability of critical systems and data. This topic is covered in detail later in this chapter.

Authenticity

Authenticity ensures that the source of a message (such as an email or text message), data, a transaction, or other exchange of information is who (or what) it claims to be. Authenticity is verified through the process of authentication, which is achieved through identity and access management (IAM; see Chapter 7), as well as by means of integrity.

Nonrepudiation

Nonrepudiation ensures that a user can't deny an action because you can irrefutably associate the user with that action. Nonrepudiation is typically achieved through IAM (discussed in Chapter 7). If a user can be positively associated with actions they perform on the network while logged in with a username and password that is known only to that user, nonrepudiation is established. But if that user's account credentials are shared with other users or compromised by an attacker, the user cannot be positively associated with the actions performed with their account. Nonrepudiation can be strengthened if the IAM platform requires multifactor authentication and/or biometric factors, both discussed in Chapter 7.

Evaluate and Apply Security Governance Principles

For the CISSP exam, you must fully understand and be able to apply security governance principles, including the following:

➤➤ Alignment of security function with business strategy, goals, mission, and objectives

➤➤ Organizational processes

➤➤ Organizational roles and responsibilities

➤➤ Security control frameworks

➤➤ Due care and due diligence

CROSS REFERENCE

This section covers Objective 1.3 of the Security and Risk Management domain in the CISSP Exam Outline (May 1, 2021).

Alignment of security function to business strategy, goals, mission, and objectives

To be effective, an information security program must be aligned with the organization's mission, strategy, goals, and objectives; thus, you must understand the differences and relationships among an organization's mission statement, strategy, goals, and objectives. You also need to know how these elements can affect the organization's information security policies and program. Proper alignment with the organization's mission, strategy, goals, and objectives also helps build business cases, secure budgets, and allocate resources for security program initiatives. With proper alignment, security projects and other activities are appropriately prioritized, and they fit better into organization policies, practices, and processes.

Mission (not-so-impossible) and strategy

This heading is corny, yes, but there's a good chance that you're humming the *Mission Impossible* theme song now. Mission accomplished!

An organization's *mission statement* expresses its reason for existence. A good mission statement is an easy-to-understand, general-purpose statement that says what the organization is, what it does, and why it exists, doing what it does in the way that it has chosen.

An organization's *strategy* describes how it accomplishes its mission and is frequently adapted to address new challenges and business realities.

Goals and objectives

A *goal* is something (or many somethings) that an organization hopes to accomplish. A goal should be consistent with the organization's mission statement or philosophy, and it should help define a vision for the organization.

An *objective* is a milestone or a specific result that is expected and, as such, helps an organization attain its goals and achieve its mission.

Security personnel should be acutely aware of their organizations' goals and objectives. Only then can security professionals ensure that security capabilities will work with and protect all the organization's current, changing, and new products, services, and endeavors.

WARNING

Organizations often use the terms *goal* and *objective* interchangeably, without distinction. Worse, some organizations refer to goals as long-term objectives and objectives as short-term goals! For the CISSP exam, an *objective* (short-term) supports a *goal* (intermediate-term), which supports a *mission* (long-term), which is accomplished with a well-defined *strategy*. All these terms fall under the umbrella of the organization's mission statement.

Organizational processes

In this section, we discuss key processes in the realm of security governance.

Acquisitions and divestitures

Organizations, particularly in private industry, continually reinvent themselves. More than ever before, it is important to be agile and competitive. Doing so results in organizations acquiring other organizations, organizations splitting themselves into two (or more) separate companies, as well as internal reorganizations that change the alignment of teams, departments, divisions, and business units.

Several security-related considerations should be taken into account when an organization acquires another organization or when two (or more) organizations merge:

>> **Security governance and management:** How is security managed in each organization, and what important differences exist?

>> **Security policy:** How do the two organizations' policies differ, and what issues will arise when the policies are merged?

>> **Security posture:** Which security controls are present in each organization, and how different are they? Do risk treatment decisions align with established risk appetite and risk tolerance?

>> **Security operations:** What security operations are in place today, and how do they operate? Considerations include vulnerability management, event monitoring, IAM, third-party risk management, and incident management.

If the security of one organization is vastly different from another organization's security, the organizations shouldn't be too hasty in connecting their networks.

Interestingly, when an organization divides itself into two (or more) separate organizations or sells off a division, the process can be trickier. Each new company probably will need to duplicate the security governance, management, controls, operations, and tools that the single organization had before the split. The two separate security functions don't always need to be the same as the old ones; it is important to fully understand the business mission in each organization, as well as the security regulations and standards that apply to each organization. Only then can information security align to the organizations.

Governance committees and executive oversight

Security management starts (or *should* start!) at the top, with executive management and board-level oversight. This oversight generally takes the form of *security governance,* which simply means that the organization's governing body has set the direction and the organization has policies and processes in place to ensure that executive management is following that direction, is fully informed, and is in control of information security strategy, policy, and operations.

A *governance committee* is a group of executives and/or managers who meet regularly to review security incidents, projects, operational metrics, and other aspects of concern to them. The governance committee will make or ratify risk treatment decisions or delegate risk treatment to another committee. Occasionally, the committee will issue mandates to security management about new business activities and shifts in priorities and strategic direction.

In practice, this type of governance is not much different from governance in IT and other departments. Governance is how executive management stays involved and controls the goings-on in various parts of the business.

Organizational roles and responsibilities

The truism that information security is everyone's responsibility is too often put into practice as "Everyone is responsible, but no one is accountable." To avoid this pitfall, specific roles and responsibilities for information security should be defined in an organization's security policy, individual job or position descriptions, and third-party contracts. These roles and responsibilities should apply to employees, consultants, contractors, interns, and vendors, and they should apply to every level of staff, from C-level executives to line employees.

Management

Senior-level management is often responsible for information security at several levels, including the role of information owner, which we discuss in the next section, "Users." In this context, however, management has a responsibility to demonstrate a strong commitment to an organization's information security program through the following actions:

>> **Monitoring the organization's security program:** Security leaders should brief executives periodically on the state of the organization's security program, noting any edicts and imperatives issued by executives.

>> **Creating, mandating, and approving a corporate information security policy:** This policy should include a statement of support from management and should also be signed by the chief executive officer, chief operating officer, chief information officer, or chairman.

>> **Leading by example:** A CEO who displays a mandatory identification badge, uses system access controls, and complies with the organization's acceptable-use policy sets a good example for the workforce.

>> **Rewarding compliance:** Management should expect proper security behavior and acknowledge, recognize, and/or reward employees accordingly.

REMEMBER

Management is ultimately responsible for an organization's overall information security and for any information security decisions that are made (or not made). Our role as information security professionals is to report security issues and make appropriate information security recommendations to management.

Users

An *end user* (or *user*) includes just about everyone within an organization. Users aren't specifically designated; they can be broadly defined as people who have authorized access to an organization's internal information, information systems, or facilities. Users include employees, contractors and other temporary employees, consultants, vendors, customer, and anyone else with access. Some organizations call these users employees, partners, associates, or what-have-you. Typical user responsibilities include

» Complying with all applicable security requirements defined in organizational policies, standards, and procedures; applicable legislative or regulatory requirements; and contractual requirements such as nondisclosure agreements and service-level agreements (SLAs)

» Exercising due care in safeguarding organizational information and information assets

» Participating in information security training and awareness efforts as required

» Reporting any suspicious activity, security violations, security problems, or security concerns to appropriate personnel

Security control frameworks

Organizations often adopt a control framework to aid in their legal and regulatory compliance efforts. Examples of relevant security frameworks include

» **COBIT 2019:** Developed by ISACA (formerly known as the Information Systems Audit and Control Association) and the IT Governance Institute (ITGI), COBIT consists of several components, including the following:

- *Framework:* Organizes IT governance objectives and best practices

- *Process descriptions:* Provides a reference model and common language

- *Maturity models:* Assesses organizational maturity/capability and addresses gaps

The COBIT framework is popular in organizations that are subject to the Sarbanes-Oxley Act (SOX; discussed later in this chapter) or ICOFR.

» **NIST (National Institute for Standards and Technology) Special Publication 800-53: Security and Privacy Controls for Federal Information Systems and Organizations:** Known among information security professionals as NIST 800-53, this very popular and comprehensive controls framework is required by U.S. government agencies. It also is widely used in private industry in the United States and throughout the world.

» **NIST Cybersecurity Framework:** Known as NIST CSF and originally written for organizations designated as critical infrastructure, this standard has become increasingly popular because it is comprehensive and provides straightforward guidance, including a high-level depiction of core information security control categories:

- Identify

- Protect

- Detect

- Respond

- Recover

» **COSO (Committee of Sponsoring Organizations of the Treadway Commission):** Developed by the Institute of Management Accountants (IMA), the American Accounting Association (AAA), the American Institute of Certified Public Accountants (AICPA), the Institute of Internal Auditors (IIA), and Financial Executives International (FEI), the COSO framework consists of five components:

- *Control environment:* Provides the foundation for all other internal control components.

- *Risk assessment:* Establishes objectives through identification and analysis of relevant risks, and determines whether anything will prevent the organization from meeting its objectives.

- *Control activities:* Policies and procedures that are created to ensure compliance with management directives. Various control activities are discussed in the other chapters of this book.

- *Information and communication:* Ensures that appropriate information systems and effective communications processes are in place throughout the organization.

- *Monitoring activities:* Monitor activities that assess performance over time and identify deficiencies and corrective actions.

» **ISO/IEC 27002 (International Organization for Standardization/ International Electrotechnical Commission):** Formally titled "Information technology — Security techniques — Code of practice for information security management," ISO/IEC 27002 documents security best practices in 14 domains:

- Information security policies

- Organization of information security

- Human resources security

- Asset management

- Access control

- Cryptography

- Physical and environmental security

- Operations security

- Communications security

- Systems acquisition, development, and maintenance

- Supplier relationships

- Information security incident management

- Information security aspects of business continuity management

- Compliance

>> **CIS CSC (Center for Internet Security Critical Security Controls):** Formerly known as the SANS Top 20 Controls, CIS CSC is a comprehensive and pragmatic security control framework used by numerous organizations. The control domains are

- Inventory and control of enterprise assets

- Inventory and control of software assets

- Data protection

- Secure configuration of enterprise assets and software

- Account management

- Access control management

- Continuous vulnerability management

- Audit log management

- Email and web browser protections

- Malware defenses

- Data recovery

- Network infrastructure management

- Network monitoring and defense

- Security awareness and skills training

- Service provider management

- Application software security

- Incident response management
- Penetration testing

» **ITIL (Information Technology Infrastructure Library):** A set of best practices for IT service management, consisting of five volumes:

- *Service Strategy:* Addresses IT services strategy management, service portfolio management, IT services financial management, demand management, and business relationship management

- *Service Design:* Addresses design coordination, service catalog management, service level management, availability management, capacity management, IT service continuity management, information security management system, and supplier management

- *Service Transition:* Addresses transition planning and support, change management, service asset and configuration management, release and deployment management, service validation and testing, change evaluation, and knowledge management

- *Service Operation:* Addresses event management, incident management, service request fulfillment, problem management, and access management

- *Continual Service Improvement:* Defines a seven-step process for improvement initiatives, including identifying the strategy, defining what will be measured, gathering the data, processing the data, analyzing the information and data, presenting and using the information, and implementing the improvement

TIP

Although ITIL is not strictly considered to be a security controls framework, cybersecurity depends on sound IT service management for many security controls to be effective.

Due care and due diligence

Due care is the conduct that a reasonable person exercises in a given situation, which provides a standard for determining negligence. In the practice of information security, due care relates to the steps that people or organizations take to perform their duties and implement security best practices.

Another important aspect of due care is the principle of *culpable negligence.* If an organization fails to follow a standard of due care in the protection of its assets (or its personnel), the organization may be held culpably negligent. In such cases, jury awards may be adjusted, and the organization's insurance company may be required to pay only a portion of any loss; the organization may get stuck paying the rest of the bill!

Due diligence is the prudent management and execution of due care. It's most often used in legal and financial circles to describe the actions that an organization takes to research the viability and merits of an investment or merger/acquisition opportunity. In the context of information security, due diligence commonly refers to risk identification and risk management practices, not only in the day-to-day operations of an organization, but also in the case of technology procurement, as well as mergers and acquisitions.

WARNING

The concepts of *due care* and *due diligence* are related but distinctly different. In practice, due care is turning on logging; due diligence is reviewing the logs regularly.

Determine Compliance and Other Requirements

Compliance is composed of the set of activities undertaken by an organization in its attempts to abide by applicable laws, regulations, standards, policies, and other legal obligations such as contract terms and conditions and SLAs.

CROSS REFERENCE

This section covers Objective 1.4 of the Security and Risk Management domain in the CISSP Exam Outline (May 1, 2021).

Because of the nature of compliance, and because many laws and standards are related to security and privacy, many organizations have adopted the fatally mistaken notion that being compliant with security regulations is the same thing as being secure. It is appropriate to say, however, that being compliant with security regulations and standards is a step in the right direction on the journey to becoming secure. The nature of threats today makes it plain that even organizations that are fully compliant with applicable security laws, regulations, and standards may be woefully unsecure.

Contractual, legal, industry standards, and regulatory requirements

A basic understanding of contractual, legal, industry standards, and regulatory compliance requirements is required for the CISSP exam.

Contractual

Contractual obligations with information security requirements may exist between organizations, for example, as part of a master services agreement, statement of work, work order, or SLA.

Common law

Common law (also known as *case law*) originated in medieval England and is derived from the decisions (precedents) of judges. Common law is based on the doctrine of *stare decisis* ("let the decision stand") and is often codified by statutes. Under the common-law system of the United States, three major categories of laws are defined at the federal and state levels: criminal, civil (or tort), and administrative (or regulatory) laws.

Criminal law

Criminal law defines crimes committed against society, even when the actual victim is a business or person. Criminal laws are enacted to protect the general public. As such, in the eyes of the court, the victim is incidental to the greater cause.

CRIMINAL PENALTIES

Penalties under criminal law have two main purposes:

- >> **Punishment:** Penalties may include jail/prison sentences, probation, fines, and/or financial restitution to the victim.
- >> **Deterrence:** Penalties must be severe enough to dissuade any further criminal activity by the offender or anyone else considering a similar crime.

BURDEN OF PROOF UNDER CRIMINAL LAW

To be convicted under criminal law, a judge or jury must believe *beyond a reasonable doubt* that the defendant is guilty. Therefore, the burden of proof in a criminal case rests firmly with the prosecution.

CLASSIFICATIONS OF CRIMINAL LAW

Criminal law has two main classifications, depending on severity, such as type of crime/attack or total loss in dollars:

>> **Felony:** More-serious crimes, normally resulting in jail/prison terms of more than one year

>> **Misdemeanor:** Less-serious crimes, normally resulting in fines or jail/prison terms of less than one year

Civil law

Civil (tort) law addresses wrongful acts committed against a person or business, either willfully or negligently, resulting in damage, loss, injury, or death.

CIVIL PENALTIES

Unlike criminal penalties, civil penalties don't include jail or prison terms. Instead, civil penalties provide financial restitution to the victim:

>> **Compensatory damages:** Actual damages to the victim, including attorney/legal fees, lost profits, investigative costs, and so on

>> **Punitive damages:** Determined by a jury and intended to punish the offender

>> **Statutory damages:** Mandatory damages determined by law and assessed for violating the law

BURDEN OF PROOF UNDER CIVIL LAW

Convictions under civil law are typically easier to obtain than under criminal law because the burden of proof is much lower. To be convicted under civil law, a jury must believe *based upon the preponderance of the evidence* that the defendant is guilty, which simply means that the available evidence leads the judge or jury to a conclusion of guilt.

LIABILITY AND DUE CARE

The concepts of liability and due care are germane to civil-law cases, but they're also applicable under administrative law.

The standard criteria for assessing the legal requirements for implementing recommended safeguards evaluate the cost of the safeguard and the estimated loss from the corresponding threat, if realized. If the cost is less than the estimated loss, and the organization doesn't implement a safeguard, a legal liability may exist. This liability is based on the principle of *proximate causation*, in which an action taken or not taken was part of a sequence of events that resulted in negative consequences.

TECHNICAL STUFF

LAWYERSPEAK

Although the information in this sidebar is not tested on the CISSP examination, it may come in handy when you're attempting to learn the various laws and regulations in this domain. You'll find it helpful to know the correct parlance (fancyspeak for *jargon*) used in the United States, including the following:

18 U.S.C. § 1030 (1986) (the Computer Fraud and Abuse Act of 1986) refers to Section 1030 in Title 18 of the 1986 edition of the U.S. Code, not "18 University of Southern California squiggly-thingy 1030 (1986)."

Federal statutes and administrative laws are usually cited in the following format:

- **Title number:** Titles are grouped by subject matter.
- **Abbreviation for the code:** *U.S.C.* is U.S. Code; *C.F.R.* is Code of Federal Regulations.
- **Section number:** *§* means "The Word Formerly Known as Section."
- **Year of publication:** The year is listed in parentheses.

Other important abbreviations to understand include

- **Fed. Reg.:** Federal Register
- **Fed. R. Evid.:** Federal Rules of Evidence
- **PL:** Public Law
- **§§:** Sections (18 U.S.C. §§ 2701–11 refers to sections 2701 through 2711)
- **v.:** versus, as in United States v. Moore

Note: The rest of the civilized world understands *vs.* to mean *versus* and *v.* to mean *version* or *volume,* but you need to remember two important points: Lawyers aren't part of the civilized world, and they apparently charge by the letter (as well as by the minute).

Under federal sentencing guidelines, senior corporate officers may be personally liable if their organization fails to comply with applicable laws. Such officers must follow the *prudent man* (or person) *rule*, which requires them to perform their duties

>> In good faith

>> In the best interests of the enterprise

>> With the care and diligence that ordinary, prudent people in a similar position would exercise under similar circumstances

International law

Given the global nature of the Internet, it's often necessary for countries to cooperate to bring a computer criminal to justice. But because practically every country has its own legal system, such cooperation is always difficult and often impossible. As a starting point, countries sometimes disagree about exactly what justice is. Other problems include the following:

>> **Lack of universal cooperation:** We can't answer the question "Why can't we all just get along?", but we can tell you that it's highly unlikely that a 14-year-old hacker in some remote corner of the world will commit some dastardly crime that unites us all in our efforts to take them down, bringing about lasting world peace.

>> **Differing interpretations of laws:** What's illegal in one country (or even in one U.S. state) isn't necessarily illegal in another.

>> **Differing rules of evidence:** This problem can encompass different rules for obtaining and collecting evidence, as well as different rules for admissibility of evidence.

>> **Priority:** Nations have different views regarding the seriousness of computer crimes, and in the realm of international relations, computer crimes are usually of minimal concern.

>> **Outdated laws and technology:** This problem is related to the priority problem. Technology varies greatly throughout the world, and many countries (not only in the Third World) are far behind others. For this reason and many others, computer crime laws are often a low priority and aren't kept current. This problem is further exacerbated by the different technical capabilities of the various law enforcement agencies that may be involved in an international case.

>> **Extradition:** Many countries don't have extradition treaties and won't extradite suspects to a country that has different or controversial practices, such as capital punishment. Although capital punishment for a computer crime may sound extreme, recent events and the threat of cyberterrorism make this possibility very real.

Besides common-law systems (which we talk about in the section "Common law" earlier in this chapter), other countries use legal systems including the following:

>> **Civil-law systems:** Not to be confused with U.S. civil law, which is based on common law. *Civil-law* systems use constitutions and statutes exclusively and aren't based on precedent. The role of a judge in a civil-law system is to interpret the law. Civil law is the most common type of law system used throughout the world.

- **Napoleonic code:** Originating in France after the French Revolution, the Napoleonic code has spread to many other countries in Europe and elsewhere. In this system, laws are developed by legislative bodies and interpreted by the courts, but there is often no formal concept of legal precedent.

- **Religious-law systems:** Derived from religious beliefs and values, common religious-law systems include Sharia in Islam, Halakha in Judaism, and canon law in Christianity.

- **Pluralistic-law systems:** These systems are combinations of various systems, such as civil and common law, civil and religious law, and common and religious law.

Administrative law

Administrative (regulatory) laws define standards of performance and conduct for major industries (including banking, energy, and health care), organizations, and government agencies. These laws are typically enforced by various government agencies, and violations may result in financial penalties and/or imprisonment.

Industry standards

Compliance with industry standards may be mandated for organizations in some cases, such as he Payment Card Industry Data Security Standard (PCI DSS), discussed later in this chapter. More commonly, organizations voluntarily or contractually comply with industry standards. An organization might choose to pursue and/or maintain ISO/IEC 27002 compliance (discussed earlier in this chapter) to attract customers, achieve a competitive advantage, or reduce its cyberinsurance premiums.

Privacy requirements

Privacy and data protection laws are enacted to protect information collected and maintained on people from unauthorized disclosure or misuse. Privacy laws are one area in which the United States lags behind many others, particularly the European Union (EU) and its General Data Protection Regulation (GDPR), which has defined increasingly restrictive privacy regulations that regulate the transfer of personal information to countries (including the United States) that don't protect such information equally. The EU GDPR privacy rules include the following requirements about personal data and records:

- The data must be collected fairly and lawfully, and only after the subject has provided explicit consent.

>> The data must be used for only the purposes for which it was collected and for only a reasonable period of time.

>> The data must be accurate and kept up to date.

>> The data must be accessible to people who request a report on personal information held about themselves.

>> People must have the right to have any errors in their personal data corrected.

>> People must have the right for their information to be expunged from an organization's information systems.

>> Personal data can't be disclosed to other organizations or people unless disclosure is authorized by law or personal consent.

>> Transmission of personal data to locations where equivalent privacy protection cannot be ensured is prohibited.

Specific privacy and data protection laws are discussed later in this chapter.

Understand Legal and Regulatory Issues That Pertain to Information Security

CISSP candidates are expected to be familiar with the laws and regulations that are relevant to information security throughout the world and in various industries, which could include national laws, local laws, and any laws that pertain to the types of activities performed by organizations.

CROSS REFERENCE

This section covers Objective 1.5 of the Security and Risk Management domain in the CISSP Exam Outline (May 1, 2021).

Cybercrimes and data breaches

Cybercrime consists of any criminal activity in which computer systems or networks are used as tools. Cybercrime also includes crimes in which computer systems are targeted or in which computers are the scene of the crime committed. That's a pretty wide spectrum.

The real world, however, has difficulty dealing with computer crimes. Several reasons why cybercrimes are hard to cope with include

>> **Lack of understanding:** In general, legislators, judges, attorneys, law enforcement officials, and jurors don't understand the many technologies and issues involved in a cybercrime.

>> **Inadequate laws:** Laws are slow to change and fail to keep pace with rapidly evolving new technology.

>> **Encryption:** Increasingly, law enforcement organizations are hindered in their criminal investigations because of advanced encryption techniques in mobile devices.

>> **Multiple roles of computers in crime:** These roles include crimes committed *against* a computer (such as hacking into a system and stealing information) and crimes committed *by using* a computer (such as using a system to launch a distributed denial of service [DDoS] attack). Computers may also *support* criminal enterprises when criminals use computers for crime-related record-keeping or communications.

Cybercrimes are often difficult to prosecute for the reasons we just listed and also because of the following issues:

>> **Lack of tangible assets:** Traditional rules of property doesn't always clearly apply in a computer crime case. But property rules have been extended in many countries to include electronic information. Computing resources, bandwidth, and data (in the form of magnetic particles) are often the only assets at issue, and these assets can be very difficult to quantify and assign a value to. The asset valuation process, which we discuss later in this chapter, can provide vital information for valuing electronic information.

>> **Rules of evidence:** Often, original documents aren't available in a cybercrime case. Most evidence in such a case is considered to be hearsay evidence and must meet certain requirements to be admissible in court. Often, evidence is a computer itself or data on its hard drive.

>> **Lack of evidence:** Many cybercrimes are difficult to prosecute because law enforcement agencies lack the skills or resources even to identify the perpetrator, much less gather sufficient evidence to bring charges and prosecute successfully. Frequently, skilled computer criminals use a long trail of compromised computers in different countries to make it as difficult as possible for even diligent law enforcement agencies to identify them. Further, encryption techniques sometimes prevent law enforcement from being able to search computers and mobile devices for evidence.

>> **Definition of loss:** A loss of confidentiality or integrity of data goes far beyond the normal definition of loss in a criminal or civil case.

>> **Location of perpetrators:** Often, the people who commit cybercrimes against specific organizations do so from locations outside the victim's country. Cybercriminals know that even if they make a mistake and create discoverable evidence that identifies them, law enforcement agencies in the victim's country will have difficulty apprehending them.

>> **Criminal profiles:** Cybercriminals aren't necessarily hardened criminals and may include the following:

- *Juveniles:* Juvenile laws in many countries aren't taken seriously and are inadequate to deter crime. A busy prosecutor is unlikely to pursue a low-profile crime committed by a juvenile that results in a three-year probation sentence for the offender.

- *Trusted people:* Many cybercriminals hold a position of trust within a company and have no criminal record. Such a person likely can afford a dream team for legal defense, and a judge may be inclined to levy a more-lenient sentence for the first-time offender. Recent corporate scandals in the United States, however, have set a strong precedent for punishment at the highest levels.

Cybercrimes are often classified in one of the following seven major categories:

>> **Industrial espionage:** Businesses are increasingly the targets of industrial espionage. These attacks include competitive intelligence gathering; theft of product specifications, plans, and schematics; and business information such as marketing and customer information. Businesses can be inviting targets for an attacker due to

- *Lack of expertise:* Despite heightened security awareness, a shortage of qualified security professionals exists and is getting worse. As a result, organizations don't have adequate preventive, detective, and response capabilities.

- *Lack of resources:* Businesses often lack the resources to prevent, or even detect, attacks against their systems.

- *Lack of concern:* Executive management and boards of directors in many organizations still turn a blind eye to requests for security resources.

- *Lack of reporting or prosecution:* Because of public relations concerns and the inability to prosecute cybercriminals because of a lack of evidence or a lack of properly handled evidence, the majority of business attacks still go unreported. Further, few jurisdictions require organizations to disclose break-ins involving intellectual property.

The cost to businesses can be significant, including loss of trade secrets or proprietary information, loss of revenue, and loss of reputation when intrusions are made public.

» **Financial attacks:** Banks, large corporations, and e-commerce sites are the targets of financial attacks, many of which are motivated by greed. Financial attacks may seek to steal or embezzle funds, gain access to online financial information, extort people or businesses, or obtain the personal credit card numbers of customers. Ransomware attacks are immensely successful forms of financial attacks that encrypt information and demand a cryptocurrency ransom for the key to decrypt the information. Destructware attacks are similar to ransomware in that they often demand ransoms but do not provide keys for recovering the encrypted information.

» **"Fun" attacks:** Fun attacks are perpetrated by thrill-seekers who are motivated by curiosity or excitement. Although these attackers may not intend to do any harm or use any of the information they access, they're still dangerous, and their activities are still illegal.

These attacks can be relatively easy to detect and prosecute. Because the perpetrators are often *script kiddies* (hackers who use scripts or programs written by other hackers because they don't have programming skills themselves) or otherwise-inexperienced hackers, they may not know how to cover their tracks effectively.

Also, because no real harm is normally done or intended against the system, it may be tempting (although ill-advised) for a business to prosecute the attacker and put a positive public relations spin on the incident. You've seen the film at 11: "We quickly detected the attack, prevented any harm to our network, and prosecuted the responsible individual; our security is *unbreakable!*" Such action, however, will likely motivate others to launch more-serious and concerted grudge attacks against the business.

Many cybercriminals in this category seek only notoriety. Although it's one thing to brag to a small circle of friends about defacing a public website, the wily hacker who appears on CNN reaches the next level of hacker celebrity. These twisted individuals want to be caught to revel in their 15 minutes of fame.

» **Grudge attacks:** Grudge attacks target people or businesses, and the attacker is motivated by a desire to take revenge. A disgruntled employee, for example, may steal trade secrets, delete valuable data, or plant a logic bomb in a critical system or application.

Fortunately, these attacks (at least in the case of a disgruntled employee) can be easier to prevent or prosecute than many other types of attacks because

- The attacker is often known to the victim.
- The attack has a visible impact that produces a viable evidence trail.

- Most businesses (already sensitive to the possibility of wrongful-termination suits) have well-established termination procedures.

- Specific laws (such as the U.S. Economic Espionage Act of 1996, which we discuss later in this chapter) provide very severe penalties for such crimes.

>> **Hacktivism:** Ideological attacks, commonly known as *hacktivism,* have become increasingly common in recent years. Hacktivists typically target businesses or organizations to protest a controversial position that does not agree with their own ideology. These attacks typically take the form of DDoS attacks but can include data theft.

>> **Military and political intelligence attacks:** Military and political intelligence attacks are perpetrated by criminals, traitors, or foreign military and intelligence agents seeking classified government, law enforcement, or military information. Such attacks are often carried out by governments during times of war and conflict.

>> **Terrorist attacks:** Terrorism exists at many levels on the Internet. Following the terrorist attacks against the United States on September 11, 2001, the general public became painfully aware of the extent of terrorism on the Internet. Terrorist organizations and cells use online capabilities to coordinate attacks; transfer funds; harm international commerce; disrupt critical systems; disseminate propaganda; recruit new members; and gain useful information about developing techniques and instruments of terror, including nuclear, biological, and chemical weapons.

In an effort to combat identity theft, many U.S. states have passed disclosure laws that compel organizations to publicly disclose security breaches that may result in the compromise of personal data.

Although these laws typically include statutory penalties, the damage to an organization's reputation and the potential loss of business — caused by the public disclosure requirement of these laws — can be the most significant and damaging aspect for affected organizations. Thus, public disclosure laws shame organizations into implementing effective information security policies and practices to lessen the risk that a data breach will occur in the first place.

By requiring organizations to notify the public of a data breach, disclosure laws enable potential victims to take defensive or corrective action to avoid or minimize the damage resulting from identity theft.

Important international computer crime and information security laws and standards that the CISSP candidate should be familiar with include

>> U.S. Computer Fraud and Abuse Act of 1986

>> U.S. Electronic Communications Privacy Act (ECPA) of 1986

- U.S. Computer Security Act of 1987
- The Computer Misuse Act of 1990 (UK)
- U.S. Federal Sentencing Guidelines of 1991 (not necessarily specific to computer crime but certainly relevant)
- U.S. Communications Assistance for Law Enforcement Act of 1994
- U.S. Economic Espionage Act of 1996
- U.S. Child Pornography Prevention Act of 1996
- Safe Harbor (1998)
- Information Technology Act 2000 (India)
- USA PATRIOT Act of 2001
- The Council of Europe's Convention on Cybercrime (2001)
- Cybercrime Act of 2001 (Australia)
- U.S. Sarbanes-Oxley Act of 2002
- U.S. Homeland Security Act of 2002
- California Security Breach Information Act (2002)
- U.S. Controlling the Assault of Non-Solicited Pornography and Marketing (CAN-SPAM) Act of 2003
- U.S. Identity Theft and Assumption Deterrence Act of 2003
- Privacy and Electronic Communications Regulations of 2003 (UK)
- U.S. Intelligence Reform and Terrorism Prevention Act of 2004
- U.S. Federal Information Systems Modernization Act of 2014
- General Data Protection Regulation (GDPR) (EU) (2018)
- Payment Card Industry Data Security Standard (PCI DSS)

It is important to understand that cybersecurity and privacy laws change from time to time. The list of such laws in this book should not be considered to be complete or up to date. Instead, consider the following to be a sampling of laws in the United States and elsewhere.

U.S. Computer Fraud and Abuse Act of 1986, 18 U.S.C. § 1030 (as amended)

In 1986, the first U.S. federal computer crime law, the U.S. Computer Fraud and Abuse Act, was passed. This intermediate act was narrowly defined and somewhat ambiguous. The law covered

>> Classified national defense or foreign relations information

>> Records of financial institutions or credit reporting agencies

>> Government computers

The U.S. Computer Fraud and Abuse Act of 1986 enhanced and strengthened the 1984 law, clarifying definitions of criminal fraud and abuse for federal computer crimes and removing obstacles to prosecution.

The act established two new felony offenses for the unauthorized access of federal interest computers and a misdemeanor for unauthorized trafficking in computer passwords:

>> **Felony 1:** Unauthorized access, or access that exceeds authorization, of a federal interest computer to further an intended fraud, shall be punishable as a felony [Subsection (a)(4)].

>> **Felony 2:** Altering, damaging, or destroying information in a federal interest computer or preventing authorized use of the computer or information, that causes an aggregate loss of $1,000 or more during a one-year period or potentially impairs medical treatment, shall be punishable as a felony [Subsection (a)(5)].

   Note: This provision was stricken in its entirety and replaced by a more general provision, which we discuss later in this section.

>> **Misdemeanor:** Trafficking in computer passwords or similar information if it affects interstate or foreign commerce or permits unauthorized access to computers used by or for the U.S. government [Subsection (a)(6)].

TIP

The act defines a *federal interest computer* (the term was changed to *protected computer* in the 1996 amendments) as a computer

>> "[E]xclusively for the use of a financial institution or the United States government, or, in the case of a computer not exclusively for such use, used by or for a financial institution or the United States government and the conduct constituting the offense affect that use by or for the financial institution or the government"

>> "[W]hich is used in interstate or foreign commerce or communication"

Several minor amendments to the U.S. Computer Fraud and Abuse Act were made in 1988, 1989, and 1990, and more significant amendments were made in 1994, 1996 (by the Economic Espionage Act of 1996), and 2001 (by the USA PATRIOT Act of 2001). The act in its present form establishes eight specific computer crimes. In addition to the three that we discuss in the preceding list, these crimes include the following five provisions (we discuss subsection [a][5] in its current form in the following list):

>> Unauthorized access, or access that exceeds authorization, to a computer that results in *disclosure of U.S. national defense or foreign relations information* (emphasis added) [Subsection (a)(1)].

>> Unauthorized access, or access that exceeds authorization, to a protected computer to *obtain any information on that computer* (emphasis added) [Subsection (a)(2)].

>> Unauthorized access to a protected computer, or access that exceeds authorization, to a protected computer that *affects the use* (emphasis added) of that computer by or for the U.S. government [Subsection (a)(3)].

>> Unauthorized access to a protected computer causing damage or reckless damage, or *intentionally transmitting malicious code* (emphasis added) which causes damage to a protected computer [Subsection (a)(5), as amended].

>> Transmission of interstate or foreign commerce communication *threatening to cause damage* (emphasis added) to a protected computer for the purpose of extortion [Subsection (a)(7)].

In the section "USA PATRIOT Act of 2001" later in this chapter, we discuss major amendments to the U.S. Computer Fraud and Abuse Act of 1986 (as amended) that Congress introduced in 2001.

The U.S. Computer Fraud and Abuse Act of 1986 is *the* major computer crime law currently in effect. The CISSP exam likely tests your knowledge of the act in its original 1986 form, but you should also be prepared for revisions to the exam that may cover the more recent amendments.

U.S. Electronic Communications Privacy Act (ECPA) of 1986

The U.S. Electronic Communications Privacy Act (ECPA) provides the legal basis for network monitoring. It complements the U.S. Computer Fraud and Abuse Act of 1986 and prohibits eavesdropping, interception, or unauthorized monitoring of wire, oral, and electronic communications. The act does provide specific statutory exceptions, however, allowing network providers to monitor their networks for legitimate business purposes if they notify the network users of the monitoring process.

The ECPA was amended extensively by the USA PATRIOT Act of 2001. These changes are discussed in the upcoming "USA PATRIOT Act of 2001" section.

U.S. Computer Security Act of 1987

The U.S. Computer Security Act of 1987 requires federal agencies to take extra security measures to prevent unauthorized access to computers that hold sensitive information. In addition to identifying and developing security plans for sensitive systems, the act requires those agencies to provide security-related awareness training for their employees. The act also assigns formal government responsibility for computer security to the National Institute of Standards and Technology (NIST) for information security standards in general and to the National Security Agency (NSA) for cryptography in classified government/military systems and applications.

U.S. Communications Assistance for Law Enforcement Act of 1994

The U.S. Communications Assistance for Law Enforcement Act of 1994 (CALEA) provides for the lawful interception of electronic communications through wiretaps and other means with telecommunications providers. Nowadays, CALEA includes landline phones, wireless (mobile) phones, and text messaging. Law enforcement organizations can obtain such information from telecommunications companies through a search warrant or subpoena.

U.S. Federal Sentencing Guidelines of 1991

In November 1991, the U.S. Sentencing Commission published Chapter 8, "Federal Sentencing Guidelines for Organizations," of the U.S. Federal Sentencing Guidelines. These guidelines establish written standards of conduct for organizations, provide relief in sentencing for organizations that have demonstrated due diligence, and place responsibility for due care on senior management officials, with penalties for negligence including fines of up to $290 million.

U.S. Economic Espionage Act of 1996

The U.S. Economic Espionage Act (EEA) of 1996 was enacted to curtail industrial espionage, particularly when such activity benefits a foreign entity. The EEA makes it a criminal offense to take, download, receive, or possess trade secret information that's been obtained without the owner's authorization. Penalties include fines of up to $10 million, up to 15 years in prison, and forfeiture of any property used to commit the crime. The EEA also enacted the 1996 amendments to the U.S. Computer Fraud and Abuse Act of 1986, which we talk about earlier in this chapter.

U.S. Child Pornography Prevention Act of 1996

The U.S. Child Pornography Prevention Act of 1996 was enacted to combat the use of computer technology to produce and distribute pornography involving children, including adults portraying children.

USA PATRIOT Act of 2001

Following the terrorist attacks against the United States on September 11, 2001, the USA PATRIOT (Uniting and Strengthening America by Providing Appropriate Tools Required to Intercept and Obstruct Terrorism) Act of 2001 was enacted in October 2001 and renewed in March 2006. Many provisions originally set to expire have since been made permanent under the renewed act. This act takes great strides to strengthen and amend existing computer crime laws, including the U.S. Computer Fraud and Abuse Act and the ECPA, as well as to empower U.S. law enforcement agencies, if only temporarily. Federal courts have subsequently declared some of the act's provisions to be unconstitutional.

The sections of the act that are relevant to the CISSP exam include

>> **Section 202, Authority to Intercept Wire, Oral, and Electronic Communications Relating to Computer Fraud and Abuse Offenses:** Under previous law, investigators couldn't obtain a wiretap order for violations of the U.S. Computer Fraud and Abuse Act. This amendment authorizes such action for felony violations of that act.

>> **Section 209, Seizure of Voice-Mail Messages Pursuant to Warrants:** Under previous law, investigators could obtain access to email under the ECPA but not voicemail, which was covered by the more restrictive wiretap statute. This amendment authorizes access to voicemail with a search warrant rather than a wiretap order.

>> **Section 210, Scope of Subpoenas for Records of Electronic Communications:** Under previous law, subpoenas of electronic records were restricted to very limited information. This amendment expands the list of records that can be obtained and updates technology-specific terminology.

>> **Section 211, Clarification of Scope:** This amendment governs privacy protection and disclosure to law enforcement of cable TV, telephone, and Internet service provider records.

>> **Section 212, Emergency Disclosure of Electronic Communications to Protect Life and Limb:** Before this amendment, no special provisions allowed a communications provider to disclose customer information to law enforcement officials in emergency situations, such as an imminent crime or terrorist attack, without exposing the provider to civil liability suits from the customer.

TECHNICAL STUFF

>> **Section 214, Pen Register and Trap and Trace Authority under FISA (Foreign Intelligence Surveillance Act):** This amendment clarifies law enforcement authority to trace communications on the Internet and other computer networks, and authorizes the use of a pen/trap device nationwide instead of limiting it to the jurisdiction of the court.

A *pen/trap device* refers to a *pen register,* which shows outgoing numbers called from a phone, and a *trap and trace device,* which shows incoming numbers that called a phone. Pen registers and trap and trace devices are collectively referred to as pen/trap devices because most technologies allow the same device to perform both types of traces (incoming and outgoing numbers).

>> **Section 217, Interception of Computer Trespasser Communications:** Under previous law, it was permissible for organizations to monitor activity on their own networks but not necessarily for law enforcement to assist these organizations in monitoring, even when such help was specifically requested. This amendment allows organizations to authorize people "acting under color (pretense or appearance) of law" to monitor trespassers on their computer systems.

>> **Section 220, Nationwide Service of Search Warrants for Electronic Evidence:** This amendment removes jurisdictional issues in obtaining search warrants for e-mail. For an excellent example of this problem, read *The Cuckoo's Egg: Tracking a Spy Through the Maze of Computer Espionage,* by Clifford Stoll (Doubleday).

>> **Section 814, Deterrence and Prevention of Cyberterrorism:** This amendment greatly strengthens the U.S. Computer Fraud and Abuse Act, including raising the maximum prison sentence from 10 years to 20 years.

>> **Section 815, Additional Defense to Civil Actions Relating to Preserving Records in Response to Government Requests:** This amendment clarifies the "statutory authorization" (government authority) defense for violations of the ECPA.

>> **Section 816, Development and Support of Cybersecurity Forensic Capabilities:** This amendment requires the attorney general to establish regional computer forensic laboratories; maintain existing laboratories; and provide forensic and training capabilities to federal, state, and local law enforcement personnel and prosecutors.

WARNING

The USA PATRIOT Act of 2001 changes many of the provisions in the computer crime laws, particularly the U.S. Computer Fraud and Abuse Act, which we discuss earlier in this chapter, and the ECPA, also discussed earlier in this chapter. As a security professional, you must keep abreast of current laws and affairs to perform your job effectively.

U.S. Sarbanes-Oxley Act of 2002 (SOX)

In the wake of several major corporate and accounting scandals, the U.S. Sarbanes-Oxley (SOX) Act was passed in 2002 to restore public trust in publicly held corporations and public accounting firms by establishing new standards and strengthening existing standards for these entities, including auditing, governance, and financial disclosures.

SOX established the Public Company Accounting Oversight Board (PCAOB), which is a private-sector, not-for-profit corporation responsible for overseeing auditors in the implementation of SOX. PCAOB's Accounting Standard 2 recognizes the role of information technology as it relates to a company's internal controls and financial reporting. The standard identifies the responsibility of chief information officers for the security of information systems that process and store financial data, and it has many implications for information technology security and governance.

U.S. Homeland Security Act of 2002

This law consolidated 22 U.S. government agencies to form the Department of Homeland Security (DHS). The law also provided for the creation of a privacy official to enforce the Privacy Act of 1974.

U.S. Federal Information Systems Modernization Act (FISMA) of 2014

The U.S. Federal Information Systems Modernization Act (FISMA) of 2014 extends the U.S. Computer Security Act of 1987 by requiring regular audits of both U.S. government information systems, and organizations providing information services to the U.S. federal government. It supersedes the U.S. Federal Information Systems Management Act of 2002.

U.S. Controlling the Assault of Non-Solicited Pornography and Marketing (CAN-SPAM) Act of 2003

The U.S. Controlling the Assault of Non-Solicited Pornography and Marketing (CAN-SPAM) Act of 2003 establishes standards for sending commercial e-mail messages, charges the U.S. Federal Trade Commission with enforcement of the provision, and provides penalties that include fines and imprisonment for violations of the act.

U.S. Identity Theft and Assumption Deterrence Act of 2003

This law updates earlier U.S. laws on identity theft.

Safe Harbor (1998)

In a 1998 agreement, the European Union and the U.S. Department of Commerce developed a certification program called Safe Harbor, which permits U.S.-based organizations to certify themselves as properly handling private data belonging to European citizens.

U.S. Intelligence Reform and Terrorism Prevention Act of 2004

This law facilitates the sharing of intelligence information among various U.S. government agencies, as well as protections of privacy and civil liberties.

California Security Breach Information Act

Passed in 2003, the California Security Breach Information Act (SB-1386) was the first U.S. state law to require organizations to notify all affected people "in the most expedient time possible and without unreasonable delay, consistent with the legitimate needs of law enforcement," if their confidential or personal data is lost, stolen, or compromised, unless that data is encrypted.

The law applies to any organization that does business in California — even a single customer or employee. An organization is subject to the law even if it doesn't directly do business in California (such as if it stores personal information about California residents for another company).

Other states quickly followed suit; now 46 states, the District of Columbia, Puerto Rico, and the U.S. Virgin Islands have public disclosure laws. These laws aren't necessarily consistent from one state to another, however, and are not without flaws and critics.

Until early 2008, for example, Indiana's Security Breach Disclosure and Identity Deception law (HEA 1101) did not require an organization to disclose a security breach "if access to the [lost or stolen] device is protected by a *password* [emphasis added] that has not been disclosed." Indiana's law has since been amended and is now one of the toughest state disclosure laws in effect, requiring public disclosure unless "all personal information . . . is protected by encryption."

Finally, a provision in California's and Indiana's disclosure laws, as well as in most other states' laws, allows an organization to avoid much of the cost of disclosure if the cost of providing such notice would exceed $250,000 or if more than 500,000 people would need to be notified. Instead, a substitute notice — consisting of email notifications, conspicuous posting on the organization's website, and notification of major statewide media — is permitted.

The Council of Europe's Convention on Cybercrime (2001)

The Convention on Cybercrime is an international treaty, signed by more than 40 countries (the United States ratified the treaty in 2006), requiring criminal laws to be established in signatory nations for computer hacking activities, child pornography, and intellectual property violations. The treaty also attempts to improve international cooperation with respect to monitoring, investigations, and prosecution.

The Computer Misuse Act of 1990 (UK)

The Computer Misuse Act 1990 defines three criminal offenses related to computer crime: unauthorized access (whether successful or unsuccessful), unauthorized modification, and hindering of authorized access (DoS).

Privacy and Electronic Communications Regulations of 2003 (UK)

Similar to U.S. "do not call" laws, this law makes it illegal to use equipment to make automated telephone calls that play recorded messages.

Information Technology Act 2000 (India)

This law modernizes computer crimes and defines activities such as data theft, creation and spreading of malware, identity theft, pornography, child pornography, and cyberterrorism. This law also validates electronic contracts and electronic signatures.

Cybercrime Act of 2001 (Australia)

The Cybercrime Act 2001 (Australia) establishes criminal penalties, including fines and imprisonment, for people who commit computer crimes (including unauthorized access, unauthorized modification, or DoS) with intent to commit a serious offense.

General Data Protection Regulation (GDPR)

Adopted in 2016 and effective in 2018 European General Data Protection Regulation (GDPR) has introduced sweeping changes in privacy and the requirements on organizations that store or process personal information on European residents. GDPR embodies numerous privacy concepts, including transparency, a shift from "opt out" to "opt in," and "the right to be forgotten." Many newer privacy laws in various countries and U.S. states have adopted some of the terms defined in GDPR such as "processor" and "controller."

Payment Card Industry Data Security Standard (PCI DSS)

Although not a legal mandate, the Payment Card Industry Data Security Standard (PCI DSS) is one example of an industry initiative for mandating and enforcing security standards. PCI DSS applies to any business worldwide that transmits, processes, or stores payment card (meaning credit card) transactions to conduct business with customers — whether that business handles thousands of credit card transactions a day or a single transaction a year. Compliance is mandated and enforced by the payment card brands (American Express, MasterCard, Visa, and so on), and each payment card brand manages its own compliance program.

TIP

Although PCI DSS is an industry standard rather than a legal mandate, many states are beginning to introduce legislation that would make PCI compliance (or at least compliance with certain provisions) mandatory for organizations that do business in that state.

PCI DSS requires organizations to submit an annual assessment and network scan or to complete onsite PCI data security assessments and quarterly network scans. The actual requirements depend on the number of payment card transactions handled by an organization and other factors, such as previous data loss incidents.

PCI DSS version 3.2 consists of 6 core principles, supported by 12 requirements, and more than 200 specific procedures for compliance, including the following:

>> **Principle 1:** Build and maintain a secure network:

- *Requirement 1:* Install and maintain a firewall configuration to protect cardholder data.

- *Requirement 2:* Don't use vendor-supplied defaults for system passwords and other security parameters.

- **Principle 2:** Protect cardholder data:
  - *Requirement 3:* Protect stored cardholder data.
  - *Requirement 4:* Encrypt transmission of cardholder data across open, public networks.
- **Principle 3:** Maintain a vulnerability management program:
  - *Requirement 5:* Use and regularly update antivirus software.
  - *Requirement 6:* Develop and maintain secure systems and applications.
- **Principle 4:** Implement strong access control measures:
  - *Requirement 7:* Restrict access to cardholder data by business need to know.
  - *Requirement 8:* Assign a unique ID to each person who has computer access.
  - *Requirement 9:* Restrict physical access to cardholder data.
- **Principle 5:** Regularly monitor and test networks:
  - *Requirement 10:* Track and monitor all access to network resources and cardholder data.
  - *Requirement 11:* Regularly test security systems and processes.
- **Principle 6:** Maintain an information security policy:
  - *Requirement 12:* Maintain a policy that addresses information security.

Penalties for noncompliance are levied by the payment card brands and include not being allowed to process credit card transactions, fines up to $25,000 per month for minor violations, and fines up to $500,000 for violations that result in actual lost or stolen financial data.

Licensing and intellectual property requirements

Given the difficulties in defining and prosecuting computer crimes, many prosecutors seek to convict computer criminals on more traditional criminal statutes, such as theft, fraud, extortion, and embezzlement. Intellectual property rights and privacy laws, in addition to specific computer crime laws, also exist to protect the general public and assist prosecutors.

The CISSP candidate should understand that because of the difficulty of prosecuting computer crimes, prosecutors often use more traditional criminal statutes, intellectual property rights, and privacy laws to convict criminals. In addition, you should realize that specific computer crime laws exist.

Four categories of intellectual property are protected by U.S. law:

>> Patents

>> Trademarks

>> Copyrights

>> Trade secrets

Intellectual property rights worldwide are agreed on, defined, and enforced by various organizations and treaties, including the World Intellectual Property Organization, World Customs Organization, World Trade Organization, United Nations Commission on International Trade Law, European Union, and Trade-Related Aspects of Intellectual Property Rights.

Licensing violations are among the most prevalent examples of intellectual property rights infringement. Other examples include plagiarism, software piracy, and corporate espionage.

Digital rights management attempts to protect intellectual property rights by using access control technologies to prevent unauthorized copying or distribution of protected digital media.

Patents

A *patent*, as defined by the U.S. Patent and Trademark Office (PTO), is "the grant of a property right to the inventor." A patent grant confers upon the owner (either a person or a company) "the right to exclude others from making, using, offering for sale, selling, or importing the invention." To qualify for a patent, an invention must be novel, useful, and not obvious. An invention must also be tangible; an idea cannot be patented. Examples of computer-related objects that may be protected by patents are computer hardware and physical devices in firmware.

The PTO grants a patent for an invention that has been sufficiently documented by the applicant and that the agency has been verified as being original. A U.S. patent is generally valid for 20 years from the date of application and is effective only within the United States, including territories and possessions. Patent applications must be filed with the appropriate patent office in various countries world to receive patent protection in those countries. The owner of the patent may grant a license to others for use of the invention or its design, often for a fee.

U.S. patent (and trademark) laws and rules are covered in 35 U.S.C. and 37 C.F.R., respectively. The Patent Cooperation Treaty, which more than 130 countries have adopted, provides some international protection for patents. Patent infringements are not prosecuted by the PTO. Instead, the holder of a patent must enforce their patent rights through the appropriate legal system.

REMEMBER

Patent grants were previously valid for only 17 years; now newly granted patents are valid for 20 years.

Trademarks

A *trademark,* as defined by the PTO, is "any word, name, symbol, or device, or any combination, used, or intended to be used, in commerce to identify and distinguish the goods of one manufacturer or seller from goods manufactured or sold by others." Computer-related objects that may be protected by trademarks include corporate brands and operating system logos. U.S. Public Law 105–330, the Trademark Law Treaty Implementation Act, provides some international protection for U.S. registered trademarks.

Copyrights

A *copyright* is a form of protection granted to the authors of "original works of authorship," both published and unpublished. A copyright protects a tangible form of expression rather than the idea or subject matter itself. Under the original Copyright Act of 1909, publication was generally the key to obtaining a federal copyright. The Copyright Act of 1976 changed this requirement, however; now copyright protection applies to any original work of authorship immediately, from the time that it's created in tangible form. Object code or documentation are examples of computer-related objects that may be protected by copyrights.

Copyrights can be registered through the Copyright Office of the Library of Congress, but a work doesn't need to be registered to be protected by copyright. Copyright protection generally lasts for the lifetime of the author plus 70 years.

Trade secrets

A *trade secret* is proprietary or business-related information that a company or person uses and has exclusive rights to. To be considered a trade secret, the information must meet the following requirements:

>> **Is genuine and not obvious:** Any unique method of accomplishing a task would constitute a trade secret, especially if it is backed up by copyrighted, patented, or proprietary software or methods that give that organization a competitive advantage.

>> **Gives the owner a competitive or economic advantage and therefore has value to the owner:** Google's search algorithms, for example — the "secret sauce" that makes it popular with users (and advertisers) — aren't universally known. Some secrets are protected.

>> **Is reasonably protected from disclosure:** The information doesn't have to be kept absolutely and exclusively secret, but the owner must exercise due care in its protection.

Software source code and firmware code are examples of computer-related objects that an organization may protect as trade secrets.

Import/export controls

International import and export controls exist among countries to protect both intellectual property rights and certain sensitive technologies (such as encryption).

Information security professionals need to be aware of relevant import/export controls for any countries in which their organization operates or to which their employees travel. It is not uncommon for laptops to be searched, and possibly confiscated, at airports to enforce various import/export controls, for example.

In the United States, the International Traffic in Arms Regulations (ITAR) law restricts the export of military and related technologies (including encryption algorithms) to other countries.

Transborder data flow

Related to import/export controls is the issue of transborder data flow. As discussed earlier in this chapter, data privacy and breach disclosure laws vary greatly across regions, countries, and U.S. states. Australia and EU countries, for example, have far more stringent data privacy regulations than the United States does. Many countries restrict or prevent export of their citizens' personal data.

Issues of transborder data flow and data residency (where data is physically stored) are particularly germane for organizations operating in the public cloud. For these organizations, it is important to know — and have control of — where their data is stored. Issues of data residency and transborder data flow should be addressed in any agreements or contracts with cloud service providers.

The trend in outsourcing operations to service providers — including Software as a Service (SaaS), Platform as a Service (PaaS), and Infrastructure as a Service (IaaS) — have clouded the issue of transborder data flow. Security and privacy professionals need to understand where their data physically resides to ensure compliance with privacy laws.

Privacy

The concept of *privacy* is closely related to confidentiality (discussed earlier in this chapter) but focuses more specifically on preventing the unauthorized use or disclosure of personal data.

Personal data, commonly referred to as personally identifiable information (PII), may include

>> Name

>> Addresses

>> Contact information

>> Social Security number

>> IP address

>> Mobile device serial number

>> Financial account number

>> Birthdate and birthplace

>> Race

>> Marital status

>> Sexual orientation or gender identity

>> Credit history and other financial information

>> Criminal records

>> Education

>> Employment records and history

>> Health records and medical data, known as protected health information (PHI) or electronic protected health information (ePHI)

>> Religious preference

>> Political affiliation

>> Other unique personal characteristics or traits

Every organization that collects any personal data about anyone (including employees, customers, and patients) must have a well-defined, published, and distributed privacy policy that explains why the data is being collected, how it is being used, how it will be protected, and what each person's rights are regarding the personal data that is being collected.

Finally, certain employee privacy issues often arise within an organization regarding employee rights concerning monitoring, search, drug testing, and other policies.

Monitoring commonly occurs in many forms within an organization, including Internet, email, and general computer use, as well as through surveillance cameras, access badges or keys, and time clocks, among other devices. Mandatory and random drug testing and searches of desks, lockers, work areas, and even personally owned vehicles are other common policies that can evoke employee privacy concerns.

To reduce or eliminate employee privacy concerns, organizational policies should clearly define (and require written acknowledgment of) acceptable use policies for computer, Internet, and email use. Additional policies should explain monitoring purposes, acceptable use or behavior, and potential disciplinary actions resulting from violations. Finally, organizational policies should clearly state that the employee has no expectation of privacy concerning the organization's monitoring and search policies.

Privacy in the context of electronic information about citizens is not well understood. Simply put, privacy has two main components:

>> **Data protection:** Here, we mean the usual data security measures discussed in most of this book.

>> **Appropriate handling and use:** This term refers to the ways in which information owners choose to process and distribute personal data.

Several important pieces of privacy and data protection legislation include the Federal Privacy Act, the Health Insurance Portability and Accountability Act (HIPAA), the Health Information Technology for Economic and Clinical Health Act (HITECH), and the Gramm-Leach-Bliley Act (GLBA) in the United States, and the Data Protection Act (DPA) in the United Kingdom. Finally, the PCI DSS is an example of an industry policing itself without the need for government laws or regulations.

Several privacy related laws that CISSP candidates should be familiar with include

>> U.S. Federal Privacy Act of 1974

>> U.S. Health Insurance Portability and Accountability Act (HIPAA) of 1996

>> U.S. Children's Online Privacy Protection Act (COPPA) of 1998

- » U.S. Gramm-Leach-Bliley Financial Services Modernization Act (GLBA) of 1999
- » U.S. Health Information Technology for Economic and Clinical Health Act (HITECH) of 2009
- » California Consumer Privacy Act (CCPA) of 2018
- » California Privacy Rights Act (CPRA) of 2020
- » UK Data Protection Act of 1998
- » European Union General Data Protection Regulation (GDPR)

U.S. Federal Privacy Act of 1974, 5 U.S.C. § 552A

The U.S. Federal Privacy Act of 1974 protects records and information maintained by government agencies about citizens and lawful permanent residents. Except under certain specific conditions, no agency may disclose any record about a person "except pursuant to a written request by, or with the prior written consent of, the individual to whom the record pertains." The act also has provisions for access and amendment of records by the person to whom they belong, except in cases of "information compiled in reasonable anticipation of a civil action or proceeding." The act provides individual penalties for violations, including a misdemeanor charge and fines up to $5,000.

WARNING

Although the Federal Privacy Act of 1974 predates the Internet as we know it today, don't dismiss its relevance. The provisions of the act are as important as ever and remain in full force and effect today.

U.S. Health Insurance Portability and Accountability Act (HIPAA) of 1996, PL 104–191

The U.S. Health Insurance Portability and Accountability Act (HIPAA) was signed into law effective August 1996. The HIPAA legislation provided Congress three years from that date to pass comprehensive health privacy legislation. When Congress failed to pass legislation by the deadline, the U.S. Department of Health and Human Services (HHS) received the authority to develop the privacy and security regulations for HIPAA. In October 1999, HHS released proposed HIPAA privacy regulations titled "Privacy Standards for Individually Identifiable Health Information," which took effect in April 2003. HIPAA security standards were published in February 2003 and took effect in April 2003.

Organizations that must comply with HIPAA regulations are referred to as *covered entities* and include

- » **Payer (or health plan):** A person or group health plan that provides — or pays the cost of — medical care, such as an insurer

- » **Health-care clearinghouse:** A public or private entity that processes or facilitates the processing of nonstandard data elements of health information into standard data elements, such as data warehouses

- » **Health-care provider:** A provider of medical or other health services, such as hospitals, health maintenance organizations, doctors, specialists, dentists, and counselors

Civil penalties for HIPAA violations include fines of $100 per incident and up to $25,000 per provision per calendar year. Criminal penalties include fines up to $250,000 and potential imprisonment of corporate officers for up to ten years. Additional state penalties may also apply.

In 2009, Congress passed additional HIPAA provisions as part of the American Recovery and Reinvestment Act of 2009, requiring covered entities to publicly disclose security breaches involving personal information.

U.S. Children's Online Privacy Protection Act (COPPA) of 1998

This law provides for protection of online information about children under the age of 13. The law defines rules for the collection of information from children and means for obtaining consent from parents. Organizations are also restricted from marketing to children under the age of 13.

U.S. Gramm-Leach-Bliley Financial Services Modernization Act (GLBA) of 1999, PL 106-102

Gramm-Leach-Bliley (known as GLBA) opened competition among banks, insurance companies, and securities companies. GLBA also requires financial institutions to better protect their customers' PII with three rules:

- » **Financial Privacy Rule:** Requires each financial institution to provide information to each customer regarding the protection of their private information

- » **Safeguards Rule:** Requires each financial institution to develop a formal written security plan that describes how the institution will protect its customers' PII

- » **Pretexting Protection:** Requires each financial institution to take precautions to prevent attempts by social engineers to acquire private information about institutions' customers

Civil penalties for GLBA violations are up to $100,000 for each violation. Furthermore, officers and directors of financial institutions are personally liable for civil penalties of not more than $10,000 for each violation.

U.S. Health Information Technology for Economic and Clinical Health Act (HITECH) of 2009

The U.S. Health Information Technology for Economic and Clinical Health (HITECH) Act, passed as part of the American Recovery and Reinvestment Act of 2009, broadens the scope of HIPAA compliance to include the business associates of HIPAA-covered entities, including third-party administrators; pharmacy benefit managers for health plans; claims processing, billing, and/or transcription companies; and people who perform legal, accounting, and administrative work.

Another highly important provision of the act promotes and, in many cases, funds the adoption of electronic health records to increase the effectiveness of individual medical treatment, improve efficiency in the U.S. health-care system, and reduce the overall cost of health care. Anticipating that the widespread adoption of electronic health records will increase privacy and security risks, the act introduces new security and privacy-related requirements.

In the event of a breach of "unsecured protected health information," the act requires covered entities to notify the affected people and the secretary of HHS. The regulation defines unsecured protected health information as PHI that is not secured through the use of a technology or methodology to render it unusable, unreadable, or indecipherable by unauthorized people.

The notification requirements vary according to the amount of data breached:

>> A data breach affecting more than 500 people must be reported immediately to the HHS, major media outlets, and people affected by the breach, and information must be posted on the official HHS website.

>> A data breach affecting fewer than 500 people must be reported to the people affected by the breach and to the HHS secretary.

Finally, the act requires the issuance of technical guidance on the technologies and methodologies "that render protected health information unusable, unreadable, or indecipherable to unauthorized individuals." The guidance specifies data destruction and encryption as actions that render PHI unusable if it is lost or stolen. PHI that is encrypted and whose encryption keys are properly secured provides a "safe harbor" to covered entities and does not require them to issue data-breach notifications.

California Consumer Privacy Act of 2018 (CCPA)

The California Consumer Privacy Act (CCPA) of 2018 gives California consumers greater control of the personal information that businesses collect about them, including the right to know, the right to delete, the right to opt out, and the right to nondiscrimination. The law is applicable to any company that does business with a California resident, regardless of where the company itself is located.

California Privacy Rights Act of 2020 (CPRA)

Passed by ballot initiative in 2020, the California Privacy Rights Act (CPRA) amends the CCPA by providing additional privacy rights. The act also creates the California Privacy Protection Agency as an enforcement arm of CCPA and CPRA.

UK Data Protection Act of 1998

Passed by Parliament in 1998, the UK Data Protection Act applies to any organization that handles sensitive personal data about living people. Such data includes

>> Names

>> Birth and anniversary dates

>> Addresses, phone numbers, and email addresses

>> Racial or ethnic origins

>> Political opinions and religious (or similar) beliefs

>> Trade or labor union membership

>> Physical or mental condition

>> Sexual orientation or gender identity

>> Criminal or civil records or allegations

The act applies to electronically stored information, but certain paper records used for commercial purposes may also be covered. The act consists of eight privacy and disclosure principles:

>> "Personal data shall be processed fairly and lawfully and [shall not be processed unless certain other conditions (set forth in the Act) are met]."

>> "Personal data shall be obtained only for one or more specified and lawful purposes, and shall not be further processed in any manner incompatible with that purpose or those purposes."

>> "Personal data shall be adequate, relevant, and not excessive in relation to the purpose or purposes for which they are processed."

>> "Personal data shall be accurate and, where necessary, kept up-to-date."

>> "Personal data processed for any purpose or purposes shall not be kept for longer than is necessary for that purpose or those purposes."

>> "Personal data shall be processed in accordance with the rights of data subjects under this Act."

>> "Appropriate technical and organizational measures shall be taken against unauthorized or unlawful processing of personal data and against accidental loss or destruction of, or damage to, personal data."

>> "Personal data shall not be transferred to a country or territory outside the European Economic Area unless that country or territory ensures an adequate level of protection for the rights and freedoms of data subjects in relation to the processing of personal data."

Compliance is enforced by the Information Commissioner's Office, an independent official body. Penalties generally include fines, which may also be imposed on the officers of a company.

European Union General Data Protection Regulation (GDPR)

The European Union General Data Protection Regulation, known as the GDPR, represents a significant revision of the 1995 privacy directive. Highlights of GDPR include the following:

>> The law requires the enactment of a formal, documented data privacy program, which must direct all relevant business activities to be designed with privacy by default and privacy by design.

>> The law requires organizations that collect PII from any European resident to obtain explicit consent for the collection and use of such information. Organizations' collection of PII must be opt-in as opposed to opt-out. In other words, users must choose to opt in to data collection and use.

>> Data subjects must be able to review information about themselves, request that corrections be made, and request that their data be expunged.

>> The law defines a *data controller* (an organization that stores and processes PII) and a *data processor* (an organization that stores and processes PII as directed by a data controller.

>> The law requires the appointment of a data protection officer, who oversees the creation and operation of an organization's data privacy program.

>> The law requires all affected data subjects to be notified within 72 hours of a data breach.

>> The law permits European authorities to levy fines on organizations that violate its terms, with those fines being as high as €20 million or 4 percent of the organization's annual revenue.

Understand Requirements for Investigation Types

The purpose of an investigation is to determine what happened and who is responsible, and to collect evidence that supports this hypothesis. Closely related to, but distinctly different from, investigations is incident management (discussed in detail later in this chapter). Incident management determines what happened, contains and assesses damage, and restores normal operations.

CROSS REFERENCE

This section covers Objective 1.6 of the Security and Risk Management domain in the CISSP Exam Outline (May 1, 2021).

Investigations and incident management must often be conducted simultaneously in a well-coordinated and controlled manner to ensure that the initial actions of either activity don't destroy evidence or cause further damage to the organization's assets. For this reason, it's important that computer incident (or emergency) response teams (CIRT or CERT) or computer security incident response teams (CSIRT) be properly trained and qualified to secure a computer-related crime scene or incident while preserving evidence. Ideally, the CIRT includes the people who will be conducting the investigation.

An example is a police officer who discovers a murder victim. It's important for the officer to assesses the safety of the situation quickly and secure the crime scene, but at the same time, they must be careful not to disturb or destroy any evidence. The homicide detective's job is to gather and analyze the evidence. Ideally, but rarely, the detective is the person who discovers the murder victim, allowing them to assess the safety of the situation, secure the crime scene, and begin collecting evidence. Think of yourself as a *CSI-SSP*!

Requirements for various investigation types include

>> **Operational:** After any damage from a security incident has been contained, operational investigations typically focus on root-cause analysis, lessons learned, and management reporting.

>> **Criminal:** Criminal investigations require strict adherence to proper evidence collection and handling procedures. The investigation focuses on discovering and preserving evidence for possible prosecution of any culpable parties.

>> **Civil:** A civil investigation may result from a data breach or regulatory violation. It typically focuses on quantifying any damage and establishing due diligence or negligence.

>> **Regulatory:** Regulatory investigations often take the form of external, mandatory audits and focus on evaluating security controls and compliance.

Various industry standards and guidelines provide guidance for conducting investigations. These include the American Bar Association's *Best Practices in Internal Investigations*, various best-practices guidelines and tool kits published by the U.S. Department of Justice, and ASTM International's *Standard Practice for Computer Forensics (ASTM E2763)*.

Develop, Document, and Implement Security Policies, Standards, Procedures, and Guidelines

Policies, standards, procedures, and guidelines are different, but they interact in a variety of ways. It's important to understand these differences and relationships, and to recognize the different types of policies and their applications. To develop and implement information security policies, standards, guidelines, and procedures successfully, you must ensure that your efforts are consistent with the organization's mission, goals, and objectives (discussed earlier in this chapter).

CROSS REFERENCE

This section covers Objective 1.7 of the Security and Risk Management domain in the CISSP Exam Outline (May 1, 2021).

Policies, standards, procedures, and guidelines work together as the blueprints for a successful information security program. They do all the following things:

>> Establish governance

>> Provide valuable guidance and decision support

>> Help establish legal authority

>> Ensure that risks are kept to acceptable levels

Too often, technical security solutions are implemented without these important blueprints. The results are often expensive and ineffective controls that aren't uniformly applied and don't support an overall security strategy.

Governance collectively represents the system of policies, standards, guidelines, and procedures — together with management oversight — that help steer an organization's day-to-day operations and decisions.

Policies

A *security policy* forms the basis of an organization's information security program. RFC 2196, *The Site Security Handbook*, defines a security policy as "a formal statement of rules by which people who are given access to an organization's technology and information assets must abide."

The four main types of policies are

>> **Senior management:** A high-level management statement of an organization's security objectives, organizational and individual responsibilities, ethics and beliefs, and general requirements and controls.

>> **Regulatory:** Highly detailed and concise policies usually mandated by federal, state, industry, or other legal requirements.

>> **Advisory:** Not mandatory, but highly recommended, often with specific penalties or consequences for failure to comply. Most policies fall into this category.

>> **Informative:** Informs only, with no explicit requirements for compliance.

REMEMBER

Standards, procedures, and guidelines are supporting elements of a policy and provide specific implementation details of the policy.

TIP

ISO/IEC 27002, *Information Technology — Security Techniques — Code of Practice for Information Security Management,* is an international standard for information security policy. ISO/IEC is the International Organization for Standardization and International Electrotechnical Commission. ISO/IEC 27002 consists of 12 sections that largely (but not completely) overlap the eight (ISC)² security domains.

Standards (and baselines)

Standards are specific, mandatory requirements that further define and support higher-level policies. A standard might require the use of a specific technology, such as a minimum requirement for encryption of sensitive data using the Advanced Encryption Standard (AES). A standard may go so far as to specify the exact brand, product, or protocol to be implemented. A device or system *hardening standard* would define specific security configuration settings for applicable systems.

Baselines are similar to and related to standards. A baseline can be useful for identifying a consistent basis for an organization's security architecture, taking into account system-specific parameters, such as operating systems. After consistent baselines are established, appropriate standards can be defined across the organization.

Some organizations call their configuration documents *standards* (and still others call them *standard operating environments*) instead of *baselines*. This practice is common and acceptable.

Procedures

Procedures provide detailed instructions on implementing specific policies and meeting the criteria defined in standards. Procedures may include standard operating procedures, run books, and user guides. A procedure might be a step-by-step guide for encrypting sensitive files by using a specific software encryption product.

Guidelines

Guidelines are similar to standards but function as recommendations rather than as requirements. A guideline might provide tips or recommendations for determining the sensitivity of a file and whether encryption is required.

In a discussion of policies, standards, guidelines, and procedures, we cannot forget *requirements*: statements of required characteristics of a business process or information system. Requirements are typically derived from policies, standards, and controls.

Identify, Analyze, and Prioritize Business Continuity (BC) Requirements

Business continuity and disaster recovery (discussed in detail in Chapter 9) work hand in hand to provide an organization the means to continue and recover business operations when a disaster strikes. Business continuity and disaster recovery are two sides of the same coin. Each process springs into action when a disaster strikes. But both have different goals:

>> **Business continuity** deals with keeping business operations running — perhaps in another location or by using different tools and processes — after a disaster has struck. Business continuity is sometimes called *continuity of operations* (see the nearby sidebar "Cooperation is the key").

>> **Disaster recovery** deals with restoring normal business operations after the disaster takes place.

CROSS REFERENCE

This section covers Objective 1.8 of the Security and Risk Management domain in the CISSP Exam Outline (May 1, 2021).

Although the business continuity team is busy keeping business operations running via one of possibly several contingency plans, the disaster recovery team members are busy restoring the original facilities and equipment so that they can resume normal operations.

Here's an analogy. Two boys kick a big anthill — a disaster for the ant colony. Some of the ants scramble to save the eggs and the food supply; that's Ant City business continuity. Other ants work on rebuilding the anthill; that's Ant City disaster recovery. Both teams work to ensure the anthill's survival, but each team has its own role to play.

Business continuity and disaster recovery planning have these common elements:

>> **Identification of critical business functions:** The business impact analysis and risk assessment (discussed in "Business impact analysis" later in this chapter) identify these functions.

>> **Identification of possible scenarios:** The planning team identifies all the likely human-made and natural-disaster activation scenarios, ranked by event probability and impact on the organization.

>> **Experts:** *Experts* in this context are people who understand the organization's critical business processes.

The similarities end with this list. Business continuity planning concentrates on *continuing* business operations, whereas disaster recovery planning focuses on *recovering* the original business functions. Although both plans deal with the long-term survival of the business, they involve different activities. When a significant disaster occurs, both activities kick into gear at the same time, keeping vital business functions running (business continuity) and getting things back to normal as soon as possible (disaster recovery).

BUSINESS CONTINUITY AND DISASTER RECOVERY: A SIMPLE ILLUSTRATION

Here's a scenario: A business is a delivery service that has one delivery truck, which delivers goods around the city.

Business continuity deals with keeping the delivery service running in case something happens to the truck, presumably with a backup truck, substitute drivers, maps that show ways to get around traffic jams, and other contingencies to keep the delivery function running.

Disaster recovery, on the other hand, deals with fixing (or replacing) the original delivery truck, which might involve making repairs or even buying or leasing a new truck if the original truck is damaged beyond repair.

Business continuity (and disaster recovery) planning exist because bad things happen. Organizations that want to survive a disastrous event need to make formal and extensive plans — contingency plans to keep the business running and recovery plans to return operations to normal.

Keeping a business operating during a disaster can be like juggling with one arm tied behind your back. (We first thought of using plate-spinning and one-armed paper hangers as similes, but most of our readers are probably too young to understand these concepts.) You'd better plan how you're going to do it, and practice! A disaster could happen at night, you know (and one-handed juggling in the dark is a lot harder).

COOPERATION IS THE KEY

Like many other disciplines based in technology, business continuity and disaster recovery planning are changing rapidly. One new approach is continuity of operations (COOP), which blends business continuity and disaster recovery into a single mission: keeping the organization running after a disaster.

If you're interested in learning more, an excellent reference is U.S. Federal Emergency Management Agency Guide IS-1300, *Introduction to Continuity of Operations,* which is available at https://training.fema.gov/is/courseoverview.aspx?code=IS-547.a.

Before business continuity planning can begin, everyone on the project team has to make and understand some basic definitions and assumptions. These critical items include

» **Senior management support:** Developing a business continuity plan is time-consuming, with no immediate or tangible return on investment. To ensure a successful business continuity planning project, you need the support of the organization's senior management, with adequate budget, manpower, and visible statements backing the project. Senior management needs to make explicit statements identifying the responsible parties, as well as the importance of the business continuity planning project, budget, priorities, urgency, and timing.

» **Senior management involvement:** Senior management can't just bless the project. Because senior managers and directors may have implicit and explicit responsibility for the organization's ability to recover from a disaster, senior management needs to have a degree of direct involvement in the business continuity planning effort. The careers that these people save may be their own.

» **Project team membership:** Which people do you want to put on the business continuity planning project team? The team must represent all relevant functions and business units. Many of the team members probably have their usual jobs, too, so the team needs to develop a realistic timeline for how quickly the business continuity planning project can make progress.

» **Who brings the doughnuts:** Because it's critical that business continuity planning meetings be well attended, quality doughnuts or other tasty fare are essential.

Business impact analysis

A *business impact analysis* (BIA) describes the impact that a disaster is expected to have on business operations. This important early step in business continuity planning helps an organization figure out which business processes are more resilient and which are more fragile.

A disaster's impact includes quantitative and qualitative effects. The *quantitative impact* is generally financial, such as loss of revenue or output of production. The *qualitative impact* has more to do with the quality of goods and/or services.

Any BIA worth its salt needs to perform the following tasks well:

» **Perform a vulnerability assessment.** This assessment is not so much an application/infrastructure vulnerability assessment, but a big-picture, business process vulnerability assessment.

>> **Carry out a criticality assessment.** This assessment determines how critically important a particular business function is to the ongoing viability of the organization.

>> **Determine the maximum tolerable downtime.** This is a measure of the longest business interruption that the organization can withstand before its survival is at risk.

>> **Establish recovery targets.** These measures, including recovery time objective and recovery point objective, are basic parameters that contribute to the development of business continuity and disaster recovery plans.

>> **Determine resource requirements.** This describes the personnel and funding required to develop business continuity and disaster recovery plans, and to sustain them through training, testing, and periodic updates to keep them relevant.

You can get the scoop on these activities in the following sections.

Assessing vulnerability

Often, a BIA includes a vulnerability assessment that gets a handle on obvious and not-so-obvious weaknesses in business critical systems. A vulnerability assessment has quantitative (financial) and qualitative (operational) sections, similar to those in a risk assessment, which is covered later in this chapter.

The purpose of a vulnerability assessment is to determine the impact — both quantitative and qualitative — of the loss of a critical business function.

Quantitative losses include

>> Revenue

>> Operating capital

>> Market share

>> Personal liabilities

>> Increased expenses

>> Penalties due to violations of business contracts, laws, and regulations (which can result in fines and civil penalties)

Qualitative losses include

>> Service quality

>> Competitive advantages

>> Customer satisfaction

>> Trust

>> Prestige and reputation

The vulnerability assessment identifies critical support areas, which are business functions that, if lost, would cause significant harm to the business by jeopardizing critical business processes or the lives and safety of personnel. The vulnerability assessment should carefully study critical support areas to identify the resources that those areas require to continue functioning.

Quantitative losses include an increase in operating expenses because of any higher costs associated with executing the contingency plan. In other words, planners need to remember to consider operating costs that may be higher during a disaster situation.

Assessing criticality

The business continuity planning team should conduct a criticality assessment (CA), an inventory all high-level business functions (such as customer support, order processing, returns, cash management, accounts receivable, and payroll), and rank them in order of criticality. The team should also describe the impact of a disruption to each function on overall business operations.

The team members need to estimate the duration of a disaster event to prepare an effective criticality assessment. Project team members need to consider the impact of a disruption based on the length of time that a disaster impairs specific critical business functions. You can see the vast difference in business impact of a disruption that lasts one minute, compared with one hour, one day, one week, or longer. Generally, the criticality of a business function depends on the degree of impact that its impairment has on the business.

REMEMBER

Planners need to consider disasters that occur at different times in the business cycle, whatever that might be for an organization. Response to a disaster at the busiest time of the month (or year) may vary quite a bit from response at other times.

Identifying key players

Although you can consider a variety of angles when evaluating vulnerability and criticality, commonly, you start with a high-level organization chart. (Hip people call it an *org chart.*) In most companies, the major functions pretty much follow the structure of the organization.

REMEMBERING PAYROLL

Organizations that inventory and categorize their business processes usually look outward to the goods and services that they provide their customers. During a disaster-related crisis, organizations that survive have effective contingency plans for these processes.

But some organizations overlook internal services that support ongoing operations. An important example is payroll. Some disasters can last weeks or even months while organizations rebuild their goods and services delivery. If you don't have payroll high on the list of processes to recover, employees could find themselves going without a paycheck for quite a while. An organization in this position may find itself losing the people it needs to get normal operations running again, which could precipitate a secondary disaster that has long-term consequences.

A retail organization that we're familiar with has an interesting contingency plan for paying its branch-office employees. Branch managers are authorized to pay their employees a fixed amount of cash each week if the organization's payroll system stops functioning. When automated payroll systems are restored, the cash payments are entered into the system so that payroll records for each employee are accurate.

Following an org chart helps the business continuity planning project team consider all the steps in a critical process. Walk through the org chart, stopping at each manager's or director's position, and asking "What do they do?" This mental stroll can jog your memory and help you see all the parts of the organization's big picture.

TIP

When you're cruising an org chart to make sure that it covers all areas of the organization, you may easily overlook outsourced functions that may not show up in the org chart. If your organization outsources accounts payable (A/P) functions, for example, you might miss this detail if you don't see it on an org chart. Okay, you'd probably notice the absence of *all* A/P. But if your organization outsources only part of A/P — say, a group that detects and investigates A/P fraud, looking for payment patterns that suggest the presence of phony payment requests, your org chart probably doesn't include that vital function.

Establishing maximum tolerable downtime

An extension of the criticality assessment is a statement of maximum tolerable downtime (MTD) or maximum tolerable period of disruption (MTPD) for each critical business function. *MTD* is the maximum period that a critical business function can be inoperative before the company incurs significant and long-lasting damage.

Suppose that your favorite online merchant — a bookseller, an auction house, or an online trading company — goes down for an hour, a day, or a week. At some point, you have to figure that a prolonged disruption will sink the ship, meaning that the business can't survive. Determining MTD involves figuring out at what point the organization will suffer permanent, measurable loss as a result of a disaster. Online retailers know that even short outages may mean some customers will switch brands and take their business elsewhere.

Make the MTD assessment a major factor in determining the criticality — and priority — of business functions. A function that can withstand only two hours of downtime obviously has higher priority than a function that can withstand several days of downtime.

MTD is a measure of the longest period of time that a critical business function can be disrupted without suffering unacceptable consequences, perhaps threatening the survival of the organization.

Determining maximum tolerable outage

During the criticality assessment, you establish a statement of maximum tolerable outage (MTO) for each critical business function. *MTO* is the maximum period of time that a critical business function can be operating in emergency or alternative processing mode. This statement matters because in many cases, emergency or alternative processing mode performs at a lower level of throughput or quality or at a higher cost. Although an organization's survival can be ensured for an interim period in alternative processing mode, the long-term business model may not be able to sustain the differences in throughput, quality, cost, or whatever aspects of alternative processing mode are different from normal processing.

Establishing recovery targets

When you establish the criticality assessment, MTD, and MTO for each business process (which we talk about in the preceding sections), the planning team can establish recovery targets. These targets represent the period from the onset of a disaster to the time when critical processes resume functioning.

Two primary recovery targets are usually established for each business process: a recovery time objective and a recovery point objective. We discuss these targets in the following sections.

HOW BAD DOES IT HAVE TO BE?

Establishing reasonable MTD and MTO values can be difficult. The issue is similar to pain threshold and the actual effects of a disaster. Years ago, we used to say that an MTD value was valid when its magnitude was sufficient to cause the complete failure of a business. Now we believe that this threshold is too high. After all, some organizations won't fail even in a huge disaster. Local governments and religious institutions, for example, won't go out of business and disappear from the landscape.

So what's a reasonable measure of MTD? The answer depends on your particular organization and situation, but here are some ideas:

- Threshold of public outcry
- Loss of a certain number of market-share points
- Loss of a certain percentage of constituents
- Loss of life

You need to identify a reasonable threshold of MTD — short of your organization's ceasing to exist, but something more reasonable, such as a significant loss of business or loss of confidence in your organization.

Similarly, determining a reasonable MTO is far from simple. A disaster may result in an organization's relying on emergency operations for an extended period, which may put financial strain on the organization. Higher costs or lower quality will take a toll, and organizations and their investors have to decide how much is enough before they throw in the towel.

RECOVERY TIME OBJECTIVE

A *recovery time objective* (RTO) is the maximum period in which a business process must be restored after a disaster. An organization without a business continuity plan that suffers a serious disaster, such as an earthquake or hurricane, could experience recovery time of one to two weeks or more. An organization could need this length of time to select a new location for processing data, purchase new systems, load application software and data, and resume processing. An organization that can't tolerate such a long outage needs to establish a shorter RTO and determine the level of investments required to meet that target.

RECOVERY POINT OBJECTIVE

A *recovery point objective* (RPO) is the maximum period in which data might be lost if a disaster strikes. A typical schedule for backing up data is once per day. If a disaster occurs before backups are done, the organization can lose an entire day's worth of information because system and data recovery are often performed with the last good set of backups. An organization that requires a shorter RPO needs to figure out a way to make copies of transaction data more frequently than once per day.

Here are some examples of how organizations might establish RPOs:

>> **Keyed invoices:** An accounts payable department opens the mail and manually keys in the invoices that it receives from its suppliers. Data entry clerks spend their entire day inputting invoices. If a disaster occurs before backups are run at the end of the business day (and if that disaster requires the organization to rebuild systems from backup tapes), those clerks have to redo that whole day's worth of data entry.

>> **Online orders:** A small business develops an online web application that customers can use to place orders. At the end of each day, the orders department runs a program that prints out all the day's orders, and the shipping department fills those orders on the following day. If a disaster occurs at any time during the day, the business loses all online orders placed since the previous day's backup.

If you establish the MTD for processes such as the ones in the preceding list as less than one business day, the organization needs to take some steps to save online data more than once per day.

Many organizations consider offsite backup media storage, in which backup tapes are transported offsite as frequently as every day, or electronic vaulting to an off-site location several times each day. An event such as a fire can destroy computers as well as nearby backup media.

HOW RTO AND RPO WORK TOGETHER

RPO and RTO targets are different measures of recovery for a system, but they work together. When the team establishes proposed targets, the team members need to understand how each target works.

At first glance, you might think that RPO should be shorter than RTO, or the other way around. Different businesses and applications present different business requirements that might make RPO less than RTO, equal to RTO, or greater than RTO. Here are some examples:

>> **RPO greater than RTO:** A business can recover an application in 4 hours (RTO), and it has a maximum data loss (RPO) of 24 hours. So if a disaster occurs, the business can get the application running again in 4 hours, but data recovered in the system consists of data entered more than 24 hours before the incident took place.

THE HIGH COST OF RAPID RECOVERY

Business continuity planning teams often establish ambitious RPOs and RTOs for systems. Teams working on recovery objectives need to understand that the speed of recovery is directly proportional to its cost.

Suppose that an RPO for an application is established at two hours. To meet that goal, the organization has to purchase new storage systems, plus an expensive data connection from the main processing center to the backup processing center. But the cost of so short an RPO may not be warranted. The project team needs to understand the cost of downtime (in dollars per hour or per day) versus the cost of recovery. If the cost of downtime for an application is $40,000 per hour, and a two-hour RPO requires a $500,000 investment in equipment and a $20,000-per-month expense, the investment may be warranted. If, however, the cost of downtime for the application is $500 per hour, the organization may not need this level of investment and should establish a longer RPO.

>> **RPO equal to RTO:** A business can recover an application in 12 hours (RTO), with a maximum data loss of 12 hours (RPO). You can imagine this scenario: An application mirrors (or replicates) data to a backup system in real time. If a disaster occurs, the disaster recovery team requires 12 hours to start the backup system. After the team gets the system running, the business has data from until 12 hours in the past — the time when the primary system failed.

>> **RPO less than RTO:** The disaster recovery team can recover an application in 4 hours (RTO), with a maximum data loss of 1 hour (RPO). How can this situation happen? Maybe a back-office transaction-posting application, which receives and processes data from order-processing applications, fails. If the back-office application is down for 4 hours, data coming from the order-processing applications may be buffered someplace else, and when the back-office application resumes processing, it can receive and process the waiting input data.

Defining resource requirements

The *resource requirements* portion of the BIA is a list of the resources that an organization needs to continue operating each critical business function. In an organization that has finite resources (which is pretty much every organization), the most critical functions get first pick, and the lower-priority functions get the leftovers.

Understanding what resources are required to support a business process helps the project team figure out what the contingency plan for that process needs to contain and how the process can be operated in emergency mode and then recovered.

Examples of required resources include

>> **Systems and applications:** For a business process to continue operating, it may require one or more IT systems or applications — not only the primary supporting application, but also other systems and applications that the primary application requires to continue functioning. Depending on the nature of the organization's primary and alternative processing resources, these systems may be physical, virtual, or cloud-based.

>> **Suppliers and partners:** Many business processes require a supply of materials or services from outside organizations, without which the business process can't continue operating.

>> **Key personnel:** Most business processes require a number of specifically trained or equipped staff members — or contingency workers such as contractors or personnel from another company — to run business processes and operate systems.

>> **Business equipment:** This category includes anything from PBXs to copiers, postage machines, point-of-sale machines, red staplers, and any other machinery required to support critical business processes.

TIP

When you identify required resources for complex business processes, you may want to identify additional information about each resource, including resource owners, criticality, and dependencies.

Develop and document the scope and the plan

The success and effectiveness of a business continuity planning project depends greatly on whether senior management and the project team define its scope properly. Business processes and technology can muddy the waters and make this task difficult. Distributed systems dependence on at least some desktop systems for vital business functions, for example, expands the scope beyond core functions. Geographically dispersed companies — often the result of mergers — complicate matters as well.

Also, large companies are understandably more complex. The boundaries between where a function begins and ends are often fuzzy, and sometimes poorly documented and not well understood.

Political pressures can influence the scope of the business continuity planning project as well. A department that thinks it's vital but that falls outside the scope of the business continuity planning project may lobby to be included. Everybody wants to be important (and some people just want to *appear* to be important). You need senior management support of scope — what the project team really needs to include and what it doesn't — to put a stop to political games.

Scope creep (what happens when a project's scope grows beyond the original intent) can become *scope leap* if you have a weak or inexperienced business continuity planning project team. For the success of the project, strong leaders must make rational decisions about the scope of the project. You can always change the scope of the business continuity planning project in later iterations of the project.

The project team needs to find a balance between a too-narrow scope, which makes the plan ineffective, and a too-wide scope, which makes the plan too cumbersome.

A complete plan consists of components that handle not only the continuation of critical business functions, but also all the functions and resources that support those critical functions. The various elements of a business continuity plan are described in the following sections.

Emergency response

Emergency response teams must be identified for every possible type of disaster. These response teams need playbooks (detailed written procedures and checklists) to keep critical business functions operating.

Written procedures are vital for two reasons:

» The people who perform critical functions after a disaster may not be familiar with them; they may not usually perform those functions. During a disaster, the people who ordinarily perform the function may be unavailable.

» The team probably needs to use different procedures and processes for performing the critical functions during a disaster than they would under normal conditions. Also, the circumstances surrounding a disaster might have people feeling out of sorts; having a written procedure guides them into action (like the "break glass" instructions on some fire alarms in case you forget what to do).

Damage assessment

When a disaster strikes, experts need to be called in to inspect the premises and determine the extent of the damage. Typically, you need experts who can assess building damage, as well as damage to any special equipment and machinery.

Depending on the nature of the disaster, you may have to perform damage assessment in stages. A first assessment may involve a quick walk-through to look for obvious damage, followed by a more time-consuming and detailed assessment to look for problems that you don't see right away.

Damage assessments determine whether an organization can still use buildings and equipment, whether it can use those items after some repairs, or whether it must abandon those items.

Salvage

When the damage to facilities and equipment is known and understood, an organization can begin the work of organizing the repair of facilities and equipment that can be salvaged and removing or destroying those that cannot. Salvage is generally performed by outside organizations.

Personnel safety

In any kind of disaster, the safety of personnel is the highest priority, ahead of buildings, equipment, computers, backup tapes, and so on. Personnel safety is critical not only because of the intrinsic value of human life, but also because people — not physical assets — make the business run.

Personnel notification

The business continuity plan must have some provisions for notifying all affected personnel that a disaster has occurred. An organization needs to establish multiple methods for notifying key business continuity personnel in case public communications infrastructures are interrupted.

Not all disasters are obvious: A fire or broken water main is a local event, not a regional one. And in an event such as a tornado or flood, employees who live even a few miles away may not know the condition of the business. Consequently, the organization needs a plan for communicating with employees no matter what the situation is.

Throughout a disaster and recovery, managers must be given regular status reports as well as updates on crucial tactical issues so that they can align resources to support critical business operations that function on a contingency basis. A manager of a corporate facilities department, for example, can loan equipment that critical departments need to keep functioning.

Backups and media storage

Things go wrong with hardware and software, resulting in wrecked or unreachable data. When data is gone, it's gone! Thus, IT departments everywhere make copies of their critical data on tapes or removable discs, in external storage systems, or in the cloud.

These backups must be performed regularly — usually, once per day. For organizations with on-premises systems, backup media must also be stored offsite in the event that the facility housing the original systems is damaged. Having backup tapes *in* the data center may be convenient for doing a quick data restore but of little value if backup tapes are destroyed along with their respective systems. For organizations with cloud-based systems, the problem is the same, but the technology differs a bit: It is imperative that data be backed up (or replicated) to a different geographic location so that it can be recovered no matter what happens.

For systems with large amounts of data, that data must be well understood to determine what kinds of backups need to be performed (real-time replication, full, differential, and incremental) and how frequently. Consider these factors:

» The time it takes to perform backups

» The effort required to restore data

» The procedures for restoring data from backups, compared with other methods of recovering the data

Consider, for example, whether you can restore application software from backup faster than by installing them from their release media (the original CD-ROMs or downloaded install files). Just make sure that you can recover your configuration settings if you reinstall software from release media. Also, if a large part of the database is static, do you really need to back it all up every day?

You must choose of-site storage of backup media and other materials (documentation and the like) carefully. Factors to consider include survivability of the off-site storage facility, as well as the distance from the offsite facility to the data center, media transportation, and alternative processing sites. The facility needs to be close enough so that media retrieval doesn't take too long (how long depends on the organization's recovery needs), but not so close that the facility becomes involved in the same natural disaster as the business.

Cloud-based data replication and backup services are viable alternatives to offsite backup media storage. Today's Internet speeds make it possible to back up critical data to a cloud-based storage provider often faster than magnetic tapes can be returned from an offsite facility and data recovered from them.

TIP

Some organizations have one or more databases so large that they can't (or don't) back them up to tape. Instead, they keep one or more replicated copies of their databases on other computers in other cities. Business continuity planners need to consider this possibility when developing continuity plans.

The purpose of offsite media storage is to ensure that up-to-date data is available in the event that systems in the primary data center are damaged.

THE END OF MAGNETIC TAPE?

Magnetic tape has been the backup medium of choice since the 1960s. Gradually improving in reliability, capacity, and throughput, magnetic tape has hung in there as the mainstay of backup. But, the era of magnetic tape may be nearing its end.

The linear access property of magnetic tape means that you have to read all the way through a tape to know its contents and to restore data that may be near the end. In addition, magnetic tape is somewhat fragile, and it is less tolerant of defects at higher storage densities.

Commercially viable alternatives to magnetic tape are emerging, including the following:

- **Virtual tape library (VTL):** Really just disk-based storage, a VTL has the appearance of magnetic tape to back up programs. In a hot-pluggable RAID array, you could send these disks offsite.

- **Replication:** An organization with two or more processing centers can consider replicating data from one location to another.

- **Cloud backup (e-vaulting):** If data sets aren't too large and Internet bandwidth is sufficient, data can be backed up to a cloud-based storage provider.

Proven but linear and relatively fragile magnetic tape may soon be part of the great data-processing museum in the sky.

Software escrow agreements

Your organization should consider software escrow agreements (wherein the software vendor sends a copy of its software code to a third-party escrow organization for safekeeping) with the software vendors whose applications support critical business functions. In the event that an insurmountable disaster (which could include bankruptcy) strikes the software vendor, your organization must consider all options for the continued maintenance of those critical applications, including in-house support.

External communications

The corporate communications, external affairs, and (if applicable) investor relations departments should have plans in place for communicating the facts about a disaster to the press, customers, and public. You need contingency plans for these functions if you want the organization to continue communicating to the outside world. Open communication during a disaster is vital so that customers, suppliers, and investors don't panic (which they might do if they don't know the true extent of the disaster).

WHO SAYS EXTERNAL AFFAIRS IS NONESSENTIAL?

Suppose that the headquarters building of a large company burns to the ground. (Such an event is very unlikely for a modern building, but stay with us.) All personnel escape unharmed. In fact, the organization is very well off because it duplicated all the information in the building and stored those duplicates in an offsite facility. Nice work! But the external affairs department, which was housed in that building, loses everything; it needs two days to recover the capability of communicating to the outside world. Because of this time lag, the company loses many of its customers, who fear the worst. This situation is especially unfortunate and ironic because the company was in pretty good shape after the conflagration, all things considered. Sometimes, a lack of credible information causes people to fear the worst.

The emergency communications plan needs to take into account the possibility that some corporate facilities or personnel may be unavailable. Thus, you need to keep the data and procedures related to the communications plan safe so that they're available in any situation.

Utilities

Data-processing facilities that support time-critical business functions must keep running in the event of a power failure. Although every situation is different, the principle remains the same: The business continuity planning team must determine how long the data-processing facility must be able to continue operating without utility power. A power engineer can find out the length of typical (we don't want to say *routine*) power outages in your area and crunch the numbers to arrive at the mean time of outages. By using that information, as well as an inventory of the data center's equipment and environmental equipment, you can determine whether the organization needs an uninterruptible power supply (UPS) alone or a UPS *and* an electric generator.

A business can use UPSs and emergency electric generators to provide electric power during prolonged power outages. A UPS is also good for a controlled shutdown if the organization is better off having its systems powered off during a disaster. A business can also use a stand-alone power system, an off-the-grid system that generates power with solar, wind, water, or employees madly pedaling stationary bicycles. (We're kidding about that last one.)

In a really long power outage (more than a day or two), it is essential to have a plan for the replenishment of generator fuel.

Logistics and supplies

The business continuity planning team needs to study every aspect of critical functions that must be made to continue in a disaster. Every resource that's needed to sustain the critical operation must be identified and then considered in every possible disaster scenario to determine what special plans must be made. If a business operation relies on a just-in-time shipment of materials for its operation, and an earthquake closes the region's only highway (or airport, seaport, or lake port), alternative means for acquiring those materials must be determined in advance. Or perhaps an emergency ration of those materials needs to be stockpiled so that the business function can continue uninterrupted.

Fire and water protection

Many natural disasters disrupt public utilities, including water supplies and delivery. In the event that a disaster has interrupted water delivery, new problems arise. Your facility may not be allowed to operate without the means for fighting a fire, should one occur.

In many places, businesses could be ordered to close if they can't prove that they can effectively fight a fire using other means, such as FM-200 inert gas. Then again, if water supplies have been interrupted, you have other issues to contend with, such as drinking water and water for restrooms. Without water, you're hosed!

We discuss fire protection in more detail in Chapter 5.

Documentation

Any critical business function must be able to continue operating after a disaster strikes. And to make sure that you can sustain operations, you need to make available all relevant documentation for every critical piece of equipment, as well as every critical process and procedure that the organization performs in a given location.

Don't be lulled into taking for granted the emerging trend of hardware and software products that don't come with any documentation. Many vendors deliver their documentation only over the Internet or charge extra for a hard copy. But many types of disasters may disrupt Internet communications, leaving an operation high and dry, with no instructions for using and managing tools and applications.

At least one set of hard copy (or CD-ROM soft copy) documentation, including your business continuity and disaster recovery plans, should be stored at the same offsite storage facility that stores the organization's backup tapes. It would also be smart to issue electronic copies of the documentation to all relevant personnel on USB storage devices (with encryption) or to store them on responders' mobile devices.

If the preceding sounds like the ancient past to you, your organization may be fully in the cloud today. In such a case, you may be more inclined to maintain multiple soft copies of all required documentation so that employees can use it as needed.

Continuity and recovery documentation must exist in hard copy in the event that the documents are unavailable via electronic means.

Data processing continuity planning

Data processing facilities are so vital to businesses today that a great deal of emphasis is placed on them. Generally, planning comes down to these variables: where and how the business will continue to sustain its data processing functions.

Because data centers are so expensive and time-consuming to build, good business sense dictates having an alternative processing site available. The types of sites are

>> **Cold site:** A *cold site* is an empty computer room with environmental facilities (UPS; heating, ventilation, and air conditioning; and so on) but no computing equipment. This option is the least costly one, but more time is required to assume a workload because computers need to be brought in from somewhere and set up, and data and applications need to be loaded. Connectivity to other locations also needs to be installed.

>> **Warm site:** A *warm site* is a cold site with computers and communications links already in place. To take over production operations, you must load the computers with application software and business data.

>> **Hot site:** Indisputably the most expensive option, a *hot site* uses the same computers as the production system, with application changes, operating-system changes, and even patches kept in sync with their live production-system counterparts. You even keep business data up to date at the hot site by using some sort of mirroring or transaction replication. Because the organization trains its staff in how to operate the organization's business applications (and staff members have documentation), the operations staff knows what to do to take over data processing operations at a moment's notice. Hot sites may be cloud-based or in a colocation center.

>> **Reciprocal site:** Your organization and another organization sign a reciprocal agreement in which you both pledge the availability of your organization's data center in the event of a disaster. Back when data centers were rare, many organizations made this sort of arrangement, but it's fallen out of favor in recent years.

>> **Multiple data centers:** Larger organizations can consider the option of running daily operations out of two or more regional data centers that are hundreds of miles (or more) apart. The advantage of this arrangement is that

the organization doesn't have to make arrangements with outside vendors for hot, warm, or cold sites, and the organization's staff is already onsite and familiar with business and computer operations.

>> **Cloud site:** Organizations with primary information processing in the cloud are likely to employ cloud assets in multiple regions, possibly from more than one cloud provider. Many organizations that have primary processing on-premises employ hybrid cloud infrastructure for disaster recovery purposes — a common way for companies to ease their way into the cloud. Depending on the degree of readiness required, a cloud site can be as ready as a hot, warm, or cold site, as determined by the resources devoted to keeping the cloud site ready for production operations.

A hot site provides the most rapid recovery capability, but it also costs the most because of the effort required to maintain its readiness. Table 3-1 compares these options side by side.

TABLE 3-1 Data Processing Continuity Planning Site Comparison

Feature	Hot Site	Warm Site	Cold Site	Multiple Data Centers	Cloud Site
Cost	Highest	Medium	Low	No additional	Variable
Computer-equipped	Yes	Yes	No	Yes	Yes
Connectivity-equipped	Yes	Yes	No	Yes	Yes
Data-equipped	Yes	No	No	Yes	Variable
Staffed	Yes	No	No	Yes	No
Typical lead time to readiness	Minutes to hours	Hours to days	Days to weeks	Minutes to hours or longer	Minutes to hours

After you complete the BIA and develop and document the scope of the business continuity plan, you know

>> What portion of the organization is included in the plan.

>> Of this portion of the organization, which business functions are so critical that the business would fail if these functions were interrupted for long (or even short) periods.

>> The general degree of impact on the business when one of the critical functions fails. This idea comes from quantitative and qualitative data.

Now the hard part of the business continuity planning project begins: developing the strategy for continuing each critical business function when disasters occur, which is known as the *continuity strategy*.

When you develop a continuity strategy, you must set politics aside and look at the excruciating details of critical business functions. You need lots of strong coffee, several pizzas, buckets of antacids, and cool heads.

Making your business continuity planning project a success

For the important and time–consuming continuity strategy phase of the project, you need to follow these guidelines:

>> **Call things as you see them.** No biases. No angles. No politics. No favorites. No favors. You're trying to ensure survival of the business before the disaster strikes.

>> **Build smaller teams of experts.** Each critical business function should have teams dedicated to just that function. That team's job is to analyze just one critical business function and figure out how you can keep it functioning despite a disaster of some sort. Pick the right people for each team — people who really understand the details of the business process that they're examining.

>> **Brainstorm.** Proper brainstorming considers all ideas, even silly ones (up to a point). Even a silly-sounding idea can lead to a good idea.

>> **Have teams share results.** Teams working on individual continuity strategies can get ideas from one another. Each team can share highlights of its work over the past week or two. Some of the things that they say may spark ideas for other teams. You can improve the entire effort by holding these sharing sessions.

>> **Don't encourage competition or politics.** Don't pit teams against one another. Identifying success factors isn't a zero-sum game: Everyone needs to do an excellent job.

>> **Retain a business continuity planning mentor or expert.** If your organization doesn't have experienced business continuity planners on staff, you need to bring in a consultant — someone who has helped develop plans for other organizations. Even more important, make sure that the consultant you hire was on the scene when disaster struck a business they were consulting for and saw a business continuity plan in action.

Simplifying large or complex critical functions

Some critical business functions may be too large and complex to examine in one big chunk. You can break those complex functions into smaller components, perhaps like this:

» **People:** Has the team identified the critical people (or, more appropriately, the critical subfunctions) required to keep the function running?

» **Facilities:** In the event that the function's primary facilities are unavailable, where can the business perform the function?

» **Technology:** What hardware, software, and other computing/network components support the critical function? If parts or all of these components are unavailable, what other equipment can support the critical business functions? Do you need to perform the functions any differently?

» **Miscellaneous:** What supplies, equipment, and services do you need to support the critical business function?

Analyzing processes is like disassembling toy houses, in that you have to break them down to the level of their components. You need to understand each step of even the largest processes to be able to develop good continuity plans for them.

If a team that analyzes a large complex business function breaks into groups, such as the groups in the preceding list, the team members need to get together frequently to ensure that their respective strategies for each group become a cohesive whole.

GETTING AMAZING THINGS DONE

It is amazing what you can accomplish if it doesn't matter who gets the credit. Nowhere is this statement truer in business than in business continuity planning. A business continuity planning project is a setting in which people will jostle for power, influence, and credit.

These forces must be neutralized. Business continuity planning should be apolitical, with differences and personal agendas are set aside. Only then does the process have a reasonable chance of success. The business, its employees, and its customers deserve nothing less.

Documenting the strategy

Now for the part that everyone loves: documentation. The details of the continuity plans for each critical function must be described in minute detail, step by step by step. Why? The people who develop the strategy may very well not be the people who execute it. The people who develop the strategy may change roles in the company or change jobs. Or the scope of an actual disaster may be wide enough that the critical personnel aren't available. Any skeptics should consider September 11 and the impact of this disaster on companies that lost practically everyone and everything.

Best practices for documenting business continuity plans exist. For this reason, you may want to have an expert around. For $300 an hour, a consultant can spend a couple of weeks developing templates. To make sure that you get a solid consultant, do the old-fashioned things: Check their references, ask for work samples, and see whether they have a decent LinkedIn page. (We're kidding about that last one!)

Implementing the plan

It is an accomplishment indeed when the documentation has been written, reviewed, edited, placed in three-ring binders, and distributed via thumb drives or online file storage accounts. But the job isn't done. The plan needs senior management buy-in; it must be announced and socialized throughout the organization; and one or more people must be dedicated to keeping it up to date. And — oh, yeah — the plan needs to be tested.

REMEMBER

WHY HIRE AN EXPERT?

Most people don't do business continuity planning for a living. Although you may be the expert in your particular business processes, you don't necessarily know all the angles of contingency planning.

Turn this question around for a minute: What would you think if an IT shop developed a security strategy without having a security expert's help? Do you think it would have a sound, viable strategy? The same argument fits equally well with business continuity planning.

For the remaining skeptics, do yourself a favor: Hire a business continuity planning expert for a short time to help validate your framework and plan. If your expert says that your plan is great, you can consider their fee to be money well spent to confirm your suspicions. If the consultant says that your plan needs help, ask for details on where and how. Then you can decide whether to rework and improve your plan.

When disaster strikes, it's too late to wish that you had a good plan.

SECURING SENIOR MANAGEMENT APPROVAL

After the entire plan has been documented and reviewed by all stakeholders, it's time for senior management to examine and approve it. Senior management must not only approve the plan, but also *publicly* approve it. We don't mean to the *general* public; we mean that senior management should make it well known inside the business that they support the business continuity planning process.

Senior management's approval is needed so that all affected and involved employees in the organization understand the importance of emergency planning.

PROMOTING ORGANIZATIONAL AWARENESS

Everyone in the organization needs to know about the plan and their role in it. You may need to establish training for potentially large numbers of people who need to be there when a disaster strikes. All employees in the organization must know about the business continuity plan and the people who are responsible for managing it.

TESTING THE PLAN

Regularly testing the plan ensures that all essential personnel required to implement the plan understand their roles and responsibilities, and that the plan is kept up to date as the organization changes. Testing methods (similar to those for testing disaster recovery plans; see Chapter 9) include

>> Read-through

>> Walk-through

>> Simulation

>> Parallel

>> Full interruption

See Chapter 9 for explanations of these testing methods.

MAINTAINING THE PLAN

No, the plan isn't finished. It has just begun! Now the business continuity planning person (by this time, the project team members have collected their commemorative denim shirts, mugs, and mouse pads, and moved on to other projects) needs to chase The Powers That Be periodically to make sure that they know about all significant changes in the environment.

In fact, if the business continuity planning person has any leadership left at this point in the process, they need to start attending meetings of the change control board or IT steering committee (whatever the company calls them) meetings and jot down notes; some details in a plan document may need some changes.

TIP

A business continuity plan is easier to modify than it is to create out of thin air. Once or twice each year, someone knowledgeable needs to examine the detailed strategy and procedure documents in the plan to make sure that they'll still work and update them if necessary.

TIP

You can read more about business continuity and disaster recovery planning in *IT Disaster Recovery Planning For Dummies*, by Peter H. Gregory (John Wiley & Sons, Inc.).

Contribute to and Enforce Personnel Security Policies and Procedures

An organization needs clearly documented personnel security policies and procedures to facilitate the use and protection of information. Numerous essential practices can protect the business and its important information assets. All these practices have to do with how people — not technology — work together to support the business.

CROSS
REFERENCE

This section covers Objective 1.9 of the Security and Risk Management domain in the CISSP Exam Outline (May 1, 2021).

This topic is collectively known as *administrative management and control*.

Note: We tend to use the term *essential practices* versus *best practices.* The reason is simple: *Best practices* refers to the *very best* practices and technologies that can be brought to bear on a business problem, whereas *essential practices* means those activities and technologies that are considered to be essential to an organization. Best practices are nearly impossible to achieve, and few organizations attempt to do so. But essential practices are . . . well, essential and definitely achievable in many organizations.

Candidate screening and hiring

Even before posting a "Help Wanted" sign (do companies still do that?) or an ad on a job search website, an employer should ensure that the position to be filled is clearly documented and contains a complete description of the job requirements, the qualifications, and the scope of responsibilities and authority.

The job (or position) description should be created as a collaborative effort between the hiring manager — who fully understands the functional requirements of the specific position to be filled — and the human resources manager — who fully understands the applicable employment laws and organizational requirements to be addressed.

Having a clearly documented job (or position) description can benefit an organization for many reasons:

>> The hiring manager knows (and can clearly articulate) exactly what skills a certain job requires.

>> The human resources manager can prescreen job applicants quickly and accurately.

>> Potential candidates can ensure that they apply only for positions for which they're qualified and can prepare themselves for interviews properly (such as by matching their skills and experiences to the specific requirements of the position).

>> After the organization fills the position, the position description (in some cases, the employment contract) helps reduce confusion about what the organization expects from the new employee and provides objective criteria for evaluating performance.

Concise job descriptions that clearly identify an employee's responsibility and authority, particularly on information security issues, can help by

>> Reducing confusion and ambiguity

>> Providing a legal basis for an employee's authority or actions

>> Demonstrating any negligence or dereliction on the employee's part in carrying out assigned duties

An organization should conduct background checks and verify application information for all potential employees and contractors. This process can help expose any undesirable or unqualified candidates, as in these scenarios:

>> A criminal conviction may immediately disqualify a candidate from certain positions in an organization.

>> Even when a criminal record doesn't automatically disqualify a candidate, if the candidate fails to disclose this information in the job application or interview, that omission should be a clear warning sign to a potential employer.

>> Some positions require a U.S. government security clearance (available only to U.S. citizens).

>> A candidate's credit history should be examined if the position has significant financial responsibilities or handles high-value assets, or if a high opportunity for fraud exists.

It has been estimated that as many as 40 percent of job applicants exaggerate the truth on their résumés and applications. Commonly omitted, exaggerated, or misleading information includes employment dates, salary history, education, certifications, and achievements. Although the information itself may not be disqualifying, a dishonest applicant should not be given the opportunity to become a dishonest employee.

Most background checks require the written consent of the applicant and disclosure of certain private information (such as the applicant's Social Security or other retirement-system number). Private information obtained for the purposes of a background check, as well as the results of the background check, must be handled properly and safeguarded in accordance with applicable laws and the organization's records retention and destruction policies.

Basic background checks and verification might include the following information:

>> Criminal record

>> Citizenship

>> Employment history

>> Education

>> Certifications and licenses

>> Financial history, including judgments

>> Reference checks (personal and professional)

>> Union and association membership

Pre- and postemployment background checks can provide an employer valuable information about a person whom an organization is considering for a job or position within an organization. Such checks can give an immediate indication of an applicant's integrity (by providing verification of information in the employment application, for example) and can help screen out an unqualified or undesirable applicant.

Personnel who fill sensitive positions should undergo a more extensive pre-employment screening and background check, possibly including the following:

>> **Credit records:** These records minimally include bankruptcies, foreclosures, and judgments. Depending on the position, a full credit report may be required.

>> **Drug screen:** Even in countries or U.S. states where certain narcotics and other substances such as cannabis are legal, if the organization's policies prohibit their use, drug testing should be used to enforce the policy.

>> **Special background investigation:** This investigation may include government-agency records, field interviews with former associates, or a personal interview with a private investigator.

Periodic postemployment screenings (such as credit records and drug testing) may also be necessary, particularly for personnel who have access to financial data, cash, or high-value assets, or who are being considered for promotions to more sensitive or responsible positions.

Many organizations that did not perform drug screenings in the past do so today. Instead of drug-screening all employees, some take a measured approach by screening employees when they are promoted to higher levels of responsibility, such as director or vice president. Other organizations perform drug screening only on personnel who operate machinery or drive fleet vehicles.

Employment agreements and policies

Various employment agreements and policies should be signed when a person joins an organization or is promoted to a more sensitive position within an organization. Employment agreements often include noncompete agreements, non-disclosure agreements (also known as confidentiality agreements), codes of conduct, and acceptable-use policies. Typical employment policies might include Internet acceptable use, social media policy, remote access, mobile and personal device use, and sexual harassment/fraternization.

Onboarding, transfers, and termination processes

Policies and procedures need to be developed to ensure that new hires and transfers are onboarded efficiently and securely. These policies and procedures should include having the employee or contractor read and formally acknowledge by signing acceptable use policies (AUPs), assigning equipment (such as a laptop computer or mobile device), and provisioning accounts with the appropriate permissions by using role-based access controls (RBAC, discussed in Chapter 7) based on the principle of least privilege (discussed in Chapter 9). If the organization allows employees and contractors to use personal devices, a "Bring Your Own Device" (BYOD) policy should define security requirements for these devices, such as

>> Passcodes or biometric access

>> Data encryption

>> Virtual private network (VPN), mobile device management (MDM), and mobile application management (MAM) software

Formal employment termination procedures should be implemented to help protect the organization from potential lawsuits, property theft and destruction, unauthorized access, or workplace violence. Procedures should be developed for various scenarios, including resignations, termination, layoffs, accident or death, immediate departures versus prior notification, and hostile situations. Termination procedures may include

>> Having the former employee surrender keys, security badges, and parking permits

>> Conducting an exit interview

>> Requiring security to escort the former employee to collect their personal belongings and/or to leave the premises

>> Recovering company assets and materials, including laptop computers, mobile phones, tablets, and so on

>> Changing door locks and system passwords as needed

>> Formally turning over duties and responsibilities

>> Removing network and system access and disabling user accounts

>> Enforcing policies regarding retention of email, personal files, and employment records

>> Notifying customers, partners, vendors, service providers, and contractors as appropriate

Vendor, consultant, and contractor agreements and controls

Organizations commonly outsource many IT functions, particularly data center hosting, call-center or contact-center support, and application development.

Information security policies and procedures must address outsourcing security and the use of service providers, vendors, and consultants when appropriate. Access control, document exchange and review, maintenance hooks, onsite assessment, process and policy review, and SLAs are good examples of outsourcing security considerations. This topic is discussed further in "Apply Supply Chain Risk Management Concepts" later in this chapter.

Compliance policy requirements

An organization's policies must align with applicable laws, regulations, and other legal obligations. Accordingly, an organization's requirements, controls, and procedures must align with policies. This chain of compliance should align and be unbroken.

Individual responsibilities for compliance with applicable policies and regulations within the organization should be understood by all personnel within an organization. Signed statements that attest to a person's understanding, acknowledgement, and/or agreement to comply may be appropriate for certain regulations and policies.

Privacy policy requirements

Applicable policy regulations and policy requirements should be documented and understood by all personnel within the organization. Signed statements that attest to a person's understanding, acknowledgement, and/or agreement to comply may also be appropriate.

Understand and Apply Risk Management Concepts

Beyond basic security fundamentals, the concepts of risk management are perhaps the most important and complex parts of the security and risk management domain. Indeed, risk management is the process from which decisions are made to establish what security controls are necessary, implement security controls, acquire and use security tools, and hire security personnel.

This section covers Objective 1.10 of the Security and Risk Management domain in the CISSP Exam Outline (May 1, 2021).

CROSS REFERENCE

Risk can never be eliminated. Given sufficient time, resources, motivation, and money, any system or environment, no matter how secure, can eventually be compromised. Some threats or events, such as natural disasters, are entirely beyond our control and often unpredictable. Therefore, the main goal of risk management is *risk treatment:* making intentional decisions about specific risks that organizations identify. Risk management consists of three main elements (each covered in the upcoming sections):

>> Risk identification

>> Risk analysis

>> Risk treatment

Identify threats and vulnerabilities

The business of information security is all about risk management. A *risk* consists of a threat to, and a vulnerability of, an asset:

>> **Threat:** Any natural or human-made circumstance or event that could have an adverse or undesirable impact, minor or major, on an organizational asset or process.

>> **Vulnerability:** The absence or weakness of a safeguard or control in an asset or process (or an intrinsic weakness) that makes a threat potentially more harmful or costly, more likely to occur, or likely to occur more frequently.

>> **Asset:** A resource, process, product, or system that has some value to an organization and must therefore be protected. Assets may be tangible (computers, data, software, records, and so on) or intangible (privacy, access, public image, ethics, and so on), and those assets may likewise have a tangible value (purchase price) or intangible value (competitive advantage).

Remember: Risk = Asset Value × Threat Impact × Threat Probability.

The *risk management triple* consists of an asset, a threat, and vulnerability.

Risk assessment/analysis

Two key elements of risk management are risk assessment and risk analysis (discussed in the following sections).

Risk assessment

A risk assessment begins with risk identification, detecting and defining specific elements of the three components of risk: assets, threats, and vulnerabilities.

The process of risk identification occurs during a risk assessment.

ASSET VALUATION

Asset valuation is an important part of risk management, because managers and executives need to be aware of the tangible and intangible value of all assets involved in specific incidents of risk management.

Once in a while, an asset's valuation can come from the accounting department's balance sheet (for better organizations that have a good handle on asset inventory, value, and depreciation), but often, that's only part of the story. If an older server is involved in an incident and must be replaced, for example, the replacement cost will be far higher than the asset's depreciated value. Further, the time required to deploy and ready a replacement server and the cost of downtime also need to be considered.

Sometimes, there are other ways to assign values to assets. An asset's contribution to revenue might change one's perspectives on an asset's value. If an asset with a $10,000 replacement cost is key in helping the organization realize $5 million in revenue, is it still worth just $10,000?

The value of an asset to an organization can be both *quantitative* (related to its cost or the revenue it helps obtain) and *qualitative* (its relative importance). An inaccurate or hastily conducted asset valuation process can have the following consequences:

>> Poorly chosen or improperly implemented controls

>> Controls that aren't cost-effective

>> Controls that protect the wrong asset

A properly conducted asset valuation process has several benefits to an organization:

>> Supports quantitative and qualitative risk assessments, BIAs, and security auditing

>> Facilitates cost-benefit analysis and supports management decisions regarding selection of appropriate safeguards

>> Can be used to determine insurance requirements, budgeting, and replacement costs

>> Helps demonstrate due care, thus (potentially) limiting personal liability on the part of directors and officers

Three basic elements used to determine the value of an asset are

>> **Initial and maintenance costs:** Most often, a tangible dollar value that may include purchasing, licensing, development (or acquisition), maintenance, and support costs.

>> **Organizational (or internal) value:** Often a difficult and intangible value. It may include the cost of creating, acquiring, and re-creating information, and the business impact or loss if the information is lost or compromised. It can also include liability costs associated with privacy issues, personal injury, and death.

>> **Public (or external) value:** Another difficult and often intangible cost. Public value can include loss of proprietary information or processes, as well as loss of business reputation.

>> **Contribution to revenue:** The asset's role in revenue realization. An asset worth $10,000 may be instrumental to the realization of $5 million in annual revenue. Hence, risk decisions for such an asset should consider not only its cost, but also its role in generating or protecting revenue.

THREAT ANALYSIS

To perform threat analysis, follow these four basic steps:

1. **Define the threat.**
2. **Identify possible consequences to the organization if the threat event occurs.**
3. **Determine the probable frequency and impact of a threat event.**
4. **Assess the probability that a threat will materialize.**

A company that has a major distribution center located along the Gulf Coast of the United States, for example, may be concerned about hurricanes. Possible consequences include power and communications outages, wind damage, and flooding. Using climatology, the company can determine that on average, three hurricanes pass within 50 miles of its location every year between June and September and that the specific probability exists that a hurricane will affecting the company's operations during this period. During the remainder of the year, the threat of hurricanes has a low probability.

The number and types of threats that an organization must consider can be overwhelming, but you can generally categorize them as follows:

>> **Natural:** Earthquakes, floods, hurricanes, lightning, fire, and so on

>> **Human-made:** Unauthorized access, data-entry errors, strikes or labor disputes, theft, terrorism, sabotage, arson, social engineering, malicious code and viruses, and so on

WARNING

Not all threats can be easily or rigidly classified. Fires and utility losses, for example, can be both natural and human-made. See Chapter 9 for more on disaster recovery.

VULNERABILITY ASSESSMENT

A *vulnerability assessment* provides a valuable baseline for identifying vulnerabilities in an asset as well as identifying one or more potential methods for mitigating those vulnerabilities. An organization might consider a DoS threat coupled with a vulnerability found in Microsoft's implementation of the Domain Name System (DNS). If an organization's DNS servers have been patched properly or the organization uses a Unix-based DNS Security Extensions (DNSSEC) server, the specific vulnerability may be adequately addressed, and no additional safeguards may be necessary for that threat.

Risk analysis

The next element in risk management is *risk analysis* — a methodical examination that brings together all the elements of risk management (identification, analysis, and control) and is critical to an organization for developing an effective risk management strategy.

Risk analysis involves the following five steps:

1. **Identify the assets to be protected, including their relative value, sensitivity, or importance to the organization.**

This component of risk identification is asset valuation.

2. **Define specific threats, including threat frequency and impact data.**

This component of risk identification is threat analysis.

3. **Calculate annualized loss expectancy (ALE).**

The ALE calculation is a fundamental concept in risk analysis; we discuss this calculation later in this section.

4. **Select appropriate safeguards.**

    This process is a component of both risk identification (vulnerability assessment) and risk treatment (which we discuss in "Risk treatment" later in this chapter).

5. **Calculate ALE for each safeguard.**

    This process helps a risk analyst identify suitable candidates for risk mitigation.

ALE provides a standard, quantifiable measure of the impact of a realized threat on an organization's assets. Because it's the estimated annual loss for a threat or event, expressed in dollars, ALE is particularly useful for determining the cost/benefit ratio of a safeguard or control. You determine ALE by using this formula:

$$SLE \times ARO = ALE$$

Here's an explanation of the elements in this formula:

» **Single loss expectancy (SLE):** A measure of the loss incurred from a single realized threat or event, expressed in dollars. You calculate the SLE by using the formula Asset value × Exposure factor.

   Exposure factor (EF) is a measure of the negative effect or impact of a realized threat or event on a specific asset, expressed as a percentage.

» **Annualized rate of occurrence (ARO):** The estimated annual frequency of occurrence for a threat or event.

The two major types of risk analysis are qualitative and quantitative, which we discuss in the following sections.

QUALITATIVE RISK ANALYSIS

Qualitative risk analysis is more subjective than a quantitative risk analysis; unlike quantitative risk analysis, this approach to analyzing risk can be purely qualitative and avoids specific numbers. The challenge of such an approach is developing real scenarios that describe actual threats and potential losses to organizational assets.

Qualitative risk analysis has some advantages compared with quantitative risk analysis:

» No complex calculations are required.

» Time and work effort involved is relatively low.

» Volume of input data required is relatively low.

Disadvantages of qualitative risk analysis, compared with quantitative risk analysis, include

>> No financial costs are defined; therefore, cost-benefit analysis isn't possible.

>> The qualitative approach relies more on assumptions and guesswork.

>> Generally, qualitative risk analysis can't be automated.

>> Qualitative analysis is less easily communicated. (Executives seem to understand "This will cost us $3 million over 12 months" better than "This will cause an unspecified loss at an undetermined future date.")

A distinct advantage of qualitative risk analysis is that a large set of identified risks can be charted and sorted by asset value, risk, or other means. Doing so can help an organization identify and distinguish higher risks from lower risks, even though precise dollar amounts may not be known.

A qualitative risk analysis doesn't attempt to assign numeric values to the components (the assets and threats) of the risk analysis.

QUANTITATIVE RISK ANALYSIS

A fully quantitative risk analysis requires all elements of the process, including asset value, impact, threat frequency, safeguard effectiveness, safeguard costs, uncertainty, and probability, to be measured and assigned numeric values.

A quantitative risk analysis attempts to assign more objective numeric values (costs) to the components (assets and threats) of the risk analysis.

Advantages of a quantitative risk analysis, compared with qualitative risk analysis, include the following:

>> Financial costs are defined; therefore, cost-benefit analysis can be determined.

>> More concise, specific data supports analysis; thus, fewer assumptions and less guesswork are required.

>> Analysis and calculations can often be automated.

>> Specific quantifiable results are easier to communicate to executives and senior management.

Disadvantages of a quantitative risk analysis, compared with qualitative risk analysis, include the following:

>> Human biases will skew results.

>> Many complex calculations are usually required.

>> Time and work effort involved is relatively high.

>> Volume of input data required is relatively high.

>> The probability of threat events is difficult to determine.

>> Some assumptions are required.

Purely quantitative risk analysis is generally not possible or practical. primarily because it is difficult to determine a precise probability of occurrence for any given threat scenario. For this reason, many risk analyses are a blend of qualitative and quantitative risk analysis, known as a hybrid risk analysis.

HYBRID RISK ANALYSIS

A hybrid risk analysis combines elements of both a quantitative and qualitative risk analysis. The challenges of determining accurate probabilities of occurrence, as well as the true impact of an event, compel many risk managers to take a middle ground. In such cases, easily determined quantitative values (such as asset value) are used in conjunction with qualitative measures for probability of occurrence and risk level. Indeed, many so-called quantitative risk analyses are more accurately described as hybrid.

Risk appetite and risk tolerance

Business leaders who make risk treatment decisions need to be familiar with a fundamental truth: How much risk is the organization willing to accept in any given situation and how business leaders will understand when any particular risk is so high that all or part of it must be reduced or transferred. Business leaders can simply rely on gut feel or be more intentional and establish a formal risk-appetite statement that defines how much risk an organization is willing to accept.

Formal risk appetite is generally described in both qualitative and quantitative forms to guide risk managers and business leaders make better risk treatment decisions. Typically, risk appetite is defined as a policy statement, sometimes within risk management program documents.

Risk managers and business leaders also need to understand the organization's risk tolerance, defined as the variance from risk appetite that an organization is willing to accept.

Risk treatment

A properly conducted risk analysis provides the basis for the next step in the risk management process: deciding what to do about risks that have been identified. The decision-making process is known as *risk treatment* (or *risk response*). The four general methods of risk treatment are

>> **Risk mitigation (or risk reduction):** Implements one or more policies, controls, or other measures to protect an asset. Mitigation generally reduces the probability of threat realization or the impact of threat realization to an acceptable level. This risk treatment is the most common.

>> **Risk transfer (or assignment):** Transfers the potential loss associated with a risk to a third party, such as an insurance company or a service provider that explicitly agrees to accept risk.

>> **Risk avoidance:** Eliminates the risk through a cessation of the activity or condition that introduced the risk in the first place.

>> **Risk acceptance:** Accepts the risk associated with a potential threat, sometimes for convenience (not prudent) but more appropriately when the cost of other countermeasures is prohibitive, the probability or impact is low, or the benefits outweigh the costs. Generally, an organization will accept only risk when it is at or below the established risk appetite.

Countermeasure selection and implementation

As stated in the preceding section, mitigation is the most common method of risk treatment. Mitigation involves the implementation of one or more countermeasures. After appropriate countermeasures have been selected, they need to be implemented in the organization and integrated with other systems and countermeasures, when appropriate. Organizations that implement countermeasures are making planned changes to their environment in specific ways. Examples of countermeasure implementation include

>> **Change to policy, standard, process, or procedure:** An update of an official policy, technology standard, process, or procedure will require planning to ensure that the change will not have unintended effects in the organization.

Some level of review(s), analysis, and discussion will occur before the changes are accepted, published, and communicated. Changes to policy, standard, process, or procedure may also require changes to technology, and vice versa.

>> **Change to technology:** An update to something as big as network architecture or as focused as the configuration setting of an individual system is used to mitigate risk. Changes to technology usually involve business processes such as change management and configuration management and may also affect procedures or standards. Significant changes may also involve discussions or processes at the level of the IT steering committee or security committee. Disaster recovery planning and the configuration management database (CMDB) may also be affected.

>> **Change to staff:** A change in staffing could include training, reallocation of responsibilities, the addition of temporary staff (contractors or consultants), or the hiring of additional staff.

Several criteria for selecting countermeasures include cost-effectiveness, legal liability, operational impact, and technical factors.

Cost-effectiveness

The most common criterion for countermeasure selection is cost-effectiveness, which is determined through cost-benefit analysis. Cost-benefit analysis for a given countermeasure (or collection of countermeasures) can be computed as follows:

ALE before countermeasure – ALE after countermeasure – Cost of countermeasure = Value of countermeasure to the organization

If the ALE associated with a specific threat (data loss) is $1,000,000, the ALE after a countermeasure (enterprise tape backup) has been implemented is $10,000 (recovery time), and the cost of the countermeasure (purchase, installation, training, and maintenance) is $140,000, the value of the countermeasure to the organization is $850,000.

When calculating the cost of the countermeasure, you should consider the total cost of ownership, including

>> Purchase, development, and licensing

>> Architecture and design

>> Testing and installation

>> Normal operating costs

>> Resource allocation

>> Maintenance and repair

>> Production or service disruptions

The total cost of a countermeasure is normally stated as an annualized amount.

Legal liability

An organization that fails to implement a countermeasure against a threat is exposed to legal liability (also known as *compliance risk*) if the cost of implementing a countermeasure is less than the loss resulting from a realized threat. (See *due care* and *due diligence*, discussed earlier in this chapter.) The legal liability we're talking about here could encompass statutory liability (as a result of failing to obey the law) or civil liability (as a result of failing to comply with a legal contract). A cost-benefit analysis is a useful tool for determining legal liability.

Operational impact

The operational impact of a countermeasure must also be considered. If a countermeasure is too difficult to implement and operate, or if it interferes excessively with normal operations or production, it may be circumvented or ignored and thus will be ineffective. The result may be a risk that is higher than the risk before the so-called mitigation.

Technical factors

In principle, the countermeasure itself shouldn't (but often does, in practice) introduce new vulnerabilities. Improper placement, configuration, or operation of a countermeasure can cause new vulnerabilities; lack of fail-safe capabilities, insufficient auditing and accounting features, or improper reset functions can cause asset damage or destruction; and covert channel access or other unsafe conditions are technical issues that can create new vulnerabilities. Every new component in an environment, including security solutions, adds to the potential attack surface.

Applicable types of controls

A *control* is defined as a safeguard that is used to ensure a desired outcome. A control can be implemented in technology (such as a program that enforces a password-complexity policy by requiring users to employ complex passwords), in a procedure (such as a security incident response process that requires an incident responder to inform upper management), or a policy (such as a policy that requires users to report security incidents to management). Organizations typically have dozens to hundreds or even thousands of controls. So many controls exist that it sometimes makes sense to categorize controls in various ways, to help security professionals better understand the types and categories of controls used in their organization. This section discusses a few of these category groupings.

The major types of controls are

>> **Preventive:** Controls that prevent errors, unwanted events, and unauthorized actions.

>> **Detective:** Controls that detect errors, unwanted events, and unauthorized activities. An example of a detective control is a video surveillance system.

>> **Deterrent:** Controls that discourage people from carrying out an activity. A video surveillance system, for example, employs visibly placed monitors to inform and remind people that video surveillance is in effect.

Other types of controls include

>> **Corrective:** Controls that reverse or minimize the impact of errors and unauthorized events, also known as *recovery controls.* An example of a recovery control is the verification of successful data recovery after a hardware failure.

>> **Administrative:** Policies, standards, or procedures (typically, statements written down in some way).

>> **Compensating:** Controls enacted when primary controls cannot be implemented for any reason, such as excessive cost.

>> **Technical (or logical):** Controls implemented within an IT system or device.

>> **Physical:** Controls that ensure the safety and security of a work facility or processing center.

>> **Recovery:** Controls that are performed in a recovery operation to restore a system or process to its pre-event state.

Another way to think of controls is how they are enforced. Controls in this category are

>> **Automatic:** Some form of automated mechanism ensures enforcement and effectiveness. Such a system may automatically display a login page that requires a user to authenticate successfully before accessing the system.

>> **Manual:** Controls must be performed manually, such as a mandatory review of proposed changes in the change control process.

Most organizations don't attempt to create their control frameworks from scratch; instead, they adopt one of these well-known industry-standard control frameworks:

>> ISO/IEC 27002 (Code of Practice for Information Security Management)

>> NIST 800-53 (Security and Privacy Controls for Federal Information Systems and Organizations)

>> NIST 800-171 (Protecting Controlled Unclassified Information in Nonfederal Systems and Organizations)

>> COBIT 5 for Information Security

>> PCI DSS

>> HIPAA Security Rule controls

>> Center for Internet Security (CIS) Critical Security controls

Organizations typically start with one of these frameworks and make individual additions, changes, or deletions until they arrive at the precise set of controls that they deem to be sufficient.

Control assessments (security and privacy)

An organization that implemented controls but failed to assess them periodically would be negligent. The periodic assessment of controls, for both security and privacy, is a necessary part of a sound risk management system.

Control assessment approach

There are various approaches to control assessments, including:

>> **Control self assessment:** An organization examines its own controls to determine whether they are being followed and whether they are effective. Typically, a control owner is prompted to assert the effectiveness of their control by answering questions about the control and providing supporting evidence in the form of procedures and/or business records.

>> **External assessment:** An organization employs an external agency (which could be a different part of the organization or an external entity, such as an audit firm or a consulting firm) to assess its controls.

Various circumstances might compel an organization to set schedules for control assessment. Critical controls may be assessed monthly or quarterly, other controls assessed annually, and low–risk controls assessed every other year.

Organizations often take a blended approach to control assessment: some controls may be assessed internally and others externally. Some controls are assessed both internally and externally.

TIP

Laws, regulations, and standards often dictate the frequency of control assessment, as well as whether controls must be assessed internally or externally.

Control assessment methodology

It would take an entire book (a long chapter, anyway) to detail the methods used to assess controls. Most of this subject matter lies outside the realm of most CISSPs, so we'll just summarize here. If you are "fortunate" enough to work in a highly regulated environment, you may get exposure to these concepts and more.

CONTROL ASSESSMENT TECHNIQUES

Five basic techniques are used to assess the effectiveness of a control:

>> **Observation:** An auditor watches a control as it is being performed.

>> **Inquiry:** An auditor asks control owners questions about the control, such as how it is performed and how records (if any) are produced.

>> **Corroborative inquiry:** An auditor asks other people about a control to see whether their descriptions agree or conflict with those given by control owners.

>> **Inspection:** An auditor examines records and other artifacts to see whether the control is operating properly.

>> **Reperformance:** An auditor performs actions associated with the control to see whether the results indicate proper control function.

Auditors often use more than one of these techniques to test control effectiveness. The method(s) used are sometimes determined by the auditor, but a law, regulation, or standard may specify the type of control testing required.

SAMPLING TECHNIQUES

Some controls are manifested in many physical locations or are present in many separate information systems. Sometimes, an auditor elects to examine a subset of systems or locations instead of all of them. In large organizations, or those in which controls are implemented identically in all locations, it makes sense to examine a subset of the total number of instances. (Auditors call the entire collection of instances the *population*.)

Available techniques include the following:

>> **Statistical:** A random selection represents the entire population.

>> **Judgmental:** The auditor selects samples based on specific criteria.

>> **Discovery:** For high-risk controls, in which even a single exception may represent a high risk, an auditor may continue to examine a large population in search of a single exception.

Some laws, regulations, and standards have their own rules about sampling and permitted techniques.

REPORTING

Auditors typically create formal reports that include several components, including

>> Audit objectives

>> Personnel interviewed

>> Documents and records examined

>> Dates of interviews and examinations

>> Controls examined

>> Findings for each control (effective or ineffective)

Some laws, regulations, and standards specify the elements required in audit reports and sometimes even the format of a report.

Monitoring and measurement

Any safeguards or controls that are implemented need to be managed, and as you know, you can't manage what you don't measure! Monitoring and measurement not only help you manage safeguards and controls, but also help you verify and prove effectiveness (for auditing and other purposes).

Monitoring and measurement refer to active, intentional steps taken in controls and processes so that management can understand how controls and processes are operating. Depending on the control or process, one or more of the following items will be recorded for management reporting:

>> Number of events that occur

>> Outcome of each event (such as success or failure)

>> Assets involved

>> People, departments, business units, or customers involved

>> Cost

>> Duration

>> Location

For some controls, management may direct personnel (or systems, for automatic controls) to create alerts or exceptions in specific circumstances to inform management of specific events in which they may want to take action. A bank's customer representative might be required to inform a branch manager if a customer asks for change for a $100 bill, for example.

Reporting

Regular reporting is critical for ensuring that risk management is always top of mind for management. Reports should be accurate and concise. Never attempt to hide or downplay an issue, incident, or other bad news. Any changes to the organization's risk posture — whether due to a new acquisition, changing technology, new threats, or the failure of a safeguard, among others — should be reported and explained promptly.

Potentially, a lot of reporting goes on in a risk management process, including

>> Additions and changes to the risk ledger

>> Risk treatment decisions

>> Internal audits

>> External audits

>> Changes to controls

>> Controls monitoring and key metrics

>> Changes in personnel related to the risk management program

You guessed it: Some laws, regulations, and standards require these and other types of reports (and in some cases, in specific formats).

Continuous improvement

Continuous (or continual) improvement is more than a state of mind or a philosophy; it's also a way of thinking about security and risk management. Better organizations bake continual improvement into their business processes as a way of intentionally seeking opportunities to do things better.

ISO/IEC 27001 (Information Security Management Systems Requirements) specifically requires continual improvement in several ways:

>> It requires management to promote continual improvement.

>> It requires a statement of commitment to continual improvement in an organization's security policy.

>> It requires security planning to achieve continual improvement.

>> It requires that the organization provide resources to achieve continual improvement.

>> It requires management reviews to seek opportunities for continual improvement.

>> It requires a formal corrective action process that helps bring about continual improvement.

Risk frameworks

If you ask an experienced security and risk professional about risk frameworks, chances are that they'll think you're talking about either risk assessment frameworks or risk management frameworks. These frameworks are distinct but deal with the same general subject matter: identification of risk that can be treated in some way.

Risk assessment frameworks

Risk assessment frameworks are methodologies used to identify and assess risk in an organization. These methodologies are, for the most part, mature and well established.

Some common risk assessment methods include

>> **Factor Analysis of Information Risk (FAIR):** A proprietary framework for understanding, analyzing, and measuring information risk

>> **OpenFAIR:** An open-source version of FAIR

>> **Operationally Critical Threat, Asset and Vulnerability Evaluation (OCTAVE):** Developed by the CERT Coordination Center

>> **Threat Agent Risk Assessment (TARA):** Developed by Intel; a newer kid on the block

Risk management frameworks

A *risk framework* is a set of linked processes and records that work together to identify and manage risk in an organization. The activities in a typical risk management framework are

>> Creating strategies and policies

>> Establishing risk tolerance

>> Categorizing systems and information

>> Selecting a baseline set of security controls

>> Implementing security controls

>> Assessing security controls for effectiveness

>> Authorizing system operation

>> Monitoring security controls

There is no need to build a risk management framework from scratch. Instead, several excellent frameworks can be adapted for any size and type of organization. These frameworks include

>> NIST SP800-37 (Guide for Applying the Risk Management Framework to Federal Information Systems)

>> ISO/IEC 27005 (Information Security Risk Management)

>> Risk Management Framework (RMF) from the National Institute of Standards and Technology (NIST)

>> COBIT 5 from ISACA

>> Enterprise Risk Management — Integrated Framework from the Committee of Sponsoring Organizations (COSO) of the Treadway Commission

Understand and Apply Threat Modeling Concepts and Methodologies

Threat modeling is a type of risk analysis used to identify security defects in the design phase of an information system or business process. Threat modeling is most often applied to software applications, but it can be used for operating systems, devices, and business processes with equal effectiveness.

CROSS REFERENCE

This section covers Objective 1.11 of the Security and Risk Management domain in the CISSP Exam Outline (May 1, 2021).

Threat modeling is typically attack centric; threat modeling most often is used to identify vulnerabilities that can be exploited by an attacker in software applications.

Threat modeling is most effective when performed in the design phase of an information system, application, or process. When threats and their mitigation are identified in the design phase, much effort is saved by avoiding design changes and fixes in an existing system.

Although organizations take different approaches to threat modeling, the typical steps are

1. Identify threats.
2. Determine and diagram potential attacks.
3. Perform reduction analysis.
4. Remediate threats.

Identifying threats

Threat identification is the first step performed in threat modeling. *Threats* are those actions that an attacker may be able to perform successfully if vulnerabilities are present in the application or system.

For software applications, STRIDE (developed by Microsoft) is a common mnemonic used as a memory aid during threat modeling to list basic threats:

» Spoofing of user identity
» Tampering

» Repudiation

» Information disclosure

» Denial of service

» Elevation of privilege

Although the mnemonic itself doesn't contain threats, it assists the person who's performing threat modeling by reminding them of basic threat categories.

TIP

Appendices D and E in NIST SP800-30 (Guide for Conducting Risk Assessments) are an excellent general-purpose source for threats.

Determining and diagramming potential attacks

After threats have been identified, threat modeling continues through the creation of diagrams that illustrate attacks on an application or system. An *attack tree* can be developed to outline the steps required to attack a system. Figure 3-2 illustrates an attack tree for a mobile banking application.

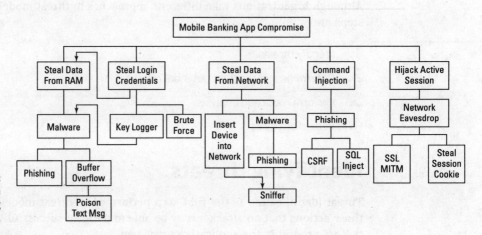

FIGURE 3-2:
Attack tree for a mobile banking application.

REMEMBER

An attack tree illustrates the steps used to attack a target system.

Performing reduction analysis

When performing a threat analysis on a complex application or a system, it is likely that many similar elements will represent duplications of technology. *Reduction analysis* is an optional step in threat modeling to avoid duplication of effort. It doesn't make sense to spend a lot of time analyzing different components in an environment if all of them use the same technology and configuration.

Here are typical examples:

>> An application contains several form fields (derived from the same source code) that request a bank account number. Because all the field input modules use the same code, detailed analysis needs to be done only once.

>> An application sends several types of messages over the same transport layer security (TLS) connection. Because the same certificate and connection are being used, detailed analysis of the TLS connection needs to be done only once.

Remediating threats

As in routine risk analysis, the next step in threat analysis is taking potential measures to mitigate the identified threat. Because the nature of threats varies widely, remediation may consist of one or more of the following for each risk:

>> Changing source code (such as adding functions to closely examine input fields and filter out injection attacks)

>> Changing configuration (such as switching to a more secure encryption algorithm or expiring passwords more frequently)

>> Changing business process (such as adding or changing steps in a process or procedure to record or examine key data)

>> Changing personnel (such as providing training or moving responsibility for a task to another person)

>> Adding a safeguard (such as adding a web application firewall to block attacks)

REMEMBER

The four options for risk treatment are mitigation, transfer, avoidance, and acceptance. In the case of threat modeling, some threats may be accepted as they are.

Apply Supply Chain Risk Management (SCRM) Concepts

Integrating security risk considerations into your supply chain risk management (also known as third-party risk management) strategy helps minimize the introduction of new or unknown risks into the organization.

CROSS REFERENCE

This section covers Objective 1.12 of the Security and Risk Management domain in the CISSP Exam Outline (May 1, 2021).

It is often said that security in an organization is only as strong as its weakest link. In the context of your supply chain, the security of all organizations in a given ecosystem will be dragged down by shoddy practices in any one of them. Connecting organizations before sufficient analysis can result in significant impairment of the security capabilities overall.

REMEMBER

The task of reconciling policies, requirements, business processes, and procedures in the supply chain is rarely straightforward. Further, no one should assume that one organization's policies, requirements, processes, and procedures are right or best for all parties in the supply chain. Each organization's policies, requirements, processes, and procedures should be assessed to identify the best solution for all organizations across the entire supply chain.

Risks associated with hardware, software, and services

Any new hardware, software, or services that an organization is considering should be evaluated appropriately to determine both how it will affect the organization's overall security and risk posture, as well as how it will affect other hardware, software, services, and processes already in place. Integration issues can have a negative impact on a system's integrity and availability.

Third-party assessment and monitoring

It's important to consider the third parties that organizations use. Organizations need to examine their third-party risk programs carefully; they also need to take a fresh look at the third parties to ensure that the risk level related to each one hasn't changed to the detriment of the organization.

Any new third-party assessments or monitoring should be considered carefully. Contracts (including privacy, nondisclosure requirements, and security requirements) and SLAs should be reviewed to ensure that all important security issues and regulatory requirements are addressed adequately.

Fourth-party risk

It's no longer considered sufficient to identify third parties: it's also important to understand *their* third parties, which are known as *fourth parties.* The assessment of a third-party organization should include its third-party risk program and the identification of its critical third parties.

Minimum security requirements

Minimum security requirements, standards, and baselines should be documented to ensure that they are fully understood and considered. Addressing security requirements across separate organizations within the supply chain is almost never as easy as simply combining them into one document. Instead, many instances of overlap, underlap, gaps, and contradiction must be reconciled. A transition period may be required so that there is ample time to adjust the security configurations, architectures, processes, and practices to meet the requirements of all organizations in the supply chain.

Service-level agreement requirements

SLAs establish minimum performance standards for a system, application, network, service, or process. An organization establishes internal SLAs and operating-level agreements to provide its users a realistic expectation of the performance of its information systems, services, and processes. A help-desk SLA might prioritize incidents as 1, 2, 3, and 4, and establish SLA response times of 10 minutes, 1 hour, 4 hours, and 24 hours, respectively. In third-party relationships, SLAs provide contractual performance requirements that an outsourcing partner or vendor must meet. An SLA with an Internet service provider might establish a maximum acceptable downtime that, if exceeded within a given period, results in invoice credits or (if desired) cancellation of the service contract.

Establish and Maintain a Security Awareness, Education, and Training Program

A CISSP candidate should be familiar with the tools and objectives of security awareness, training, and education programs. Adversaries are well aware that as organizations' technical defenses improve, the most effective way to attack an organization is through its people. Hence, all personnel in an organization need to be aware of attack techniques so that they can be on the lookout for these attacks and not be fooled by them.

CROSS REFERENCE

This section covers Objective 1.13 of the Security and Risk Management domain in the CISSP Exam Outline (May 1, 2021).

Methods and techniques to present awareness and training

Security awareness is an often-overlooked factor in an information security program. Although security is the focus of security practitioners in their day-to-day functions, it's often taken for granted that common users possess this same level of security awareness. As a result, users can unwittingly become the weakest links in an information security program's defenses. Several key factors are critical to the success of a security awareness program:

>> **Senior-level management support:** Under ideal circumstances, senior management is seen attending and actively participating in training efforts.

>> **Clear demonstration of how security supports the organization's business objectives:** Workers need to understand why security is important to the organization and how it benefits the organization as a whole.

>> **Clear demonstration of how security affects all workers and their job functions:** The awareness program needs to be relevant for all workers so that everyone understands that security is everyone's responsibility.

>> **Consideration of the audience's current level of training and understanding of security principles:** Training that's too basic will be ignored; training that's too technical will not be understood.

>> **Ensuring that training is relevant and engaging:** Training needs to be relevant and engaging for all audiences, reflecting applicable regulations, technologies in use, and the organization's culture.

>> **Action and follow-up:** A glitzy presentation that's forgotten as soon as the audience leaves the room is useless. Find ways to incorporate the security information you present into day-to-day activities and follow-up plans.

The three main components of an effective security awareness program are a general awareness program, formal training, and education.

Awareness

A general security awareness program provides basic security information and ensures that everyone understands the importance of security. Awareness programs may include the following elements:

>> **Indoctrination and orientation:** New employees and contractors should receive basic indoctrination and orientation. During the indoctrination, they may receive a copy of the corporate information security policy, be required to acknowledge and sign acceptable-use statements and nondisclosure agreements, and meet immediate supervisors and pertinent members of the security and IT staff.

>> **Presentations:** Lectures, video presentations, and interactive computer-based training are excellent tools for disseminating security training and useful information. Bonuses and performance reviews are sometimes tied to participation in these types of programs.

>> **Printed materials:** Security posters, corporate newsletters, and periodic bulletins are useful for disseminating basic information such as security tips and promoting awareness of security.

>> **Social engineering and phishing campaigns:** Many organizations are increasingly simulating the same tactics that an attacker would use to determine where improvement is needed within the organization. The help desk might randomly ask users for their passwords to see whether they actually comply (if they give up their password, they fail!) or simulate an email phishing campaign to see which users click the malicious link. In the latter

case, the link would report the user's action back to the security team, and the user would win a prize, like more security awareness training!

>> **Identification of security champions:** Identifying and educating security champions across the organization in different departments, teams, and roles is a great force multiplier that helps security become embedded throughout the organization.

>> **Gamification:** If you've ever wondered why people are so addicted to their mobile devices, the reason is all the badges, stars, and coins they earn in their favorite games. Gamification is a great way to recognize your users' good security habits and help them get excited about "winning."

Training

Formal training programs provide more in-depth information than an awareness program and may focus on specific security-related skills or tasks. Such training programs may include

>> **Classroom training:** Instructor-led or other formally facilitated training, possibly at corporate headquarters or a company training facility

>> **Self-paced training:** Usually web-based training in which students can proceed at their own pace

>> **On-the-job training:** May include one-on-one mentoring with a peer or immediate supervisor

>> **Technical or vendor training:** Training on a specific product or technology provided by a third party

>> **Apprenticeship or qualification programs:** Formal probationary status or qualification standards that must be satisfactorily completed within a specified period

Education

An education program provides the deepest level of security training, focusing on underlying principles, methodologies, and concepts. In all but the largest organizations, this training is delivered by external agencies, as well as colleges, universities, and vocational schools.

An education program may include

>> **Continuing-education requirements:** Continuing Education Units are becoming popular for maintaining high-level technical or professional certifications such as the CISSP or Certified Information Systems Auditor.

- » **Certificate programs:** Many colleges and universities offer adult education programs that have classes about current and relevant subjects for working professionals.

- » **Formal education or degree requirements:** Many companies offer tuition assistance or scholarships for employees enrolled in classes that are relevant to their profession.

Periodic content reviews

Congratulations! You've chosen a profession that is constantly and rapidly chang-ing. As a result, you must constantly review and update security education, train-ing, and awareness programs to ensure that they remain relevant — and to ensure that your own knowledge of current security concepts, trends, and technologies remains current. We suggest that you examine the content of security education and training programs at least once per year to ensure that there is no mention of obsolete or retired technologies or systems and that current topics are included.

Program effectiveness evaluation

As we say often in this book, you can't manage what you don't measure. Security awareness training is definitely included here. It is vital that security awareness training include different measurements so that security managers and company leadership know whether the effort is worthwhile. Examples include

- » **Quizzes:** Whether delivered in the classroom or via on-demand web-based training, quizzes send a clear message that workers are expected to learn and retain security awareness knowledge. When minimum passing scores are enacted, quizzes are even more effective.

- » **Participation:** It's helpful to track completion rates to ensure that as many workers as possible complete required and optional training.

- » **Other security program metrics:** It may be interesting to track security awareness training metrics with other metrics, such as security incidents, reports to ethics hotlines, and employees' reporting of security issues. It should be noted that some of these metrics may trend upward, which would represent workers' greater awareness of security-related issues and a greater likelihood that those issues will be reported.

Chapter **4**

Asset Security

The Asset Security domain addresses the collection, classification, handling, and protection of information assets throughout the information life cycle, as well as the management of physical and virtual assets such as servers, endpoints, and network devices. Essential concepts within this domain include data ownership, privacy, data security controls, and support. This domain represents 10 percent of the CISSP certification exam.

Identify and Classify Information and Assets

Information and information systems are valuable business assets that require protection from cybersecurity threats. The appropriate level of protection is determined by the value of the information and any applicable regulations with which the organization must comply.

CROSS REFERENCE

This section covers Objective 2.1 of the Asset Security domain in the CISSP Exam Outline (May 1, 2021).

Asset management is generally an IT function used to track many sorts of assets, including

>> **Servers:** From both operational and security perspectives, organizations must know the state and location of every physical server in their digital estate (including on-premises data centers, server closets, and colocation facilities) so they can be actively monitored, properly updated, and securely operated.

>> **Virtual machines and containers:** Unlike physical servers, virtual machines and containers are harder to manage as assets because they're abstract, dynamic (hundreds or thousands of virtual machines and containers can be provisioned and deprovisioned on demand, particularly in a highly orchestrated microservices architecture), and ephemeral (containers are often provisioned for a specific purpose and may be very short-lived, from a few minutes to milliseconds — just long enough to execute a service or process).

>> **User endpoints:** Most breaches today begin on a compromised end-user desktop or laptop. Keeping track of all your end-user desktops and laptops is critical to help ensure that they're properly maintained and that appropriate security controls — such as security updates, extended detection and response (XDR), virtual private network (VPN) software, and digital certificates — are properly installed and configured.

>> **Mobile devices:** The proliferation of mobile devices — including smartphones, tablets, and wearables — has created a massive attack surface that organizations must address. Maintaining an accurate inventory of mobile devices can be particularly challenging due to the number and numerous types of devices, the prevalence of "bring your own device" (BYOD) policies, and the lack of mobile device management (MDM) software in many organizations. MDM software enables organizations to keep track of their mobile devices and enforce policies, such as requiring a passcode or biometric lock. Other key MDM software capabilities include preventing rooted or jailbroken devices from connecting to the network, containerizing (or isolating) company applications and data, and remote wiping to delete company applications and data securely if a device is lost or stolen or when an employee or contractor leaves the company.

>> **Internet of Things (IoT) devices:** Devices such as surveillance cameras, smart assistants, door actuators, temperature sensors, building management systems, and security systems that don't necessarily have a human-machine interface (HMI) often run on internal data networks. Many of these devices may be easily overlooked in an organization's vulnerability management program. These devices may also be installed or deployed with default

passwords or settings by staff members who aren't focused on security, such as maintenance personnel or field technicians. As a result, attackers can often break into an organization through one of these devices. For instance, IP cameras and home routers were targeted and compromised by the Mirai botnet in 2016 and used to launch a significant DDoS attack on Dyn.

» **Network devices:** Switches, routers, firewalls, and other network devices have their own security configurations and operating systems that must be kept up to date. These devices are increasingly being deployed in public and private clouds as virtual appliances that must also be accounted for and managed properly.

» **Network information:** Network information is another asset that organizations must manage. Examples include Internet Protocol address space, autonomous system numbers (ASNs), domain names, telecommunications circuits, and direct inward dialing (DID) numbers.

» **Operating systems:** Keeping track of the various operating systems in use throughout your organization — including servers, desktops, laptops, mobile devices, hypervisors, network devices and tools, security appliances and tools, industrial control systems (ICS), and medical devices — is critical to ensure that all systems are kept current with security updates and compliant with licensing requirements and to help identify systems that may be affected by a new zero-day vulnerability.

» **Software applications:** Maintaining an accurate inventory of all your software applications is critical to ensure that all of them are kept current with security updates and compliant with licensing requirements and to help identify systems that may be affected by a new zero-day vulnerability.

» **Information:** Although ransomware and cryptomining attacks have been getting lots of media attention over the past few years, information theft remains the primary motivation for attackers, and data is still a valuable asset. Knowing what information you have — such as financial data, personally identifiable information, protected health information, and intellectual property — as well as its classification level (discussed later in the "Data classification" section of this chapter), everywhere that it's accessed and stored (including user endpoints, mobile devices, servers, databases, cloud storage, and backups), is critical to ensure that you can protect your data appropriately and completely. Many regulations specify safe-harbor provisions in the event of a data breach if an organization can prove that any exfiltrated data was encrypted properly.

Additionally, data privacy regulations such as the European Union's General Data Protection Regulation (GDPR; discussed in Chapter 3) codify a person's right to be forgotten. This right requires an organization to delete all of a person's private data except under limited circumstances, if formally requested to do so by that person. Having a complete inventory of your information assets helps ensure that you can comply with such requests in a timely and accurate manner.

>> **Personnel:** Although maintaining an inventory of your organization's personnel isn't necessarily a security responsibility, this information does provide important input for several security functions. Your incident response, business continuity, and disaster recovery plans (discussed in Chapter 9), for example, all define specific roles and responsibilities for personnel within your organization. Identity and access management (discussed in Chapter 7) requires user accounts to be bound to authorized personnel. Information security awareness and training (discussed in Chapter 3) requires personnel rosters to ensure full participation. Finally, personnel safety and security (discussed in Chapter 9) requires accurate knowledge of who is in a building, for example, in the event of a fire that requires evacuation. Sources for personnel lists might include your human resources department, company directories, organizational charts, and department rosters.

>> **User accounts:** As discussed in the personnel item, user accounts must be associated with authorized personnel to ensure effective identity and access management (see Chapter 7). Knowing which personnel have left an organization or no longer require access to systems and networks is critical so that accounts can be properly disabled, deprovisioned, and deleted.

>> **Facilities:** Security personnel should have a complete list of all facilities, work centers, business locations, and so on to facilitate activities such as ensuring that all facilities are adequately monitored and protected. Chapter 5 includes a deep dive into facility security, and Chapter 9 discusses physical security.

>> **Service providers:** Every service provider that manages or processes information of any kind must be included in your service inventory and evaluated periodically. Third-party risk management (TPRM), also known as supply-chain risk management (SCRM), is covered in Chapter 3.

Discrepancies in the tracking of any of these categories of assets can weaken your cybersecurity posture. If the complete inventory of servers or endpoint computers is unknown, for example, some systems may not be configured properly or patched with the latest security patches. Cybercriminal organizations that are intent on breaking into such an environment can identify these weak points quickly and successfully attack the organization.

REMEMBER

It takes only one unpatched system or device, or one user account, to give an attacker an easy way into the environment.

Data classification

A *data classification* scheme helps an organization assign a value to its information assets based on its *sensitivity* to loss or disclosure and its *criticality* to the organization's mission or purpose. The scheme also helps the organization determine the appropriate level of protection. Additionally, data classification schemes may be required for regulatory or other legal compliance. System and site classification schemes determine the level and type of protection for information systems and the facilities where they reside and personnel work.

Applying a single protection standard uniformly across all an organization's assets is neither practical nor desirable. In such a case, noncritical assets are over-protected, or critical assets are underprotected. In the words of former U.S. national security adviser McGeorge "Mac" Bundy, "If we guard our toothbrushes and diamonds with equal zeal, we will lose fewer toothbrushes and more diamonds."

An organization's employees also need to understand the classification schema being used, proper classification of information assets, handling and safeguarding requirements, and proper destruction or disposal procedures.

Commercial data classification

Commercial data classification schemes are typically implemented to protect information with monetary or intrinsic value, comply with applicable laws, protect privacy, and limit liability. Criteria by which commercial data is classified include

>> **Value:** The most common classification criterion in commercial organizations, based on monetary or some other value.

>> **Age/useful life:** Information that loses value over time, becomes obsolete or irrelevant, or becomes common/public knowledge.

>> **Regulatory requirements:** Sensitive or personal information, such as medical records subject to the U.S. Health Insurance Portability and Accountability Act (HIPAA) or consumer data subject to the EU's General Data Protection Regulation (GDPR), may have legal requirements for protection. Classification of such information may be based not only on compliance, but also on liability limits. Read Chapter 3 to learn more about data protection and privacy regulations.

Descriptive labels (such as "Confidential and Proprietary" and "Internal Use Only") are often applied to company information. The organizational requirements for protecting information labeled as such, however, aren't always formally defined. Organizations should formally identify standard classification levels and specific requirements for labeling, handling, storage, and destruction/disposal. Figure 4-1 shows an example of a label on a hard-copy document.

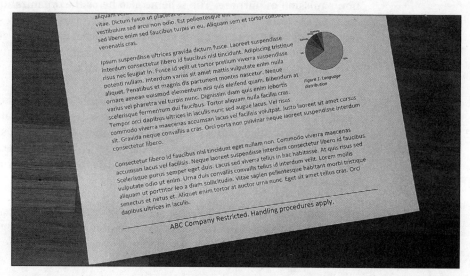

Photo courtesy of authors

FIGURE 4-1: Example document marking.

Government data classification

Government data classification schemes are generally implemented to

» Protect national interests or security

» Comply with applicable laws

» Protect privacy

Within each classification level, certain safeguards are required to access, use, handle, reproduce, transport, and destroy classified information. In addition to having an appropriate clearance level at or above the level of information being processed, people must have a need to know before they can access the information. *Need to know* is defined as requiring the information to perform an assigned job function.

REMEMBER

A common system used within the U.S. Department of Defense consists of five broad categories for information classification:

- » **Unclassified:** The lowest government data classification level is Unclassified. Unclassified information isn't sensitive, and unauthorized disclosure won't cause any harm to national security. Unclassified information may include information that was once classified at a higher level but has since been declassified by an appropriate authority. Unclassified information isn't automatically releasable to the public and may include additional modifiers such as For Official Use Only or For Internal Use Only.

- » **Sensitive but Unclassified (SBU):** Sensitive but Unclassified information is a common modifier of unclassified information. It generally includes information of a private or personal nature. Examples include test questions, disciplinary proceedings, and medical records.

- » **Confidential:** Confidential information is information that, if compromised, could damage national security. Confidential information is the lowest level of classified government information.

- » **Secret:** Secret information is information that, if compromised, could seriously damage national security. Secret information must normally be accounted for throughout its life cycle, all the way to its destruction.

- » **Top Secret:** Top Secret information is information that, if compromised, could gravely damage national security. This information may require additional safeguards, such as special designations and handling restrictions.

A person must have both the appropriate clearance level and need to know to gain access to classified information.

Data handling

Along with the classification levels described in the preceding sections, the classification of a file or database means little unless people know what is required for the different classification levels. For this reason, data classification policies should also describe how data is to be handled and protected at each level of classification. Data classification handling guidelines typically include numerous use cases, such as

- » Storing on a server
- » Storing on a desktop or laptop
- » Storing on a mobile device
- » Sending via email
- » Sending via instant message
- » Storing in a cloud service

>> Copying to a USB external storage device

>> Printing

>> Storing hard copy

>> Sending via Fax

>> Shipping hard copy or electronic copy via courier

Generally, the higher the classification level, the more restrictions are placed on any of these activities. And when the data is in electronic form, encryption is often required to further protect the most sensitive information.

Table 4-1 shows a typical data handling policy. This example is depicted in abbreviated form for illustration purposes. An actual handling policy would include more detail regarding permitted and forbidden actions.

TABLE 4-1 ## Typical Data Handling Guidelines

Action	Public	Confidential	Restricted	Secret
Server storage	Permitted	Permitted	Encryption required	Encryption required
Endpoint storage	Permitted	Encryption required	Encryption required	Not permitted
Mobile device storage	Permitted	Encryption required	Encryption required	Not permitted
Sending via email	Permitted	Permitted	Encryption required	Not permitted
Sending via instant message	Permitted	Encryption required	Encryption required	Not permitted
Sending via fax	Permitted	Attended fax only	Attended fax only	Not permitted
Shipping via courier	Permitted	Permitted	Double-sealed	Double-sealed
Cloud storage	Permitted	Permitted	Permitted	Permitted by exception only
USB external storage	Permitted	Encryption required	Not permitted	Not permitted
Printing	Permitted	Attended printing only	Attended printing only	Not permitted
Hard-copy storage	Permitted	Locked drawer or cabinet	Double-locked	Double-locked

Asset classification

Data classification is related to the identification of the sensitivity or criticality of data and proper procedures for handling that data. Asset classification, however, also applies to the information systems that store, process, and transmit this information. Better organizations go beyond data classification by implementing system classification. The two approaches are distinctly related: Systems that store or process data at higher classification levels should be protected better than systems that don't store or process data at higher classification levels.

This practice has been around for quite a long time. Under the Payment Card Industry Data Security Standards (PCI DSS), for example, systems that are in scope are required to have additional safeguards that an organization may not implement on its other systems. Organizations often place these systems in separate networks that have stricter network-level access controls.

Organizations can go still further with data and asset classification by implementing a facilities classification scheme that stipulates mandatory protection for various facilities, based on location, activities, and the presence of classified assets and data. Table 4-2 illustrates an example facilities classification policy.

TABLE 4-2   **Example Facilities Classification Policy**

Control	Sales Office	Processing or Development Center	Data Center
Fencing	Not required	Not required	6-foot chain-link fence
Video surveillance	Not required	Reception, ingress, and egress points	Reception, ingress and egress points; all internal corridors
Video recording	Not required	Motion sensor, 30-day retention	Full recording, 90-day retention
Key card control	Main entrance	All entrances	All entrances and zones
Security guard	Not required	Reception 5x10	Reception and patrol 7x24
Visitor control	Visitor log	Visitor log and proof of ID	Visitor log, relinquish ID, appointment only
Company signage	Permitted	Permitted	Not permitted
Parking	Open	Permit required	Permit required, key card access
Risk and hazards assessment	Not required	Initial	Annual

Establish Information and Asset Handling Requirements

Establishing information and system asset processes and reliable inventories is a means to an end. That end is determining how those assets are to be handled and used. Handling requirements fall into two broad categories:

>> **Information assets:** Structured and unstructured information needs to be classified according to the organization's data classification rules (discussed in the preceding section). These rules also include data retention and policies that determine what users are permitted to do with information and who has access to it in the first place.

>> **Information systems:** This broad category includes servers, endpoints, network devices, and mobile devices. From a security perspective, the imperatives here include system and device hardening (discussed in Chapter 5), patch management (also discussed in Chapter 5), and other operational processes and procedures.

CROSS REFERENCE

This section covers Objective 2.2 of the Asset Security domain in the CISSP Exam Outline (May 1, 2021).

Sensitive information such as financial records, employee data, and information about customers must be clearly marked, properly handled and stored, and appropriately destroyed per established organizational policies, standards, and procedures:

>> **Marking:** How an organization identifies sensitive information, whether electronic or hard copy. A marking might read Confidential, for example (as discussed earlier in this chapter). The method for marking will vary, depending on the type of data we're talking about. Electronic documents might have a marking in the margin at the footer of every page. When an application displays sensitive data, the application itself may inform the user of the classification of the data being displayed.

>> **Handling:** The organization should have established procedures for handling sensitive information. These procedures detail how employees can transport, transmit, and use such information, as well as any applicable restrictions.

>> **Storage and backup:** As with handling, the organization must have procedures and requirements specifying how sensitive information must be stored and backed up and how backup media must be protected.

>> **Destruction:** Sooner or later, an organization must destroy a document that contains sensitive information. The organization must have a data retention schedule and procedures detailing how to destroy sensitive information that has been previously retained, regardless of whether the data is in hard-copy form or saved as an electronic file.

Similarly, information systems and network devices must be designed, implemented, and operated securely according to *their* classifications. Activities include

>> **Architecture and design:** Overall and detailed architecture and design standards ensure that all information systems will be implemented and operated securely and that the overall environment will be resistant to attack.

>> **Patch and configuration management:** This category includes policies and procedures regarding the initial configuration of systems and schedules/procedures for applying security patches.

>> **Hardening:** This category refers to the architecture, design, and configuration of a system or device that makes it more resistant to attack and abuse.

>> **Security event monitoring:** Many systems and devices are configured to transmit security-related events to a centralized security event and information management system, so that security operations personnel are aware of events that could be signs of an attack.

>> **Resilience:** Related to business continuity and disaster recovery planning requirements, resilience is a characteristic of a system that determines its ability to continue operating despite various destructive events.

DETERMINING APPROPRIATE HANDLING REQUIREMENTS

You may be wondering, "How do I determine what constitutes appropriate handling requirements for each classification level?" You have two main ways to figure out the answer:

- **Applicable laws, regulations, and standards:** Often, regulations and standards such as the EU's GDPR, the U.S. Health Insurance Portability and Accountability Act (HIPAA), the California Consumer Privacy Act (CCPA), the Gramm-Leach-Bliley Act (GLBA), and PCI DSS contain specific requirements for handling sensitive information.

- **Risk assessment:** As described in Chapter 3, a risk assessment identifies relevant threats and vulnerabilities and establishes controls to mitigate risks. Some of these controls may take the form of data handling policies and requirements that would become a part of an organization's asset classification program.

Provision Resources Securely

The secure provisioning of resources involves intentionality. Business rules and policies stipulate how information systems and information assets are to be acquired, protected, used, and discarded when no longer needed. From an information security perspective, the intention of these policies and rules ensures the confidentiality, integrity, and availability of these assets, protecting them from accidental or intentional disclosure, harm, and destruction.

CROSS REFERENCE

This section covers Objective 2.3 of the Asset Security domain in the CISSP Exam Outline (May 1, 2021).

Better organizations have formal architecture and design standards to ensure consistency in their systems and operations. Further, formal operations procedures reduce errors and configurations that could make systems easier to attack and compromise.

Information and asset ownership

Within an organization, *owners* and *custodians* of systems, data, and the business or mission (specifically, a line of business or mission aspect) are implicitly or explicitly assigned.

TIP

Organizations should explicitly define owners and custodians of sensitive assets to avoid confusion or ambiguity regarding roles, responsibilities, and accountability.

An *owner* is typically assigned at an executive or senior management level within an organization, such as a department head, director, or vice president. An owner doesn't legally own the asset assigned to them; the owner is ultimately responsible for safeguarding designated assets and may have fiduciary responsibility or be held personally liable for negligence in protecting these assets under the concept of due care. For more on due care, read Chapter 3.

Typical responsibilities of an owner may include

>> Determining classification levels for assigned assets

>> Determining policy for access to the asset

>> Maintaining inventories and accounting for assigned assets

>> Periodically reviewing classification levels of assigned assets for possible downgrading, destruction, or disposal

>> Delegating day-to-day responsibility (but not accountability) and functions to a custodian

A *custodian* is a person who has day-to-day responsibility for protecting and managing assigned assets. IT systems administrators or network administrators often fill this role. Typical responsibilities may include

>> Performing regular backups and restoring systems and/or data, when necessary

>> Ensuring that appropriate permissions are properly implemented on systems, directories, and files, and that they provide sufficient protection for the asset

>> Ensuring that IT systems are adequately protected with system hardening and other safeguards

>> Assigning new users to appropriate permission groups and revoking user privileges when required

>> Maintaining classified documents or other materials in a vault or secure file room

REMEMBER

The distinction between owners and custodians, particularly regarding their responsibilities, is an essential concept in information security management. The data owner has ultimate responsibility for the security of the data, whereas the data custodian is responsible for day-to-day security administration.

Asset inventory

An accurate inventory of system and information assets is key to effective security. An old saying in our business is, "You cannot protect what you do not know about." Although this saying may sound almost smart-alecky, it's a fundamental truth in cybersecurity. Only identified and known information assets can be adequately and appropriately protected, including information as well as information systems and even facilities. Having an up-to-date and accurate asset inventory (for information as well as information systems) is a prerequisite for many other security activities and operations, such as patch management and data retention. An inaccurate or incomplete asset inventory will result in blind spots where assets may not be adequately protected and may provide an entry point for attackers.

Generally, many tools in use will contribute to a complete picture of asset inventory, including

>> **Security scanning tools:** Scan networks and identify vulnerabilities are useful for contributing to an asset database

>> **System management tools:** Manage the configuration of systems and devices; these tools have databases representing all of the systems they manage

>> **Data discovery tools:** Scan systems' stored data to contribute to the overall information asset database

None of the various types of tools has what is considered to be a complete master list of assets. Instead, organizations that deploy an asset management system will integrate other tools into the asset management tool and apply business rules for naming, reconciliation, and asset review to ensure that the list is as complete as possible.

Asset inventory isn't just about knowing the numbers of servers and laptops or walking into a data center to count the power cords. Although some types of inventory are easy to see and understand, others are less so. We can think of assets in two broad categories:

>> **Tangible:** Servers, desktop and laptop computers, mobile devices, IoT devices, network devices, work facilities, data centers, employees, contractors, and service providers

>> **Intangible:** Virtual machines, containers, information, intellectual property, personally identifiable information (PII), user accounts, and domains

WARNING

Organizations need to keep their inventories of all types of assets up to date. A breach is a terrible time to figure out what assets exist in a specific place or other context.

Asset management

Simply put, *asset management* is a life-cycle process that formally manages the information, software, hardware, and other types of assets and maintains an up-to-date inventory. As we stated at the start of this chapter, the accuracy of asset inventories is critical to the effectiveness of cybersecurity safeguards.

Better IT and security organizations establish formal asset management programs with tools and processes to keep track of hardware, software, and information assets. This task is usually easier said than done, however. Generally, an accurate asset inventory results from the merging of data from various tools. An accurate inventory of servers and endpoints, for example, often relies on information feeds from tools used to configure servers and endpoints, security scanning tools, and other tools that manage various aspects of those systems. Often, each of these tools has a slightly different list of assets; analysts need to compare these lists periodically to ensure that actions are taken to ensure complete coverage so that some assets don't go astray.

We often say that effective cybersecurity relies on good IT service management. Asset management is a prime example of this type of management — and perhaps the most important example of all.

Manage Data Life Cycle

Data life-cycle management is a set of processes and procedures used to track all the data in an organization, whether that data is in electronic or paper form. Like asset management, data management doesn't just happen magically; instead, data management must be established as a formal activity with roles and responsibilities assigned to key personnel.

CROSS REFERENCE

This section covers Objective 2.4 of the Asset Security domain in the CISSP Exam Outline (May 1, 2021).

The purpose of data management is to ensure that all the data residing in an organization is known, properly managed, adequately protected, properly used, and discarded when it's no longer needed — all in compliance with internal policy as well as applicable standards, laws, and regulations.

Data management generally relies on a central store of information about data in the organization, which we might call a *data catalog*. This catalog could reside in a simple worksheet or be managed in a purpose-built application. Activities to keep the data catalog up to date include interviews with application and system owners and periodic scans of data stores.

Many organizations do little or nothing regarding big-picture data management. This situation is changing, however, with the advent of sweeping privacy laws around the world. The penalties exacted by many of these laws are frighteningly large, prompting organizations to (finally) establish data management processes. These privacy protection laws and regulations exist at continental (such as the European Union), country (or federal), state, and local levels throughout the world, as well as in various industries. Privacy protection laws are among some of the most stringent laws enacted, and legal requirements vary greatly. These laws also commonly limit the collection, use, and retention of personal data and trans-border information flows (or export) of personal data. Privacy laws are discussed in Chapter 3.

Data roles

As with any other sensitive data, organizations must assign data owners and custodians (or processors) who are ultimately responsible for safeguarding personal data and for the secure collection, processing, and use of the data. Anyone within an organization who has access to personal data in any capacity must be thoroughly familiar with established procedures for collecting, handling, and safeguarding such information throughout its entire life cycle. These procedures include retention and destruction of private data and technical issues such as data remanence. Concerning the complete life cycle of data in an organization, the roles for managing data include

>> **Owner:** Usually a business owner, typically a department head, who uses data to support one or more principal business processes.

>> **Controller:** An organization (or a part of an organization) that determines the processing performed on a set of data.

>> **Custodian:** Typically, an IT department that manages data at the technology level, in information systems, applications, database management systems, storage systems, and so on.

>> **Processor:** An organization (or a part of an organization) that processes data as directed by (and on behalf of) a controller.

>> **User:** A human who accesses and perhaps manipulates data, including adding, changing, and removing it.

>> **Subject:** A more inclusive role than user, including not only human data users, but also machine users. In other words, a subject accessing a particular data set could be a human user or an autonomous program.

TIP

The terms *controller* and *processor* have gained prominence in recent years, as they are principal terms defined in the GDPR and other privacy laws.

Data collection

Data collection describes any means through which an organization acquires data, particularly about a person. Generally, data collection is concerned with human interaction when someone is typing personal information into a web form or filling out a hard-copy form.

Data collection has also attracted more attention with the passage of many new privacy laws. Often, organizations are required to disclose their privacy policy at the point of data collection and provide users an opportunity to opt in or out of the privacy policy.

Data location

Increasingly, laws, standards, and other legal mandates include stipulations regarding the locations where various types of data may be processed and stored. Some privacy laws require information about a country's citizens to reside within the country, for example. Organizations need to be aware of location-specific requirements and include steps in their asset management processes to identify and confirm the physical location of assets, as well as to conduct regular reviews to confirm that asset locations comply with regulatory requirements. Chapter 3 discusses privacy in greater detail.

Data maintenance

Data maintenance refers to any activity in which data is being reviewed, updated, corrected, or discarded. This activity can be human-driven, as when a user or specialist examines data (one record at a time or in bulk) and makes necessary changes, or it can be machine-assisted or entirely autonomous, as when a software program makes needed changes.

The concept of *integrity* should come to mind when you consider data maintenance activities. The human- and machine-driven activities concerned with maintaining and updating data should include safeguards to ensure that these activities preserve and improve the integrity and accuracy of that data.

Data retention

Most organizations are bound by various laws, regulations, and standards to collect and store certain information and keep it for specified periods. An organization must be aware of legal requirements and ensure that it complies with all applicable regulations and standards.

Records-retention policies should cover all physical records, as well as all electronic records that may be located on file servers or in document management systems, databases, email systems, archives, and records management systems. These records also include paper copies and backup media stored at offsite facilities.

Organizations that want to retain information longer than required by law should formally and firmly establish why such information should be kept longer. Nowadays, just having information can be a liability, so keeping sensitive information longer should be the exception rather than the rule. Organizations should consider the risks associated with extending retention beyond legal requirements when retention decisions are made.

Data retention applies equally to the minimum and the maximum periods that data may be retained in an organization. Retaining data longer than necessary (or permitted by law) increases an organization's liability, particularly where sensitive information is concerned. PCI DSS, for example, requires that credit card data be retained for as short a period as possible and that certain items, such as magnetic-stripe data and personal identification numbers, not be retained at all, whereas log data must be retained for at least one year to aid in possible security investigations.

At the opposite end of the records-retention spectrum, many organizations now destroy records (including backup media) as soon as legally permissible to limit the scope and cost of any future discovery requests or litigation. Before implementing any draconian retention policies that severely restrict your organization's retention periods, you should fully understand the negative implications of a policy for your disaster recovery capabilities. Also, consult your organization's legal counsel to ensure that you're in full compliance with all applicable laws and regulations.

WARNING

Although extremely short retention policies and practices may be prudent for limiting future discovery requests or litigation, they're illegal for limiting pending discovery requests or litigation. In such cases, don't destroy pertinent records; if you do, you'll go to jail. You go directly to jail, you don't pass Go, you don't collect $200, and (oh, yeah) you don't pass the CISSP exam — or even remain eligible for CISSP certification!

Data remanence

Data remanence refers to residual data that remains on storage media or in memory after a file or data has been deleted or erased. Data remanence occurs because standard delete routines mark deleted data as storage or memory space that's available to be overwritten. To eliminate data remanence, the storage media and memory must be properly wiped, degaussed, encrypted, or physically (and completely) destroyed. *Object reuse* refers to an object (such as memory space in a program or a storage block on media) that may present a risk of data remanence if it isn't properly cleared.

Data destruction

Data destruction refers to various techniques used to remove data from a system or a data storage medium. Organizations destroy data for a variety of reasons, including data retention (when specific records or even an entire database are removed according to the data retention schedule), the retirement of computers and storage systems, the discarding of older storage media, and the migration of data from on-premises systems to cloud-based systems.

Various techniques for destroying data are employed to ensure that no one can reconstruct that data. A good example is the common situation in which an organization discards older desktops and servers, and that equipment is later sold or donated to a charitable organization. IT departments need to make sure that there's no way for company information to be reconstructed later. The techniques used include overwriting, degaussing, and shredding.

Ensure Appropriate Asset Retention

IT departments pay attention to the vintage of their computing and network hardware, as well as the operating systems and applications that run on them. As a crucial part of overall asset management, tracking hardware and software versions is used for long-term planning to ensure that organizations avoid situations in which production hardware and software are no longer supported by their manufacturers. *Technical debt* is a term that describes an organization's continued use of unsupported software and hardware.

CROSS REFERENCE

This section covers Objective 2.5 of the Asset Security domain in the CISSP Exam Outline (May 1, 2021).

End of life

Hardware and software vendors typically publish a support schedule. Organizations pay close attention to end-of-life (EOL) dates and plan upgrades or migrations to newer products to safely avoid running production applications on EOL systems. But despite their best efforts, many organizations find themselves marooned on EOL hardware or software for a variety of reasons:

>> The software or hardware vendor is out of business, and no newer replacement is available.

>> The cost of upgrading to newer hardware or software is prohibitive.

>> Software applications aren't supported in newer versions of operating systems.

>> External equipment, such as medical laboratory equipment or manufacturing equipment, runs only on now-unsupported operating systems or hardware.

Organizations that find themselves in EOL situations can often enact other safeguards, such as more extreme isolation of unsupported systems or increased security monitoring.

End of support

End of support (EOS) refers to a state in which a manufacturer no longer provides technical support for specific hardware or software products. Product manufacturers generally publish an EOL or EOS schedule, often years in advance, so that organizations can include upgrades and migrations in their long-term strategic planning. The migration of a complex software application can require months of planning and months of migration work; hence, learning about EOS on short notice generally isn't helpful.

EOS is often a contentious topic between manufacturers and their customers. Sometimes, customers view vendors' announcements of EOS as being a thinly veiled scheme for selling expensive upgrades to boost their bottom line. Pragmatically, however, it can be costly for a manufacturer to support many versions of products, each with specific components and technologies. It can be impractical to support old versions of hardware when their chipsets are no longer available, and old versions of software often contain older components, themselves reaching EOS from their producers.

Determine Data Security Controls and Compliance Requirements

Information security is so-called because its mission is protecting important information. A detailed inventory of the information stored and processed by an organization is critical so that that the cybersecurity function can fulfill the task of protecting all of the information properly.

CROSS REFERENCE

This section covers Objective 2.6 of the Asset Security domain in the CISSP Exam Outline (May 1, 2021).

Determining measures to protect information is a top-down activity, beginning with the organization's determining which laws, regulations, standards, and requirements apply to specific business processes, information systems, and data. Although this undertaking may sound like a complex one, without it, the organization is flying blind at best, without a clear picture of what it should be doing to protect information.

Data states

Inventorying data requires more than just counting databases and files in file storage systems. Let's take a step back and consider that data exists in many states in an environment, including

» **Creation:** Data is created by an end-user, an incoming data feed, or an application. Data needs to be classified at this time, based on its criticality and sensitivity, and a data owner (usually but not always the creator) needs to be assigned. Data may exist in many forms such as in documents, spreadsheets, email and text messages, database records, forms, images, presentations (including videoconferences), and printed documents.

» **Distribution (data in motion):** Data may be distributed or retrieved internally within an organization or transmitted to external recipients. Distribution may be manual (such as via courier) or electronic (typically, over a network). Data in motion is vulnerable to compromise, so appropriate safeguards must be implemented based on the classification of the data. Encryption may be required to send specific sensitive data over a public network, for example. In such cases, appropriate encryption standards must be established. Data loss prevention (DLP) technologies may also be used to prevent accidental or intentional unauthorized distribution of sensitive data.

» **Use (data in use):** This stage refers to data that has been accessed by an end user or application and is being actively used (read, analyzed, modified, updated, or duplicated) by that user or application. Data in use must be accessed only on systems that are authorized for the classification level of the data and only by users and applications that have appropriate permissions (clearance) and purpose (need to know).

» **Transport (data in transit):** This stage refers to data storage media, including hard drives, backup tape, and paper records being physically transported from one location to another. Many organizations store backup media in a secure offsite storage facility.

>> **Maintenance (data at rest):** Any time between the creation and disposition of data when it isn't in motion or in use, data is maintained at rest. Maintenance includes the storage (on media such as a hard drive, removable USB thumb drive, backup magnetic tape, or paper) and filing (such as in a directory and file structure) of data. Classification levels of data should be routinely reviewed (typically, by the data owner) to determine whether a classification level needs to be upgraded (not common) or downgraded. Appropriate safeguards must be implemented and audited regularly to ensure

- *Confidentiality (and privacy):* System, directory, and file permissions and encryption, for example

- *Integrity:* Baselines, cryptographic hashes, cyclic redundancy checks, and file locking (to prevent or control modification of data by multiple simultaneous users), for example

- *Availability:* Database and file clustering (to eliminate single points of failure), backups, and real-time replication (to prevent data loss), for example

>> **Disposal:** When data no longer has any value or is no longer useful to the organization, it needs to be destroyed properly per corporate retention and destruction policies, as well as any applicable laws and regulations. Certain sensitive data may require a final disposition determination by the data owner and may require specific destruction procedures (such as witnesses, logging, shredding, or degaussing).

WARNING

Data that has merely been deleted *has not* been destroyed properly. It's merely data at rest waiting to be overwritten — or inconveniently discovered by an unauthorized and potentially malicious third party!

REMEMBER

Data remanence refers to data that still exists on storage media or in memory after the data has been deleted.

Scoping and tailoring

Because different parts of an organization and its underlying IT systems store and process various data sets, it doesn't make sense for an organization to establish a single set of controls and impose them on all systems. As in an oversimplified data classification program and its resulting overprotection or underprotection of data, organizations often divide themselves into logical zones and then specify which controls and sets of controls are applied to these zones.

Another approach is to tailor controls and sets of controls to different IT systems and parts of the organization. Controls on password strength, for example, can have categories applied to systems with varying security levels or classifications.

Both approaches for applying a complex control environment to a complex IT environment are valid because they're really just different ways of achieving the same objective: applying the right level of control to various systems and environments based on the information they store and process or on other criteria.

Standards selection

Several excellent control frameworks are available for security professionals' use. In no circumstances is it necessary to start from scratch. Instead, the best approach is to start with one of several industry-leading control frameworks and then add or remove individual controls to suit the organization's needs.

Numerous security control frameworks and standards include asset management as a critical function. Examples include

» **Center for Internet Security (CIS)Controls v8:** Controls 1 and 2 address Inventory and Control of Enterprise Assets and Inventory and Control of Software Assets, respectively.

» **International Organization for Standardization (ISO) 27002, Information technology — Security techniques — Code of practice for information security controls:** Section 8 addresses asset management, including responsibility for assets, information classification, and media handling.

» **U.S. National Institute of Standards and Technologies (NIST)Special Publication (SP) 800-53R5, Security and Privacy Controls for Information Systems and Organizations:** Section PE-20 covers asset monitoring and tracking.

» **NIST SP1800-5, IT Asset Management:** This extensive publication is devoted entirely to the topic of IT asset management, including approach, architecture, security characteristics, and how-to guides.

No matter how control frameworks are organized, they will always include asset management. Without effective asset management, few other security activities can protect an organization effectively. Again, you can't protect the assets you don't know about. Chapter 3 contains additional content on control frameworks.

Data protection methods

Information security is about data protection at its core, ensuring data confidentiality, integrity, and availability (CIA). Several methods are used to protect data, depending on the context of data storage and use. These methods are described in the following sections, and most are explored in more detail throughout this entire book.

Digital rights management (DRM)

Digital rights management (DRM) refers to a wide variety of techniques used to enforce the use, modification, and distribution of information and software. For years, software programs have employed license keys and other safeguards to prevent software piracy. More recently, various means are used to control the use and distribution of sensitive or valuable information. Mechanisms such as copy protection and integrated access controls are used to prevent the illicit distribution and use of digital files and documents.

In the context of protecting documents, the ultimate objective of DRM is to permit only specific people to open and read a document and to cause the expiration and destruction of documents that exceed their intended storage life. Such controls may be more difficult to enforce when documents are out of the physical control of the originator or owner.

Data loss prevention (DLP)

Data loss prevention (DLP) refers to various tools and techniques employed to provide visibility into various forms of data storage, transmission, and use. DLP tools can also block certain actions when those actions violate company policy. DLP generally takes two forms:

>> **Static:** Tools are used to scan data stores to detect the presence of various types of information. Reports and metrics are produced, and data analysts can take action when they determine that data is being stored improperly (such as a payroll extract residing on a widely readable file share).

>> **Dynamic:** Tools are integrated into storage systems, email programs, firewalls, and other systems and are used to detect the transmission of various types of information. Dynamic DLP systems are configured to log events that violate policy, issue warnings to users who attempt to transmit data that violates policy, and even prevent the actions that users or systems are attempting.

Organizations should initially configure dynamic DLP systems in learn mode to better understand data use. Only after valid uses of data transmissions are thoroughly understood should a DLP system be configured to prevent actions prohibited by policy.

Cloud access security broker (CASB)

A *cloud access security broker*, generally referred to as CASB (pronounced *KAS-bee*), is a network device or endpoint agent configured to control users' access to cloud services. More than a web content filter or firewall that simply blocks users from accessing certain websites, a CASB has more intelligence and directs users away from unsupported sites to company-sanctioned cloud services. If a user attempts to store company documents on a personal Box.com cloud drive, for example, a CASB will display a page that informs the user that only OneDrive may be used to store company data.

As with firewalls, web content filters, intrusion prevention systems (IPS), and DLP systems, organizations are advised to implement a CASB in learn mode first to better understand valid data use before configuring the CASB to intervene and block activities.

Cryptography

Cryptography plays a critical role in data protection, whether we're talking about data in motion through a network or at rest on a server, workstation, or storage device. Cryptography is all about hiding data in plain sight. In some situations, people may be able to access sensitive data; cryptography denies them access unless they possess an encryption key and the decryption method. Cryptography is explored in detail in Chapter 5.

Access controls

One of the first functions and still a function of prime importance, data security relies heavily on access controls that determine who and what can access important information. Access controls and access management are explored in detail in Chapter 7.

Privacy controls

Data that includes information about specific people may fall under the scope of privacy laws, which require the implementation of controls to enforce citizens' privacy preferences. Additional controls may include anonymization, pseudonymization, processing opt-in and opt-out requests, and removing data when requested by data subjects. Privacy is explored in detail in Chapter 3.

» **Understanding security models**

» **Choosing the right controls and countermeasures**

» **Using security capabilities in information systems**

» **Assessing and mitigating vulnerabilities**

» **Deciphering cryptographic concepts and fundamentals**

» **Getting physical with physical security design concepts**

Chapter **5**

Security Architecture and Engineering

S ecurity must be part of the design of information systems, as well as the facilities housing information systems and workers, which is covered in the Security Architecture and Engineering domain. This domain represents 13 percent of the CISSP certification exam.

Research, Implement, and Manage Engineering Processes Using Secure Design Principles

It is a natural human tendency to build things without first considering their design or security implications. A network engineer who is building a new network may just start plugging cables into routers and switches without thinking about the overall design — much less any security or privacy considerations. Similarly, a software engineer assigned to write a new program is apt to begin coding without planning the program's architecture or design.

CROSS REFERENCE

This section covers Objective 3.1 of the Security Architecture and Engineering domain in the CISSP Exam Outline (May 1, 2021).

If we observe the outside world and the consumer products that are available, sometimes we see egregious usability and security flaws that make us wonder how the person or organization was ever allowed to participate in its design and development.

TIP

Security professionals need to help organizations understand that security-by-design principles are vital components of the development of any system.

The engineering processes that require the inclusion of secure design principles include the following:

>> **Concept development:** From the idea stage, security considerations are vital to the success of any new IT engineering endeavor. Every project and product starts with something: a whiteboard session, sketches on cocktail napkins or pizza boxes, or a conference call. However the project starts, someone should ask how vital data, functions, and components will be protected in this new thing. We're not looking for detailed answers; we're looking for just enough confidence to know that we aren't the latest lemmings rushing toward the nearest cliff.

>> **Requirements:** Before actual design begins, one or more people will define the requirements for the new system or feature. Often, there are several categories of requirements. Security, privacy, and regulatory requirements need to be included.

>> **Design:** After all requirements have been established and agreed on, formal design of the system or component can begin. Design must incorporate all requirements established in the preceding step.

>> **Development:** Depending on what is being built, development may take many forms, including creating

- System and device configurations
- Data center equipment racking diagrams
- Data flows for management and monitoring systems

>> **Testing:** Individual components and the entire system are tested to confirm that each requirement developed earlier has been achieved. Generally, someone other than the builder/developer should perform testing.

>> **Implementation:** When the system or component is placed into service, security considerations help ensure that the new system/component and related things are not at risk. Implementation activities include

- Configuring and cabling network devices
- Installing and configuring operating systems (OSes) and subsystems, such as database management systems, web servers, and applications
- Construction of physical facilities, work areas, and data centers

>> **Maintenance and support:** After the system or facility is placed into service, all subsequent changes need to undergo similar engineering steps to ensure that new or changing security risks are quickly mitigated.

>> **Decommissioning:** When a system or facility reaches the end of its service life, it must be decommissioned without placing data, other systems, or personnel at risk.

TIP

The Building Security in Maturity Model (BSIMM) is a software security benchmarking tool that provides a framework for software security. The model is composed of 256 measurements and 113 activities. BSIMM activities consist of 12 practices organized into four domains, including governance, intelligence, SSDL touchpoints, and deployment. Go to `https://www.bsimm.com` to learn more.

The application development life cycle also includes security considerations that are nearly identical to the security engineering principles discussed here. Application development is covered in Chapter 10.

Design principles and concepts associated with security architecture and engineering include the following:

>> Threat modeling

>> Least privilege (and need to know)

>> Defense in depth

>> Secure defaults

>> Fail securely

>> Separation of duties

>> Keep it simple

>> Zero trust

>> Privacy by design

>> Trust but verify

>> Shared responsibility

These principles and concepts are discussed in detail in the remainder of this section.

Threat modeling

Threat modeling is a type of risk analysis used to identify security defects in the design phase of an information system or business process. Threat modeling is most often applied to software applications, but it can be used for OSes, devices, and business processes with equal effectiveness.

Threat modeling is typically attack-centric; threat modeling is most often used to identify vulnerabilities that can be exploited by an attacker in information systems.

Threat modeling is most effective when performed during the design phase of an information system, application, or process. When threats and their mitigation are identified during the design phase, much effort is saved by the avoidance of fixes in a completed system.

Although there are different approaches to threat modeling, the typical steps are

>> Identifying threats

>> Determining and diagramming potential attacks

>> Performing reduction analysis

>> Remediating threats

Identifying threats

Threat identification is the first step in threat modeling. *Threats* are those actions that an attacker may be able to perform successfully if corresponding vulnerabilities are present in the application, system, or process.

For software applications, two mnemonics are used as a memory aid during threat modeling:

>> STRIDE, a list of basic threats (developed by Microsoft):

- Spoofing of user identity

- Tampering

- Repudiation

- Information disclosure

- Denial of service

- Elevation of privilege

>> DREAD, an older technique used for assessing threats:

- Damage

- Reproducibility

- Exploitability

- Affected users

- Discoverability

Although these mnemonics themselves don't contain threats, they do assist the person performing threat modeling by reminding them of basic threat categories (STRIDE) and their analysis (DREAD).

TIP

Appendices D and E in NIST SP800-30, *Guide for Conducting Risk Assessments*, are excellent general-purpose sources for threats.

Determining and diagramming potential attacks

After threats have been identified, threat modeling continues through the creation of diagrams that illustrate attacks on an application or system. An attack tree can be developed, outlining the steps required to attack a system. Figure 5-1 illustrates an attack tree for a mobile banking application.

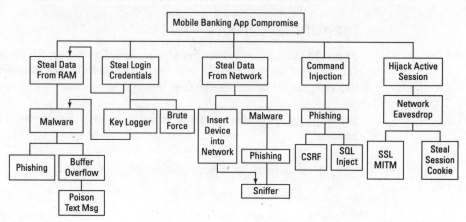

© John Wiley & Sons, Inc.

FIGURE 5-1:
Attack tree for a
mobile banking
application.

REMEMBER

An attack tree illustrates the steps used to attack a target system.

Performing reduction analysis

When you're performing a threat analysis on a complex application or a system, it is likely that many similar elements will represent duplications of technology. *Reduction analysis* is an optional step in threat modeling that prevents duplication of effort. It doesn't make sense to spend a lot of time analyzing different components in an environment if all of them have the same technology and configuration.

Here are typical examples:

>> An application contains several form fields (derived from the same source code) that request bank account numbers. Because all the field input modules use the same code, detailed analysis needs to be done only once.

>> An application sends several different types of messages over the same Transport Layer Security connection. Because the same certificate and connection are being used, a detailed analysis of the TLS connection needs to be done only once.

Remediating threats

As in routine risk analysis, the next step in threat analysis is enumerating potential measures to mitigate the identified threat. Because the nature of threats varies widely, remediation may consist of carrying out one or more of the following tasks for each risk:

>> Change source code (such as adding functions to closely examine input fields and filter out injection attacks).

>> Change configuration (such as switching to a more secure encryption algorithm or expiring passwords more frequently).

>> Change business process (such as adding or changing steps in a process or procedure to record or examine key data).

>> Change personnel such as providing training or moving responsibility for a task to another person).

REMEMBER

Recall that the four options for risk treatment are mitigation, transfer, avoidance, and acceptance. In the case of threat modeling, some threats may be accepted as they are.

IT HAS TO BE PASTA

A promising threat modeling technique called PASTA (Process for Attack Simulation and Threat Assessment) is used to detect cyber threats as a business problem, not merely a technical analysis.

The stages of PASTA methodology are

- Define objectives.
- Define scope.
- Decompose the application.
- Analyze threats.
- Analyze vulnerabilities and weaknesses.

The PASTA methodology includes business risk and business impact, which is a big leap forward in threat modeling. Risk treatment is an upper management function, and PASTA expresses risks in business language that upper management understands and can act on.

Least privilege (and need to know)

The principle of *least privilege* states that people should have the capability to perform only the tasks (or access only the data) required to perform their primary jobs— no more.

Giving a person more privileges and access than required increases risk and invites trouble. Offering the capability to perform more than the job requires may become a temptation that results, sooner or later, in an abuse of privilege.

Giving a user full permissions on a network share rather than just read and modify rights to a specific directory, for example, opens the door not only to abuse of those privileges (such as reading or copying other sensitive information on the network share), but also to costly mistakes (such as accidentally deleting a file — or the entire directory!). As a starting point, organizations should approach permissions with a "deny all" mentality and add needed permissions as required.

WARNING

Giving users excessive privileges in network shares makes them vulnerable to ransomware attacks.

The concept of *need to know* states that only people with a valid business justification should have access to specific information or functions. In addition to having a need to know, a person must have an appropriate security clearance level to be granted access. Conversely, a person with the appropriate security clearance level but without a need to know should not be granted access.

One of the most difficult challenges in managing need to know is the use of controls that enforces the concept. Information owners need to be able to distinguish genuine need from curiosity and proceed accordingly.

TIP

Least privilege is closely related to separation of duties and responsibilities, described later in this section. Distributing the duties and responsibilities of a given job function among several people means that those people require fewer privileges on a system or resource.

REMEMBER

The principle of least privilege states that people should have the fewest privileges necessary to perform their tasks.

Several important concepts associated with need to know and least privilege include

> » **Entitlement:** When a new user account is provisioned in an organization, the permissions granted to that account must be appropriate for the level of access required by the user. In too many organizations, human resources

simply instructs the IT department to give a new user "whatever so-and-so (another user in the same department) has access to." Instead, entitlement needs to be based on the principle of least privilege.

>> **Aggregation:** When people transfer between jobs and/or departments within an organization (see the section on job rotations later in this chapter), they often need different access and privileges to do their new jobs. Far too often, organizational security processes do not adequately ensure that access rights that a person no longer requires are revoked. Instead, people accumulate privileges, and over a period of many years, they can have far more access and privileges than they need. This process is known as *aggregation*, and it's the antithesis of least privilege.

Privilege creep and *accumulation of privileges* are others terms commonly used in this context.

>> **Transitive trust:** Trust relationships (in the context of security domains) are often established within and between organizations to facilitate ease of access and collaboration. A trust relationship enables subjects (such as users or processes) in one security domain to access objects (such as servers or applications) in another security domain. (See chapters 5 and 7 for more about objects and subjects.) A transitive trust extends access privileges to the subdomains of a security domain (analogous to inheriting permissions to subdirectories within a parent directory structure). Instead, a nontransitive trust should be implemented by requiring access to each subdomain to be explicitly granted based on the principle of least privilege, rather than inherited.

Defense in depth

Defense in depth is a strategy for resisting attacks. A system that employs defense in depth has two or more layers of protective controls designed to protect the system or data stored there.

An example defense-in-depth architecture would consist of a database protected by several components, such as

>> Screening router

>> Firewall

>> Intrusion prevention system

>> Hardened OS

>> OS-based network access filtering

All the layers listed here help protect the database. In fact, each by itself offers nearly complete protection. But when considered together, all these controls offer a varied (in effect, deeper) defense — hence, the term *defense in depth*.

True defense in depth employs heterogeneous, versus homogeneous, protection. Employing two back-to-back firewalls of the same make and model, for example, constitutes a poor implementation of defense in depth: a security flaw in one of the firewalls is likely to be present in the other one. A better example of defense in depth would be back-to-back firewalls of different makes (such as one made by Cisco and the other made by Palo Alto Networks); a security flaw in one is unlikely to be present in the other.

Defense in depth refers to the use of multiple layers of protection.

Secure defaults

The concept of *secure defaults* encompasses several techniques, including

>> **Secure by design:** The relationship of components in a system lends to its resilience to attack.

>> **Secure by default:** Configuration settings and other options are adjusted to secure settings.

>> **Secure by deployment:** The procedures used to implement a system don't compromise its security.

These techniques ensure that the design of new information systems includes inherent security in all phases of development and implementation. When the techniques are performed correctly, little or no retrofit to a system will be required after it is tested by security specialists who use techniques such as threat modeling and penetration testing.

Fail securely

Fail securely is a concept that describes the result of the failure of a control or safeguard. A control or safeguard is said to fail securely if its failure does not result in a reduction in protection. Consider a door that is used to control personnel access to a secure location. If the mechanism used to admit authorized personnel to the secure location fails, the door should remain locked, meaning that it is secure and continues to block unauthorized access.

Fail securely replaces the terms *fail open* and *fail closed*. These two older terms were sometimes confusing, depending on the context of a control. In some examples, failing open was secure, but in other examples, failing closed was secure. The confusion was not unlike the use of a double negative, such as a security door that is not secure in certain circumstances. Conversations that included fail open and fail closed often digressed into discussions of the meaning of the terms and whether failing open or failing closed was good or bad. Fortunately, fail securely came to the rescue, helping us better understand the context of a conversation.

Separation of duties

The concept of *separation of duties* (SoD, or *segregation of duties and responsibilities*) ensures that no single person has complete authority and control of a critical system or process. SoD is discussed further in Chapter 9.

Keep it simple

It is often said that complexity is the enemy of security and, conversely, that simplicity is the friend of security. These adages reflect the realization that more complex environments are inherently more difficult to secure, and the security posture of such an environment is harder to understand because of the higher number of components.

In information security, simplicity often calls for consistency of approach to system and data protection. Elegance of design is another way to think about simplicity. In security, less is more: Given two identical environments, the one with a simple yet effective design will be easier for engineers to understand than a complex architecture.

Security engineers and specialists often call on the KISS (Keep It Simple, Stupid) principle. No, we're not calling you or anyone stupid. We didn't make up this principle, but we do see it cited often.

Zero trust

The concept of *zero trust* has been around for a long time but is now gaining a lot of favor. Zero trust (ZT) is a popular buzzword these days, although it is not always well understood. We want you to be buzzword-compliant, so read on to find out more.

Zero trust is an about-face to the earlier notion that all devices within an organization's network were considered to be trustworthy. Organizations have been compromised countless times because of this fateful assumption, often because attackers found it way too easy to attack trusted systems and endpoints; they usually gained carte blanche access to other systems because the compromised system was considered to be trustworthy.

Zero trust is not a product, tool, or technique; it's a design principle that is implemented in different ways to ensure that systems retain their security and integrity. Here are some examples of zero trust in action:

>> An endpoint is not permitted to connect to a network (whether onsite, wired, wireless, or remote) until it can prove that its antivirus and other mechanisms are functioning properly and are up to date.

>> A user is not permitted to perform a high-risk or high-value transaction until they reauthenticate, proving that they are still in control.

>> A system is not permitted to communicate with another system until each is able to authenticate to the other.

>> A user is not permitted to access files or directories in a file share unless they demonstrate a need to know.

>> A newly acquired piece of open-source software is not considered to be secure until it can be analyzed by a source code analyzer to ensure that it is free of exploitable vulnerabilities.

>> New executable programs are not permitted to run on a system until they have been vetted by security personnel and added to a whitelist.

Privacy by design

Privacy (as we discuss more fully in Chapter 3) includes measures not only to protect information about people, but also to ensure the proper uses of personal information. Focusing on proper use here, the principle of privacy by design ensures that information systems have several capabilities, including

>> Providing mechanisms to control who has access to personal information

>> Providing visibility into the uses of personal information

>> Providing visibility into the movement of personal information as it enters, moves about, and leaves the organization

>> Providing means for performing anonymization and pseudonymization of individual records and entire databases

>> Providing means for removing business records containing personal information when they reached their retention life

>> Providing means for easily determining the uses of personal information for specific persons upon request

>> Providing means for excluding specific people from various types of processing upon request of those people (a process commonly known as *opt-out*)

>> Alerting management when new or unauthorized uses of personal information occur

Since the passage of recent privacy laws (generally starting with the European General Data Protection Regulation [GDPR]), it's not enough for organizations simply to protect personal information. Now organizations must build structures that provide visibility into and control of the uses of personal information so that organizations do not run afoul of these new laws.

We'll further explain some of the preceding terms. Organizations are realizing that the consequences of failing to use and protect personal information properly are climbing rapidly, with potential fines that can wipe out an organization's profitability. New privacy laws incentivize organizations to remove personal information from their databases as soon as that information is no longer needed. The rights of data subjects to opt out and to be forgotten can compel organizations to build mechanisms to remove them from their records. Techniques that organizations can use include

>> **Anonymization:** An organization can remove from its databases specific fields that identify a data subject. These fields might include a subject's name, address, government identifiers such as social insurance or driver's license number, email address, and phone number.

>> **Pseudonymization:** An organization can substitute pseudonyms for personal data in identifiable fields so that the records no longer relate to specific people.

Although pseudonymisation has many uses, it should be distinguished from anonymization, as it may provide only limited protection for the identity of data subjects because may allow identification using indirect means. Where a pseudonym is used, it may be possible to identify the data subject by analyzing the underlying or related data When done properly, these two techniques constitute the effective removal of a data subject from an organization's records.

WARNING

Before removing records from a database upon the request of a data subject, organizations must also consider minimum retention periods required by other laws. Generally, those other laws will prevail.

TIP

Readers who want to understand more about data privacy can pick up a copy of *Certified Information Privacy Manager All-In-One Exam Guide* (www.mhprofessional.com/cipm-certified-information-privacy-manager-all-in-one-exam-guide-2315615) or *Certified Data Privacy Solutions Engineer All-In-One Exam Guide* (www.mhprofessional.com/cdpse-certified-data-privacy-solutions-engineer-all-in-one-exam-guide-2261479).

Trust but verify

The concept *trust but verify* was made popular in the 1980s, when President Ronald Reagan enacted a treaty with the Soviet Union that included provisions for each country to not only enforce the limitation of nuclear armaments, but also inspect the other's nuclear arsenal to confirm compliance with the treaty.

SYSTEM HARDENING

Most types of information systems, including computer operating systems, have several general-purpose features that make setup easy. But systems that are exposed to the Internet should be hardened according to the following concepts:

- Remove all unnecessary components.
- Remove all unnecessary accounts.
- Close all unnecessary network listening ports.
- Change all default passwords to complex, difficult-to-guess passwords.
- All necessary programs should run at the lowest possible privilege.
- Security patches should be installed as soon as they are available.

System hardening guides can be obtained from sources such as

- The Center for Internet Security (https://www.cisecurity.org)
- Information Assurance Support Environment, from the U.S. Defense Information Security Agency (https://public.cyber.mil/stigs/)

Software and OS vendors often provide their own hardening guides, which may also be useful.

In information security, the principle means that certain controls or mechanisms should be examined or tested periodically to ensure that they comply with policies or requirements. Although examining or testing a system is an operational activity performed on a system after it has been designed and implemented, the design of a system should permit it to be examined. Here are a couple of examples:

- >> A system's source code should not be obfuscated (at least in testing phases), because doing so would inhibit the use of source-code inspection tools.
- >> It should be possible to inspect the contents of encrypted channels to verify their proper use.
- >> Password hashes should be cracked periodically to ensure that passwords comply with policy.
- >> Access logs should be checked periodically to see whether personnel attempt to access workspaces they are not authorized to enter.

Shared responsibility

This is a fundamental truth that is not universally understood: *Cloud providers do not take care of information security* — not all of it, anyway. More breaches and information leaks than we can count have occurred because organizations and people did not understand this concept (and because of lack of training and plain old sloppiness).

Better cloud service providers — and by this, we mean Infrastructure as a Service (IaaS), Platform as a Service (PaaS), and Software as a Service (SaaS) — have developed specific documents known as *shared responsibility matrices,* often in visual form, so that their customers have a clearer idea of what security controls are taken care of by the service provider and what controls are the responsibility of the customer. Sometimes, however, specific service providers don't provide clear guidance, in which case a skilled information security specialist needs to examine the characteristics of the service and discern the responsibility boundaries. However you get to this clear determination, it's critical that organizations understand *precisely* what they should be doing with regard to security and privacy and what the service provider is supposed to be doing.

Figures 5-2 and 5-3 show typical shared responsibility matrices from Amazon Web Services (AWS) and Microsoft Azure, respectively. Note that the matrices visually depict the areas in which AWS and Azure provide security and those in which customers are required to provide security.

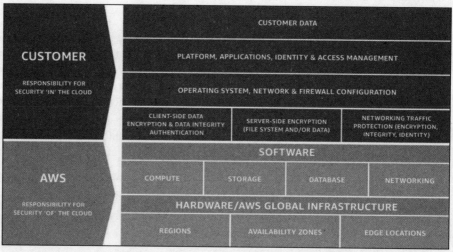

FIGURE 5-2: AWS shared responsibility matrix.

Source: AWS

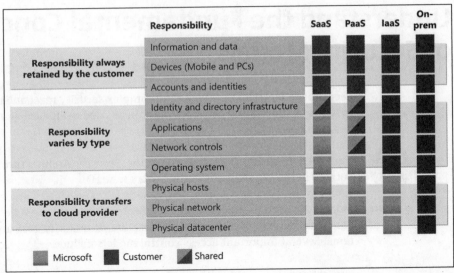

Responsibility		SaaS	PaaS	IaaS	On-prem
Responsibility always retained by the customer	Information and data				
	Devices (Mobile and PCs)				
	Accounts and identities				
Responsibility varies by type	Identity and directory infrastructure				
	Applications				
	Network controls				
	Operating system				
Responsibility transfers to cloud provider	Physical hosts				
	Physical network				
	Physical datacenter				

■ Microsoft   ■ Customer   ◩ Shared

FIGURE 5-3: Azure shared responsibility matrix.

Source: Microsoft

Examples of what shared responsibility means at various levels for different cloud services include the following:

>> **Access control:** Generally, a SaaS, IaaS, and PaaS provider is going to take care of administrative access to the system or underlying system you use. It's your responsibility, however, to create, allocate, and manage all user access to the system you run in that environment.

>> **Network security:** Generally, a PaaS and SaaS provider will employ firewalls and other capabilities to protect their environments from many kinds of attacks. Most IaaS services provide no such controls, however. If you want firewalls to protect your IaaS OSes, you have to implement them yourself.

>> **System security:** SaaS and PaaS providers will ensure that underlying OSes are current, patched, and hardened. But in an IaaS environment, you are required to configure and manage OSes with whatever level of security you want.

>> **Source-code security:** Generally, a PaaS and SaaS provider will employ means to verify that the software it provides is reasonably free of exploitable defects. But you must do all the work on your own to ensure that any software you develop and use in a PaaS or IaaS environment is free of defects.

>> **Physical security:** Virtually all SaaS, PaaS, and IaaS providers are going to take care of physical security (and environmental) concerns for all system components they provide. If part of an overall system resides in your own premises, however, you have to protect those systems with whatever measures you deem appropriate.

Understand the Fundamental Concepts of Security Models

Security models help us understand complex security mechanisms in information systems by illustrating concepts that can be used to analyze an existing system or design a new one.

CROSS REFERENCE

This section covers Objective 3.2 of the Security Architecture and Engineering domain in the CISSP Exam Outline (May 1, 2021).

Models are used to express access control requirements in a theoretical or mathematical framework and precisely describe or quantify real access control systems. Several important access control models include

>> Biba

>> Bell-LaPadula

>> Access Matrix

>> Discretionary Access Control

>> Mandatory Access Control

>> Take-Grant

>> Clark-Wilson

>> Information Flow

>> Noninterference

These models are discussed in the following sections.

REMEMBER

The Bell–LaPadula, Access Matrix, Mandatory Access Control, Discretionary Access Control, and Take–Grant models address the confidentiality of stored information. The Biba and Clark–Wilson models address the integrity of stored information.

Biba

The Biba integrity model (sometimes referred to as *Bell–LaPadula upside down*) was the first formal integrity model. Biba is a lattice-based model that addresses the first goal of integrity: ensuring that modifications to data aren't made by unauthorized users or processes. (See Chapter 3 for a complete discussion of the three goals of integrity.) Biba defines the following two properties:

>> **Simple integrity property:** A subject can't read information from an object that has a lower integrity level than the subject (also called *no read down*).

>> ***-integrity property (star integrity property):** A subject can't write information to an object that has a higher integrity level than the subject (also known as *no write up*).

Bell-LaPadula

The Bell–LaPadula model was the first formal confidentiality model of a mandatory access control system. (We discuss mandatory and discretionary access controls in Chapter 7.) It was developed for the U.S. Department of Defense (DoD) to formalize a multilevel security policy. As we discuss in Chapter 3, the DoD classifies information based on sensitivity at three basic levels: Confidential, Secret, and Top Secret. To access classified information (and systems), a person must have access (a clearance level equal to or exceeding the classification of the information or system) and need to know (legitimately need of access to perform a required job function). The Bell–LaPadula model implements the access component of this security policy.

Bell–LaPadula is a state machine model that addresses only the confidentiality of information. The basic premise of the model is that information can't flow downward — that is, that information at a higher level is not permitted to be copied or moved to a lower level. Bell–LaPadula defines the following two properties:

>> **Simple security property (ss property):** A subject can't read information from an object that has a higher sensitivity label than the subject (also known as *no read up*).

>> ***-property (star property):** A subject can't write information to an object that has a lower sensitivity label than the subject (also known as *no write down*).

Bell–LaPadula also defines two additional properties that give it the flexibility of a discretionary access control model:

>> **Discretionary security property:** This property determines access based on an access matrix (see the following section).

>> **Trusted subject:** A trusted subject is an entity that can violate the *-property but not its intent.

TIP

A *state machine* is an abstract model used to design computer programs. The state machine illustrates which state the program will be in at any time.

Access Matrix

An *Access Matrix model,* in general, provides object access rights (read/write/execute) to subjects in a discretionary access control (DAC) system. An access matrix consists of access control lists (columns) and capability lists (rows). See Table 5-1 for an example.

An Access Matrix Example

Subject/Object	Directory: H/R	File: Personnel	Process: LPD
Thomas	Read	Read/Write	Execute
Lisa	Read	Read	Execute
Harold	None	None	None

Discretionary Access Control

A *DAC* system is one in which the owners of specific objects (typically, files and/or directories) can adjust access permissions at their discretion. No central administrator is needed to adjust permissions. The underlying OS enforces these access rights by permitting or denying access to specific objects.

Mandatory Access Control

A *Mandatory Access Control* (MAC) system is one that is controlled by a central administrator who determines access rights to objects The OS enforces these access rights by permitting or denying access to specific objects.

Take-Grant

Take-Grant systems specify the rights that a subject can transfer to or from another subject or object. These rights are defined through four basic operations: create, revoke, take, and grant.

Clark-Wilson

The *Clark-Wilson* integrity model establishes a security framework for use in commercial activities, such as the banking industry. Clark-Wilson addresses all three goals of integrity and identifies special requirements for inputting data based on the following items and procedures:

>> **Unconstrained data items:** Data outside the control area, such as input data.

>> **Constrained data items (CDIs):** Data inside the control area. (Integrity must be preserved.)

>> **Integrity verification procedures:** Check validity of CDIs.

>> **Transformation procedures:** Maintain integrity of CDIs.

The Clark-Wilson integrity model is based on the concept of a *well-formed transaction,* in which a transaction is sufficiently ordered and controlled that it maintains internal and external consistency.

Information Flow

An *Information Flow* model is a type of access control model based on the flow of information rather than on imposing access controls. Objects are assigned a security class and value, and their direction of flow — from one application to another or from one system to another — is controlled by a security policy. This model type is useful for analyzing covert channels through detailed analysis of the flow of information in a system, including the sources of information and the paths of flow.

Noninterference

A *Noninterference* model ensures that the actions of different objects and subjects aren't seen by (and don't interfere with) other objects and subjects on the same system.

REMEMBER

The design of specific access control systems often borrows from one or more of the models described in this section. All access control systems, for example, are either MAC or DAC, and many can also be Noninterference.

Select Controls Based Upon Systems Security Requirements

Designing and building secure software is critical to information security, but the systems that software runs on must themselves be securely designed and built. Selecting appropriate controls is essential to designing a secure computing architecture. Numerous systems security evaluation models exist to help you select the right controls and countermeasures for your environment.

CROSS REFERENCE

This section covers Objective 3.3 of the Security Architecture and Engineering domain in the CISSP Exam Outline (May 1, 2021).

Various security controls and countermeasures that should be applied to security architecture, as appropriate, include defense in depth, system hardening, implementation of heterogeneous environments, and designing system resilience. Often, these controls are enacted based upon high-level requirements that are usually determined by the context or use of a system. When baseline controls are chosen and implemented, the risk management life cycle (discussed in Chapter 3) will, over time, determine the need for additional controls as well as changes to existing controls.

Examples of contexts and uses of information systems include

>> **Services to U.S. government agencies:** Often, systems that provide services to U.S. government agencies are required to employ controls from NIST SP800-53 (Security and Privacy Controls for Information Systems and Organizations).

>> **Processing of credit card data:** Systems that store, process, or transmit credit card data are required to comply with all the requirements in the Payment Card Industry Data Security Standard (PCI DSS).

>> **Processing of health-care information:** Systems that store, process, or transmit patient health information are required to comply with requirements enacted by the Health Insurance Portability and Accountability Act (HIPAA) and Health Information Technology for Economic and Clinical Health (HITECH).

>> **Processing of personal financial information:** Systems that store, process, or transmit personal financial information are subject to privacy requirements in the Gramm-Leach-Bliley Act.

>> **Processing of personal information:** Laws in many countries and U.S. states are strengthening requirements for the protection and proper use of personal information.

Evaluation criteria

Evaluation criteria provide a standard for quantifying the security of a computer system or network. These criteria include the Trusted Computer System Evaluation Criteria (TCSEC), Trusted Network Interpretation (TNI), European Information Technology Security Evaluation Criteria (ITSEC), and the Common Criteria.

Trusted Computer System Evaluation Criteria

The Trusted Computer System Evaluation Criteria (TCSEC), commonly known as the *Orange Book*, is part of the Rainbow Series developed for the U.S. DoD by the National Computer Security Center (NCSC). It's the formal implementation of the

Bell–LaPadula model. The evaluation criteria were developed to achieve the following objectives:

>> **Measurement:** Provide a metric for assessing comparative levels of trust between computer systems

>> **Guidance:** Identify standard security requirements that vendors must build into systems to achieve a given trust level

>> **Acquisition:** Provide customers a standard for specifying acquisition requirements and identifying systems that meet those requirements

The four basic control requirements identified in the Orange Book are

>> **Security policy:** The rules and procedures by which a trusted system operates. Specific TCSEC requirements include

- *DAC:* Owners of objects are able to assign permissions to other subjects.

- *MAC:* Permissions to objects are managed centrally by an administrator.

- *Object reuse:* The confidentiality of objects that are reassigned after initial use is protected. A deleted file still exists on storage media, for example; only the file allocation table and the first character of the file have been modified. Thus, residual data may be restored, which describes the problem of data remanence. Object-reuse requirements define procedures for erasing the data.

- *Labels:* Sensitivity labels are required in MAC-based systems. (Read more about information classification in Chapter 3.) Specific TCSEC labeling requirements include integrity, export, and subject/object labels.

>> **Assurance:** Guarantees that a security policy is implemented correctly. Specific TCSEC requirements (listed here) are classified as operational assurance requirements:

- *System architecture:* TCSEC requires features and principles of system design that implement specific security features.

- *System integrity:* Hardware and firmware operate properly and are tested to verify proper operation.

- *Covert channel analysis:* TCSEC requires covert channel analysis that detects unintended communication paths not protected by a system's normal security mechanisms. A *covert storage channel* conveys information by altering stored system data. A *covert timing channel* conveys information by altering a system resource's performance or timing.

REMEMBER

A systems or security architect must understand covert channels and how they work to prevent the use of covert channels in the system environment.

- *Trusted facility management:* A specific person is assigned to administer the security-related functions of a system. This requirement is closely related to the concepts of least privilege, separation of duties, and need to know.

- *Trusted recovery:* This requirement ensures that security isn't compromised in the event of a system crash or failure. The process involves two primary activities: failure preparation and system recovery.

- *Security testing:* This requirement specifies required testing by the developer and the NCSC.

- *Design specification and verification:* This requirement calls for mathematical and automated proof that the design description is consistent with the security policy.

- *Configuration management:* This requirement calls for identifying, controlling, accounting for, and auditing all changes made to the Trusted Computing Base during the design, development, and maintenance phases of a system's life cycle.

- *Trusted distribution:* This requirement protects a system during transport from a vendor to a customer.

» **Accountability:** The ability to associate users and processes with their actions. Specific TCSEC requirements include

- *Identification and authentication:* Systems need to track who performs what activities. We discuss this topic in Chapter 7.

- *Trusted path:* A direct communications path between the user and the Trusted Computing Base (TCB) doesn't require interaction with untrusted applications or OS layers.

- *Audit:* Security-related activities in a trusted system are recorded, examined, analyzed, and reviewed.

» **Documentation:** Specific TCSEC requirements include

- *Security Features User's Guide:* This document is a user manual for the system.

- *Trusted Facility Manual:* This document is the system administrator's and/or security administrator's manual.

- *Test documentation:* According to the TCSEC manual, this documentation must be in a position to "show how the security mechanisms were tested, and results of the security mechanisms' functional testing."

- *Design documentation:* This documentation defines system boundaries and internal components, such as the Trusted Computing Base.

REMEMBER

The Orange Book defines four major hierarchical classes of security protection and numbered subclasses (higher numbers indicate higher security):

» **D:** Minimal protection

» **C:** Discretionary protection (C1 and C2)

» **B:** Mandatory protection (B1, B2, and B3)

» **A:** Verified protection (A1)

These classes are further defined in Table 5-2.

TABLE 5-2   **TCSEC Classes**

Class	Name	Sample Requirements
D	Minimal protection	These requirements are reserved for systems that fail evaluation.
C1	Discretionary protection (DAC)	The system doesn't need to distinguish between individual users and types of access.
C2	Controlled access protection (DAC)	The system must distinguish between individual users and types of access; object reuse security features are required.
B1	Labeled security protection (MAC)	Sensitivity labels are required for all subjects and storage objects.
B2	Structured protection (MAC)	Sensitivity labels are required for all subjects and objects; trusted path requirements apply.
B3	Security domains (MAC)	Access control lists are specifically required; system must protect against covert channels.
A1	Verified design (MAC)	Formal top-level specification is required; configuration management procedures must be enforced throughout the entire system life cycle.
Beyond A1		Self-protection and reference monitors are implemented in the Trusted Computing Base, which is verified to source-code level.

TIP

You don't need to know the specific requirements of each TCSEC level for the CISSP exam, but you should know at what levels DAC and MAC are implemented and the relative trust levels of the classes, including numbered subclasses.

Major limitations of the Orange Book include the following:

>> It addresses only confidentiality issues. It doesn't include integrity and availability.

>> It isn't applicable to most commercial systems.

>> It emphasizes protection from unauthorized access despite statistical evidence that many security violations involve insiders.

>> It doesn't address networking issues.

Trusted Network Interpretation

Part of the Rainbow Series, like TCSEC (discussed in the preceding section), Trusted Network Interpretation (TNI) addresses confidentiality and integrity in trusted computer/communications network systems. Within the Rainbow Series, it's known as the *Red Book*.

Part I of the TNI is a guideline for extending the system protection standards defined in the TCSEC (the *Orange Book*) to networks. Part II of the TNI describes additional security features such as communications integrity, protection from denial of service, and transmission security.

European Information Technology Security Evaluation Criteria

Unlike TCSEC, the European Information Technology Security Evaluation Criteria (ITSEC) addresses confidentiality, integrity, and availability, as well as evaluating an entire system, defined as a target of evaluation (TOE) rather than a single computing platform.

ITSEC evaluates *functionality* (security objectives, or *why*; security-enforcing functions, or *what*; and security mechanisms, or *how*) and *assurance* (effectiveness and correctness) separately. The 10 functionality (F) classes and 7 evaluation (E) (assurance) levels are listed in Table 5-3.

You don't need to know the specific requirements of each ITSEC level for the CISSP exam, but you should know how the basic functionality levels (F-C1 through F-B3) and evaluation levels (E0 through E6) correlate to TCSEC levels.

TABLE 5-3

ITSEC Functionality (F) Classes and Evaluation (E) Levels Mapped to TCSEC Levels

(F) Class	(E) Level	Description
NA	E0	Equivalent to TCSEC level D
F-C1	E1	Equivalent to TCSEC level C1
F-C2	E2	Equivalent to TCSEC level C2
F-B1	E3	Equivalent to TCSEC level B1
F-B2	E4	Equivalent to TCSEC level B2
F-B3	E5	Equivalent to TCSEC level B3
F-B3	E6	Equivalent to TCSEC level A1
F-IN	NA	TOEs with high integrity requirements
F-AV	NA	TOEs with high availability requirements
F-DI	NA	TOEs with high integrity requirements during data communication
F-DC	NA	TOEs with high confidentiality requirements during data communication
F-DX	NA	Networks with high confidentiality and integrity requirements

Common Criteria

The Common Criteria for Information Technology Security Evaluation (usually called *Common Criteria*) is an international effort to standardize and improve existing European and North American evaluation criteria. The Common Criteria has been adopted as an international standard in ISO/IEC 15408. The Common Criteria defines eight evaluation assurance levels, which are listed in Table 5-4.

TIP

You don't need to know the specific requirements of each Common Criteria level for the CISSP exam, but you should understand the basic evaluation hierarchy (EAL0 through EAL7, in order of increasing levels of trust).

System certification and accreditation

System certification is a formal methodology for comprehensive testing and documentation of information system security safeguards, both technical and non-technical, in a given environment by using established evaluation criteria (the TCSEC).

TABLE 5-4

The Common Criteria

Level	TCSEC Equivalent	ITSEC Equivalent	Description
EAL0	N/A	N/A	Inadequate assurance
EAL1	N/A	N/A	Functionally tested
EAL2	C1	E1	Structurally tested
EAL3	C2	E2	Methodically tested and checked
EAL4	B1	E3	Methodically designed, tested, and reviewed
EAL5	B2	E4	Semiformally designed and tested
EAL6	B3	E5	Semiformally verified design and tested
EAL7	A1	E6	Formally verified design and tested

Accreditation is official, written approval of the operation of a specific system in a specific environment, as documented in the certification report. Accreditation is normally granted by a senior executive or designated approving authority (DAA), a term used in the U.S. military and government. This DAA is normally a senior official, such as a commanding officer.

System certification and accreditation must be updated when any changes are made in the system or environment, and they must be revalidated periodically, typically every three years.

The certification and accreditation process has been formally implemented in U.S. military and government organizations as the Defense Information Technology Security Certification and Accreditation Process(DITSCAP) and National Information Assurance Certification and Accreditation Process (NIACAP), respectively. U.S. government agencies that use cloud-based systems and services are required to undergo FedRAMP or Cybersecurity Maturity Model Certification (CMMC) certification and accreditation processes (described in this chapter). These important processes are used to make sure that a new or changed system has the proper design and operational characteristics and is suitable for a specific task.

DITSCAP

The Defense Information Technology Security Certification and Accreditation Process (DITSCAP) formalizes the certification and accreditation process for U.S. DoD information systems through four distinct phases:

>> **Definition:** Determines security requirements by defining the organization and system's mission, environment, and architecture.

- » **Verification:** Ensures that a system undergoing development or modification remains compliant with the System Security Authorization Agreement, which is a baseline security configuration document.

- » **Validation:** Confirms compliance with the System Security Authorization Agreement.

- » **Post accreditation:** Represents ongoing activities required to maintain compliance and address new and evolving threats throughout a system's life cycle.

NIACAP

The National Information Assurance Certification and Accreditation Process (NIACAP) formalizes the certification and accreditation process for U.S. government national security information systems. NIACAP consists of four phases — definition, verification, validation, and post accreditation — that generally correspond to the DITSCAP phases. Additionally, NIACAP defines three types of accreditation:

- » **Site accreditation:** All applications and systems at a specific location are evaluated.

- » **Type accreditation:** A specific application or system for multiple locations is evaluated.

- » **System accreditation:** A specific application or system at a specific location is evaluated.

FedRAMP

The Federal Risk and Authorization Management Program (FedRAMP) is a standardized approach to assessments, authorization, and continuous monitoring of cloud-based service providers. This program represents a change from controls-based security to risk-based security.

CMMC

The Cybersecurity Maturity Model Certification (CMMC) is an assessment program used to evaluate the security of service providers that provide information system-related services to U.S. government agencies. CMMC is aligned with the NIST SP-171 ("Protecting Controlled Unclassified Information in Nonfederal Systems and Organizations") standard.

DCID 6/3

The Director of Central Intelligence Directive 6/3 is the process used to protect sensitive information that's stored on computers used by the U.S. Central Intelligence Agency.

Understand Security Capabilities of Information Systems

Basic concepts related to security architecture include the Trusted Computing Base, Trusted Platform Module, secure modes of operation, open and closed systems, protection rings, security modes, and recovery procedures.

CROSS REFERENCE

This section covers Objective 3.4 of the Security Architecture and Engineering domain in the CISSP Exam Outline (May 1, 2021).

Trusted Computing Base

A *Trusted Computing Base* (TCB) is the entire complement of protection mechanisms within a computer system (including hardware, firmware, and software) that's responsible for enforcing a security policy. A *security perimeter* is the boundary that separates the TCB from the rest of the system.

REMEMBER

A TCB is the total combination of protection mechanisms within a computer system (including hardware, firmware, and software) that's responsible for enforcing a security policy.

Access control is the ability to permit or deny the use of an object (a passive entity, such as a system or file) by a subject (an active entity, such as a person or process).

A *reference monitor* is a system component that enforces access controls on an object. Stated another way, a reference monitor is an abstract machine that mediates all access to an object by a subject.

REMEMBER

A *security kernel* is the combination of hardware, firmware, and software elements in a TCB that implements the reference monitor concept. A security kernel must

>> Mediate all access

>> Be protected from modification

>> Be verified as correct

Trusted Platform Module

A Trusted Platform Module (TPM) performs sensitive cryptographic functions on a physically separate, dedicated microprocessor. The TPM specification was written by the Trusted Computing Group and is an international standard (ISO/IEC 11889 Series).

A TPM generates and stores cryptographic keys and performs the following functions:

>> **Attestation:** Enables third-party verification of the system state, using a cryptographic hash of the known-good hardware and software configuration.

>> **Binding:** Binds a unique cryptographic key to specific hardware.

>> **Sealing:** Encrypts data with a unique cryptographic key and ensures that ciphertext can be decrypted only if the hardware is in a known-good state.

Common TPM uses include ensuring platform integrity, full disk encryption, password and cryptographic key protection, and digital rights management.

Secure modes of operation

Security modes are used in MAC systems to enforce different levels of security. Techniques and concepts related to secure modes of operation include

>> **Abstraction:** The process of viewing an application from its highest-level functions, which makes all lower-level functions into abstractions. Lower-level functions are treated as black boxes — known to work, even if we don't know how.

>> **Data hiding:** An object-oriented term that refers to the practice of encapsulating an object within another to hide the first object's functioning details.

>> **System high mode:** A system that operates at the highest level of information classification. Any user who wants to access such a system must have clearance at or above the information classification level.

>> **Security kernel:** Composed of hardware, software, and firmware components that mediate access and functions between subjects and objects. The security kernel is part of the protection rings model, in which the OS kernel occupies the innermost ring, and rings farther from the innermost ring represent fewer access rights. The *security kernel* is the innermost ring and has full access to all system hardware and data. User programs occupy outer rings and have fewer access privileges.

>> **Reference monitor:** A component implemented by the security kernel that enforces access controls on data and devices on a system. In other words, when a user tries to access a file, the reference monitor ultimately performs the "Is this person allowed to access this file?" function.

REMEMBER

The system's reference monitor enforces access controls on a system.

Open and closed systems

An *open system* is a vendor-independent system that complies with a published and accepted standard. This compliance with open standards promotes interoperability between systems and components made by different vendors. Additionally, open systems can be independently reviewed and evaluated, which facilitates the identification of bugs and vulnerabilities and the rapid development of solutions and updates. Examples of open systems include the Linux OS, the Open Office desktop productivity system, and the Apache web server.

A *closed system* uses proprietary hardware and/or software that may not be compatible with other systems or components. Source code for software in a closed system normally isn't available to customers or researchers. Examples of closed systems include the Microsoft Windows OS, the Oracle database management system, and Apple's iTunes.

TECHNICAL STUFF

The terms *open systems* and *closed systems* also refer to a system's access model. A closed system does not allow access by default, whereas an open system does.

Memory protection

Virtually all of today's OSes are multiprocessing — that is, several processes can occupy system memory and be processing simultaneously. OSes employ a means of process isolation so that each process is prevented from accessing memory allocated to all other processes. Although process isolation is automatic and usually considered to be effective, some species of malware have been able to exploit OS kernel weaknesses and access memory allocated to other processes. For this reason, it's often wise to employ obfuscation techniques or encryption to hide sensitive data in memory, such as encryption keys, and to deallocate or overwrite those memory locations when such data is no longer needed.

Encryption and decryption

Encryption and decryption can be thought of as being forms of access control, wherein data is converted to ciphertext with an encryption key; any person who is in possession of the correct decryption key may access the plaintext form of this information, but any person who lacks the decryption key may not access it. Encryption and decryption concepts are discussed later in this chapter.

Protection rings

The concept of protection rings implements multiple concentric domains with increasing levels of trust near the center. The most privileged ring is identified as Ring 0 and normally includes the OS security kernel. Additional system components are placed in the appropriate concentric ring according to the principle of least privilege and to provide isolation, so that a breach of a component in one protection ring does not automatically provide access to components in more privileged rings. The MIT MULTICS OS (whose ashes gave rise to Unix) implements the concept of protection rings in its architecture, as did Novell Netware. Figure 5-4 depicts an operating system protection ring model.

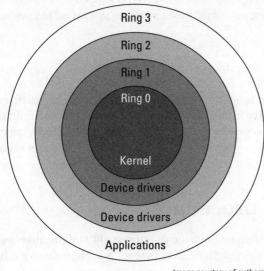

FIGURE 5-4: Protection rings provide layers of defense in a system.

Image courtesy of authors

Security modes

A system's security mode of operation describes how a system handles stored information at various classification levels. Several security modes of operation, based on the classification level of information being processed on a system and the clearance level of authorized users, have been defined. These designations, typically used for U.S. military and government systems, include

>> **Dedicated:** All authorized users must have a clearance level equal to or higher than the highest level of information processed on the system and a valid need to know.

>> **System High:** All authorized users must have a clearance level equal to or higher than the highest level of information processed on the system, but a valid need to know isn't necessarily required.

>> **Multilevel:** Information at different classification levels is stored or processed on a *trusted computer system* (a system that employs all necessary hardware and software assurance measures and meets the specified requirements for reliability and security). Authorized users must have an appropriate clearance level, and access restrictions are enforced by the system accordingly.

>> **Limited access:** Authorized users aren't required to have a security clearance, but the highest level of information on the system is Sensitive but Unclassified.

REMEMBER

A trusted computer system is a system with a TCB.

Security modes of operation generally come into play in environments that contain highly sensitive information, such as government and military environments. Most private and education systems run in multilevel mode, meaning they contain information at all sensitivity levels. See Chapter 3 for more on security clearance levels.

Recovery procedures

A hardware or software failure can potentially compromise a system's security mechanisms. Security designs that protect a system during a hardware or software failure include

>> **Fault-tolerant systems:** These systems continue to operate after the failure of a computer or network component. The system must be capable of detecting and correcting — or circumventing — a fault.

>> **Fail-safe systems:** When a hardware or software failure is detected, program execution is terminated, and the system is protected from compromise.

>> **Fail-soft (resilient) systems:** When a hardware or software failure is detected, certain noncritical processing is terminated, and the computer or network continues to function in a degraded mode.

>> **Failover systems:** When a hardware or software failure is detected, the system automatically transfers processing to a component, such as a clustered server.

Assess and Mitigate the Vulnerabilities of Security Architectures, Designs, and Solution Elements

In this section, we discuss the techniques used to identify and fix vulnerabilities in systems. We will also briefly discuss techniques for security assessments and testing, which are fully explored in Chapter 8.

CROSS REFERENCE

This section covers Objective 3.5 of the Security Architecture and Engineering domain in the CISSP Exam Outline (May 1, 2021).

Unless detected (and corrected) by an experienced security analyst, many weaknesses may be present in a system and permit exploitation, attack, or malfunction. These vulnerabilities include

>> **Covert channels:** *Covert channels* are unknown, hidden communications that take place within the medium of a legitimate communications channel.

>> **Rootkits:** By their very nature, rootkits are designed to subvert system architecture by inserting themselves into an environment in a way that makes it difficult or impossible to detect. Some rootkits run as a hypervisor and change the computer's OS into a guest, which changes the basic nature of the system in a powerful but subtle way. We wouldn't normally discuss malware in a chapter on computer and security architecture, but rootkits are game-changers that warrant mention because they use various techniques to hide themselves from the target system.

>> **Race conditions:** Software code in multiprocessing and multiuser systems, unless it's very carefully designed and tested, can result in critical errors that are difficult to find. A *race condition* is a flaw in a system where the output or result of an activity in the system is unexpectedly tied to the timing of other events. The term *race condition* comes from the idea of two events or signals racing to influence an activity.

The most common race condition is the time-of-check-to-time-of-use bug, caused by changes in a system between the checking of a condition and the use of the results of that check. Two programs that both try to open a file for exclusive use may open the file, even though only one should be able to do so.

>> **State attacks:** Web-based applications use session management to distinguish users from one another. The mechanisms used by the web application to establish sessions must be able to resist attack. Primarily, the algorithms used to create session identifiers must not permit an attacker to steal session identifiers or guess other users' session identifiers. A successful attack would result in an attacker's taking over another user's session, which can lead to the compromise of confidential data, fraud, and monetary theft.

>> **Emanations:** The unintentional emissions of electromagnetic or acoustic energy from a system can be intercepted by others and possibly used to illicitly obtain information from the system. A common form of undesired emanations is radiated energy from CRT computer monitors. (Yes, cathode-ray tubes are still out there, and not just in old movies!) A third party can discover what data is being displayed on a CRT by intercepting radiation emanating from the display adapter or monitor from as far as several hundred meters. Also, a third party can eavesdrop on a network if it has one or more unterminated coaxial cables in its cable plant.

Client-based systems

The design vulnerabilities often found on endpoints involve defects in client–side code in browsers and applications. The defects most often found include the following:

>> **Sensitive data left behind in the file system:** Generally, this data consists of temporary files and cache files, which may be accessible by other users and processes on the system.

>> **Sensitive data residing in memory:** Although OS process isolation is supposed to protect data in a program's memory space, some exploits are able to access protected memory.

>> **Unprotected local data:** Local data stores may have loose permissions and lack encryption.

>> **Vulnerable applets:** Many browsers and other client applications employ applets for viewing documents and video files. Often, the applets themselves may have exploitable weaknesses.

>> **Unprotected or weakly protected communications:** Data transmitted between the client and other systems may use weak encryption or no encryption at all.

>> **Weak or nonexistent authentication:** Authentication methods on the client, or between the client and server systems, may be unnecessarily weak. This weakness permits an adversary to access the application, local data, or server data without authenticating.

Other weaknesses may be present in client systems. For a complete understanding of application weaknesses, consult `https://owasp.org`.

Identifying weaknesses like the preceding examples requires using one or more of the following techniques:

>> OS examination

>> Network sniffing

>> Code review

>> Manual testing and observation

Server-based systems

Design vulnerabilities found on servers are the same as for client–based systems, discussed in the preceding section. The terms *client* and *server* have to do only with perspective. In both cases, software is running on a system.

Database systems

Database management systems are nearly as complex as the OSes on which they reside. Vulnerabilities in database management systems include

>> **Loose access permissions:** Like applications and OSes, database management systems have schemes of access controls that are often designed far too loosely, which permits more access to critical and sensitive information than is appropriate. Another aspect of loose access permissions is an excessive number of people who have privileged access. Finally, there can be failures to implement cryptography as an access control when appropriate.

>> **Excessive retention of sensitive data:** Keeping sensitive data longer than necessary increases the effect of a security breach.

>> **Aggregation of personally identifiable information:** The practice known as *aggregation* of data about citizens is a potentially risky undertaking that can result in an organization's possessing sensitive personal information. Sometimes, this aggregation happens when an organization deposits historic data from various sources into a data warehouse, bringing disparate sensitive data is together for the first time. The result is a gold mine or a time bomb, depending on how you look at it.

Database security defects can be identified through manual examination or automated tools. Mitigation may be as easy as changing access permissions or as complex as redesigning the database schema and related application software programs.

Cryptographic systems

Cryptographic systems are especially apt to contain vulnerabilities, for the simple reason that people focus on the cryptographic algorithm but fail to implement it properly. Like any powerful tool, a cryptographic system is useless at best and dangerous at worst if the operator doesn't know how to use it.

Following are some ways in which a cryptographic system may be vulnerable:

>> **Use of outdated algorithms:** Developers and engineers must be careful to select robust encryption algorithms. Furthermore, algorithms in use should be reviewed at least once per year to ensure that they continue to be sufficient.

>> **Use of untested algorithms:** Engineers sometimes make the mistake of either home-brewing a cryptographic system or using one that is clearly insufficient. It's best to use one of many publicly available cryptosystems that have stood the test of repeated scrutiny.

>> **Failure to encrypt encryption keys:** A proper cryptosystem sometimes requires encryption keys themselves to be encrypted.

>> **Weak cryptographic keys:** A great algorithm is all but undone if the initialization vector is too small or if the keys are too short or too simple.

>> **Insufficient protection of cryptographic keys:** A cryptographic system is only as strong as the protection of its encryption keys. If too many people have access to keys, or if the keys are not sufficiently protected, an intruder may be able to compromise the system simply by stealing and using the keys. Separate encryption keys should be used for the data encryption key (DEK) used to encrypt/decrypt data and the key encryption key (KEK) used to encrypt/decrypt the data encryption key.

These and other vulnerabilities in cryptographic systems can be detected and mitigated through peer reviews of cryptosystems, assessments by qualified external parties, and application of corrective actions to defects.

Industrial control systems

Industrial control systems (ICSes) represent a wide variety of means for monitoring and controlling machinery of various kinds, including power generation, distribution, and consumption; natural gas and petroleum pipelines; municipal water, irrigation, and waste systems; traffic signals; manufacturing; and package distribution.

Other related terms in common use include

>> **Supervisory Control and Data Acquisition (SCADA):** This somewhat-antiquated term refers to networks and systems used to monitor and control industrial systems, often related to public utilities.

>> **Operational technology(OT):** Organizations generally represent operational technology as the network and systems infrastructure that is not part of the corporate IT infrastructure. In many organizations, separate teams manage operational and information technology.

Weaknesses in industrial control systems include the following:

>> **Loose access permissions:** Access to monitoring or controls of ICSes is often set too loosely, thereby enabling some users or systems access to more data and control than they need.

>> **Failure to change default access credentials:** All too often, organizations implement ICS components and fail to change the default administrative credentials on those components, making it far too easy for intruders to take over the ICS.

>> **Access from personally owned devices:** In the name of convenience, some organizations permit personnel to control machinery from personally owned smartphones and tablets. This weakness vastly increases the system's attack surface and provides opportunities for intruders to access and control critical machinery.

>> **Lack of malware control:** Many ICSes lack security components that detect and block malware and other malicious activity, making it too easy for intruders to get in.

>> **Failure to air-gap the ICS:** Many organizations fail to *air-gap* (isolate) the ICS from the rest of its corporate network (or the Internet — gasp!), thereby enabling excessive opportunities for malware and intruders to access the ICS via a corporate network where users invite malware through phishing and other means.

>> **Failure to update ICS components:** Although the manufacturers of ICS components are notorious for failing to issue security patches, organizations are equally culpable in their failure to install these patches when they do arrive.

These vulnerabilities can be mitigated through a systematic process of establishing good controls, testing control effectiveness, and applying corrective action when controls are found to be ineffective.

Cloud-based systems

The U.S. National Institute of Standards and Technology (NIST) defines three cloud computing service models as follows:

>> **SaaS:** Customers are provided access to an application running on a cloud infrastructure. The application is accessible from various client devices and interfaces, but customers have no knowledge of, and do not manage or control, the underlying cloud infrastructure. Customers may have access to limited user-specific application settings.

>> **PaaS:** Customers can deploy supported applications to the provider's cloud infrastructure, but they have no knowledge of, and do not manage or control, the underlying cloud infrastructure. Customers have control of the deployed applications and limited configuration settings for the application-hosting environment.

>> **IaaS:** Customers can provision processing, storage, networks, and other computing resources, and they can deploy and run OSes and applications, but they have no knowledge of, and do not manage or control, the underlying cloud infrastructure. Customer have control of OSes, storage, and deployed applications, as well as some networking components (such as host firewalls).

NIST defines four cloud computing deployment models:

>> **Public:** A cloud infrastructure that is open to use by the general public. It's owned, managed, and operated by a third party (or parties) and exists on the cloud provider's premises.

>> **Community:** A cloud infrastructure that is used exclusively by a specific group of organizations.

>> **Private:** A cloud infrastructure that is used exclusively by a single organization. It may be owned, managed, and operated by the organization or a third party (or a combination of both), and may exist on- or off-premises.

>> **Hybrid:** A cloud infrastructure that is composed of two or more of the aforementioned deployment models, bound together by standardized or proprietary technology that enables data and application portability (such as failover to a secondary data center for disaster recovery or content delivery networks across multiple clouds).

Major public cloud service providers such as AWS, Azure, Google Cloud Platform, and Oracle Cloud Platform provide customers not only virtually unlimited compute and storage at scale, but also security capabilities that often exceed the capabilities of the customers themselves. These services do not mean that cloud-based systems are inherently secure, however. The shared responsibility model is used by public cloud service providers to clearly define which aspects of security the provider is responsible for and which aspects the customer is responsible for. SaaS models place the most responsibility on the cloud service provider, typically including securing the following:

>> Applications and data

>> Run-time ware and middleware

>> Servers, virtualization, and OSes

>> Storage and networking

>> Physical data center

The customer is always ultimately responsible for the security and privacy of its data. Additionally, identity and access management (IAM) is typically the customer's responsibility.

In a PaaS model, the customer is typically responsible for the security of its applications and data, as well as identity and access management.

In an IaaS model, the customer is typically responsible for the security of its applications and data, run-time ware and middleware, and OSes. The cloud service provider is typically responsible for the security of networking and the data center (although cloud service providers generally do not provide firewalls). Virtualization, server, and storage security may be managed by either the cloud service provider or the customer.

TIP

The Cloud Security Alliance publishes the Cloud Controls Matrix, which provides a framework for information security that is specifically designed for the cloud industry. The CCM is available at `https://cloudsecurityalliance.org/research/cloud-controls-matrix`.

Distributed systems

Distributed systems are systems with components scattered throughout physical and logical space. Often, these components are owned and/or managed by different groups or organizations, sometimes in different countries. Some components may be privately used, and others represent services available to the public (such as Google Maps). Vulnerabilities in distributed systems include

>> **Loose access permissions:** Individual components in a distributed system may have individual, separate access control systems, or there may be one overarching access control system for all the distributed system's components. Either way, there are too many opportunities for access permissions to be too loose, thereby giving some subjects access to more data and functions than they need.

>> **Unprotected or weakly protected communications:** Data transmitted between the server and other systems (including clients) may be using either weak encryption or no encryption at all.

>> **Weak security inheritance:** What we mean here is that in a distributed system, one component with weak security may compromise the security of the entire system. A publicly accessible component may have direct open access to other components, for example, bypassing local controls in those other components.

>> **Lack of centralized security and control:** Distributed systems that are controlled by more than one organization may lack overall oversight of security management and security operations, especially peer-to-peer systems, which are often run by end users on lightly managed or unmanaged endpoints.

>> **Critical paths:** A critical-path weakness occurs when a system's continued operation depends on the availability of a single component.

All these weaknesses can be present in simpler environments. These weaknesses and other defects can be detected through the use of security scanning tools or manual techniques, and corrective actions can be taken to mitigate those defects.

High-quality standards for cloud computing — for cloud service providers as well as organizations that use cloud services — are available on the websites of the Cloud Security Alliance (`https://cloudsecurityalliance.org`) and the European Network and Information Security Agency (`https://www.enisa.europa.eu`).

Internet of Things

The security of Internet of Things (IoT) devices and systems is a rapidly evolving area of information security. IoT sensors and devices collect large amounts of both potentially sensitive data and seemingly innocuous data, and are used to control physical systems and environments. Under certain circumstances, however, practically any data that is collected can be used for nefarious purposes, and devices can be subverted to affect physical environments. As a result, security must be a critical design consideration for IoT devices and systems, includes not only securing the data stored on the systems, but also how the data is collected, transmitted, processed, and used.

Many networking and communications protocols are commonly used in IoT devices, including the following:

>> IPv6 over low-power wireless personal area networks

>> 5G

>> Wi-Fi

>> Bluetooth, Bluetooth Mesh, and Bluetooth Low-Energy

>> Thread

>> Zigbee

The security of these various protocols and their implementations must be carefully considered in the design of secure IoT devices and systems.

Microservices

Microservices represent a variety of software-based services running on systems in a distributed environment. Using older technology terms, you could consider microservices to be like software program subroutines that run on various systems and are written in various computer languages. Put another way, you could think of microservices as being a more mature form of *mashups*, which are web applications that use content from various sources displayed through a single user interface.

Microservices are generally developed and deployed with a DevOps or DevSecOps model; typically, they communicate by using standard message–based protocols such as HTTP.

Vulnerabilities in microservices appear in several ways, including the following:

>> **Application software:** All the techniques used to ensure that software is free of security defects also apply to application software. Chapter 10 fully explores this topic.

>> **Server subsystem vulnerabilities:** The underlying subsystems that host microservices such as web servers and deeper layers such as database management systems require the usual hardening, configuration management, patching, and life-cycle management operations to ensure that they're reasonably free of exploitable defects. This classic vulnerability-management challenge faces every organization today.

>> **OS vulnerabilities:** The OSes supporting microservices must be actively managed through hardening, configuration management, and vulnerability management tools and processes to ensure that they don't have exploitable vulnerabilities that attackers could use to gain a foothold into the environment.

>> **Access management:** All layers of a microservices environment must employ hardened authentication controls to prevent intruders from attacking the environment.

It's imperative that microservices environments be fully included in all traditional IT service management processes so that they are actively managed, protected, and monitored, just like all other types of server and endpoint OSes, subsystems, software, and source code.

Containerization

Containers are relatively new innovations in virtualization environments. Instead of running multiple instantiations of software programs in their own virtual OS machines, programs are run in isolated containers within a single OS instance. The practice of building and managing containers is known as *containerization*.

For the purposes of information security, you can think of containerization as being like virtualization. Vulnerabilities can exist in several layers, including

>> **Application software:** The software running in a container must be managed like software in all other forms to ensure that it's free of exploitable defects. See Chapter 10 for comprehensive coverage of this topic.

>> **Container engines:** The subsystem that runs the containerization environment must be managed like any software subsystem, requiring active techniques such as hardening, configuration management, access management, patching, and life-cycle management to ensure that no exploitable defects or configurations could permit a successful attack.

>> **OS:** Like OSes used for every other purpose, OSes in containerization environments must be actively managed by the usual IT and security operations processes to prevent attacks.

>> **Access management:** All layers of a container environment must employ hardened authentication controls to prevent intruders from successfully attacking the environment.

>> **Operations:** Like every kind of IT environment, a containerization environment must be actively operated, monitored, and managed to ensure that it is running properly, and that all signs of malfunction and intrusion are detected and acted on.

Serverless

Serverless computing is a cloud-native development model in which virtual infrastructure (such as virtual machines or containers) is abstracted from developers, allowing them to build and run applications without having to manage the underlying infrastructure. Serverless applications are deployed using container services such as Kubernetes that automatically launch on demand when required. When an event triggers code to run, the cloud service provider dynamically allocates resources, and when it finishes executing, these resources are released. This system brings cost and resource efficiencies while also releasing developers from routine tasks, such as application scaling and server provisioning. The term serverless computing is something of a misnomer, as a server OS and infrastructure indeed exist, but they are abstracted away from the customer and provisioned, scaled, and managed by the service provider.

Using serverless applications requires a paradigm shift in how organizations approach security. Instead of building security around the application infrastructure, the developers need to build security around the functions within the applications hosted by the cloud service provider. There are two major security areas of serverless cloud infrastructure that require special attention: secure coding and identity and access management.

Vulnerabilities in a serverless environment are the same as those in software of every other type. Software developers must be trained in secure software development, and tooling must be used to identify source-code defects that must be fixed before the software is placed into production. The serverless environment must be actively monitored for security events that could be signs of intrusion.

A serverless environment must include hardened authentication controls to prevent successful intrusions by attackers. The security of the most hardened software is all for naught if the administrative interface is exposed to the Internet with simple authentication and credentials such as admin/admin.

Embedded systems

Embedded systems encompass the wide variety of systems and devices that are Internet-connected. Mainly, we're talking about devices that are not human-connected in the computing sense. Examples of such devices include

>> Automobiles and other vehicles

>> Home appliances, such as washers and dryers, ranges and ovens, refrigerators, thermostats, televisions, videogame players, video surveillance systems, and home automation systems

>> Medical care devices, such as IV pumps and monitors

>> Heating, ventilation, and air conditioning systems

>> Commercial video surveillance and key card systems

>> Automated payment kiosks, fuel pumps, and automated teller machines

>> Network devices such as routers, switches, modems, and firewalls

These devices often run embedded systems, which are specialized OSes designed to run on devices that lacking computerlike human interaction through a keyboard or display. The devices still have an OS that is very similar to that on endpoints such as laptops and mobile devices.

Design defects in this class of devices include

>> **Lack of a security patching mechanism:** Most of these devices lack any means of remediating security defects that are found after manufacture.

>> **Lack of antimalware mechanisms:** Most of these devices have no built-in defenses against attack.

>> **Lack of robust authentication:** Many of these devices have simple, easily-guessed default login credentials that cannot be changed (or, at best, are rarely changed) by their owners.

>> **Lack of monitoring capabilities:** Many of these devices lack any means of sending security and event alerts.

Because the majority of these devices cannot be altered, mitigation of these defects typically involves isolating these devices on separate, heavily guarded networks that have tools in place to detect and block attacks.

TIP

Many manufacturers of embedded, network-enabled devices do not permit customers to alter their configuration or apply security settings. As a result, organizations are compelled to place these devices on separate, guarded networks.

High-performance computing systems

High-performance computing (HPC) refers to the use of supercomputers or grid computing to solve problems that require computationally intensive processing. Topics addressed by HPC include weather forecasting and climatology, quantum mechanics, oil and gas exploration, seismology, and cryptanalysis.

HPC systems are generally characterized by having large numbers of CPUs and large amounts of memory, facilitating a high number of floating-point operations per second. Historically, HPC systems used specialized operating systems, but increasingly, Linux is used.

HPC environments use some form of parallel processing, in which computational tasks are distributed across large numbers of processors. Either a single program will execute across multiple threads, or several programs communicate by using some form of inter-process communication.

Edge computing systems

Edge computing refers to the architecture of a highly distributed environment in which computing resources are deployed near the edges of the environment, close to where data is acquired from outside. Edge computing is all about server placement in a network to reduce latency and improve performance.

The vulnerabilities in an edge computing environment are virtually the same as in any other, including

>> **Application software:** All application software running in an edge computing environment must be managed like application software in any other environment. These management activities include all those that are part of the systems development life cycle. Chapter 10 explores this topic more fully.

>> **Subsystems:** Subsystems such as web servers and database management systems must be actively managed like those in other environments. Typical activities include access management, configuration management, monitoring, patching, and life-cycle management.

>> **OS:** OSes in edge computing environments should be managed with techniques used on servers in other types of environments, including access management, hardening, configuration management, patching, and monitoring.

>> **Network devices:** All network devices in an edge computing environment must be hardened, patched, and actively managed in typical IT life-cycle management processes.

>> **Architecture:** The relationship between systems, networks, and any network or computing zones, segmentation, data flow, and other aspects of architecture must be closely examined by security specialists or engineers to ensure that the architecture is free of design flaws that could aid an intruder.

>> **Operations:** As in every other type of IT environment, active IT service management and security operations must be in place to monitor the performance, health, and security of an edge computing environment.

Virtualized systems

Virtualization is the practice of implementing multiple instances of OSes in a single hardware platform. Virtualization makes the use of computing hardware more efficient and flexible. But organizations must be mindful of certain risks associated with virtualization, including

>> **Hypervisor management and protection:** The hypervisor manages various OS instances and uses isolation and other protection, not unlike process isolation within an OS. Like OSes, hypervisors must be hardened, patched, and managed carefully to prevent various types of attacks.

>> **Virtual-machine sprawl:** In the old days, you could not implement a new OS without going through the corporate purchasing process to buy a server. In a virtual-machine environment, a new OS can be built with a few clicks. Thus, more discipline is required to control the creation and use of virtual machines.

Virtual desktop infrastructure is the practice of implementing centrally stored and managed desktop OSes that execute on individual endpoints. This practice can reduce the cost of endpoint management, as well as prevent information leakage by keeping sensitive data on central servers. Endpoints assume the role of terminals, and all processing and data manipulation is performed on servers.

Web-based systems

Web-based systems contain many components, including application code, database management systems, OSes, middleware, and the web-server software itself.

These components may, individually and collectively, have security design or implementation defects. Some of those defects are

>> **Failure to block injection attacks:** Attacks such as JavaScript injection and SQL injection can permit an attacker to cause a web application to malfunction and expose sensitive internally stored data.

>> **Defective authentication:** A website has many ways to implement authentication — too many to list here. Authentication is essential to get right, but many sites fail to do so.

>> **Defective session management:** Web servers create logical sessions to keep track of individual users. Many websites' session management mechanisms are vulnerable to abuse, notably that permit an attacker to take over another user's session.

>> **Failure to block cross-site scripting attacks:** Websites may fail to examine and sanitize input data. As a result, attackers can create attacks that send malicious content to users.

>> **Failure to block cross-site request forgery attacks:** Websites that fail to employ proper session and session context management can be vulnerable to attacks in which users are tricked into sending commands to websites that may cause them harm. Here's an example we like to use: An attacker tricks a user into clicking a link that actually takes the user to a URL like `http://bank.com/transfer?tohackeraccount:amount=99999.99`.

>> **Failure to protect direct object references:** Websites can be tricked into accessing and sending data to a user who is not authorized to view or modify it.

These vulnerabilities can be mitigated in three main ways:

>> Training developers in the techniques of safe software development

>> Including security in the development life cycle

>> Using dynamic and static application scanning tools

TIP

For an in-depth review of vulnerabilities in web-based systems, read the "Top 10" list at `https://www.owasp.org`.

Mobile systems

Mobile systems include OSes and applications on smartphones, tablets, phablets, smart watches, and wearables. The most popular OS platforms for mobile systems are Apple iOS, Android, and Windows 10.

The vulnerabilities of mobile systems include

» **Lack of robust resource access controls:** History has shown us that some mobile OSes lack robust controls that govern which apps are permitted to access resources on the mobile device, including

- Locally stored data

- Contact lists

- Camera and photo library

- Email messages

- Location services

- Microphone

» **Insufficient security screening of applications:** Some mobile platform environments are quite good at screening out applications that contain security flaws or outright break the rules, but other platforms apparently have more of an "anything goes" policy. Beware: Your mobile app may be doing more than advertised.

» **Lax default security settings:** Many mobile platforms lack enforcement of basic security. Some platforms don't require devices to lock automatically or have lock codes, for example.

In a managed corporate environment, the use of an MDM system can mitigate many or all of these risks. But individual users must do the right thing by using strong security settings.

Select and Determine Cryptographic Solutions

Cryptography (from the Greek *kryptos*, meaning *hidden*, and *graphia*, meaning *writing*) is the science of encrypting and decrypting communications to make them incomprehensible to everyone all but the intended recipient.

CROSS
REFERENCE

This section covers Objective 3.6 of the Security Architecture and Engineering domain in the CISSP Exam Outline (May 1, 2021).

Cryptography can be used to achieve several goals of information security:

» **Confidentiality:** Cryptography protects the confidentiality or secrecy of information. Even when the transmission or storage medium has been compromised, the encrypted information is practically useless to unauthorized people who don't have the proper encryption keys.

» **Integrity:** Cryptography can be used to ensure the integrity or accuracy of information through the use of hashing algorithms and message digests.

» **Authentication:** Cryptography can be used for authentication and nonrepudiation services through digital signatures, digital certificates, or a public key infrastructure (PKI).

REMEMBER

The CISSP exam tests your ability to apply general cryptographic concepts to real-world issues and problems. You don't have to memorize cryptographic algorithms or the step-by-step operation of various cryptographic systems. But you should have a firm grasp of cryptographic concepts and technologies, as well as their specific strengths, weaknesses, uses, and applications.

WARNING

Don't confuse these three points with the C-I-A triad, which we discuss in Chapter 3: The C-I-A triad deals with confidentiality, integrity, and availability; cryptography does nothing to ensure availability.

Cryptography has evolved into a complex science (some people say an art), presenting many great promises and challenges in the field of information security. The basics of cryptography include various terms and concepts, the individual components of the cryptosystem, and the classes and types of ciphers.

THE SCIENCE OF CRYPTO

Cryptography is the science of encrypting and decrypting information, such as private messages, to protect confidentiality, integrity, and/or authenticity. Practitioners of cryptography are known as *cryptographers*.

Cryptanalysis is the science of deciphering (or breaking) ciphertext without the cryptographic key. Practitioners of cryptanalysis are known as *cryptanalysts*.

Cryptology is a science that encompasses both cryptography and cryptanalysis. Practitioners of cryptology are known as *cryptologists*.

Plaintext and ciphertext

A *plaintext* message is a message in its original readable format or a ciphertext message that has been properly decrypted (unscrambled) to produce the original readable plaintext message.

A *ciphertext* message is a plaintext message that has been transformed (encrypted) into a scrambled message that's unintelligible. This term doesn't apply to messages from your boss, which may also happen to be unintelligible.

Encryption and decryption

Encryption (or *enciphering*) is the process of converting plaintext communications to ciphertext. *Decryption* (or *deciphering*) reverses that process, converting ciphertext to plaintext. (See Figure 5-5.)

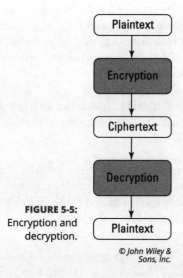

FIGURE 5-5: Encryption and decryption.

© *John Wiley & Sons, Inc.*

Traffic on a network can be encrypted via end-to-end or link encryption.

End-to-end encryption

With *end-to-end encryption*, packets are encrypted once at the original encryption source and then decrypted only at the final decryption destination. The advantages of end-to-end encryption are speed and overall security. For the packets to be routed properly, however, only the data is encrypted, not the routing information.

Link encryption

Link encryption requires each node (such as a router) to have separate key pairs for its upstream and downstream neighbors. Packets are encrypted and decrypted, and then re-encrypted at every node along the network path.

The following example, as shown in Figure 5-6, illustrates link encryption:

1. Computer 1 encrypts a message by using Secret Key A and then transmits the message to Router 1.

2. Router 1 decrypts the message by using Secret Key A, reencrypts the message by using Secret Key B, and then transmits the message to Router 2.

3. Router 2 decrypts the message by using Secret Key B, reencrypts the message by using Secret Key C, and then transmits the message to Computer 2.

4. Computer 2 decrypts the message by using Secret Key C.

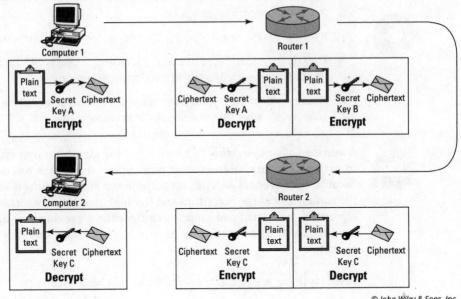

FIGURE 5-6: Link encryption.

© *John Wiley & Sons, Inc.*

The advantage of using link encryption is that the entire packet (including routing information) is encrypted. But link encryption has the two disadvantages:

>> **Latency:** Packets must be encrypted/decrypted at every node, which creates *latency* (delay) in the transmission of those packets.

>> **Inherent vulnerability:** If a node is compromised or a packet's decrypted contents are cached in a node, the message can be compromised.

Putting it all together: The cryptosystem

A *cryptosystem* is the hardware or software implementation that transforms plaintext to ciphertext (encrypting it) and back to plaintext (decrypting it).

An effective cryptosystem must have the following properties:

>> The encryption and decryption process is efficient for all possible keys within the cryptosystem's keyspace.

A *keyspace* is the range of all possible values for a key in a cryptosystem.

>> The cryptosystem is easy to use. A cryptosystem that is difficult to use might be used improperly, leading to data loss or compromise.

>> The strength of the cryptosystem depends on the secrecy of the *cryptovariables* (keys) rather than the secrecy of the algorithm.

A *restricted algorithm* refers to a cryptographic algorithm that must be kept secret to provide security. Restricted or proprietary algorithms are not very effective, because effectiveness depends on keeping the algorithm itself secret, rather than the complexity of the algorithm and the high number of variable solutions of the algorithm. Restricted and proprietary algorithms are therefore are not commonly used today. They are generally used only for applications that require minimal security.

Cryptosystems are typically composed of two basic elements:

>> **Cryptographic algorithm:** Also called a *cipher,* a cryptographic algorithm details the step-by-step mathematical function used to produce ciphertext (encipher) and plaintext (decipher).

>> **Cryptovariable:** Also called a *key,* the cryptovariable is a secret value applied to the algorithm. The strength and effectiveness of the cryptosystem largely depend on the secrecy and strength of the cryptovariable.

Key clustering occurs when identical ciphertext messages are generated from a plaintext message with the same encryption algorithm but different encryption keys. Key clustering indicates a weakness in a cryptographic algorithm because it statistically reduces the number of key combinations that must be attempted in a brute-force attack.

A *cryptosystem* consists of the cryptographic algorithm (cipher) and the crypto-variable (key), as well as all the possible plaintexts and ciphertexts produced by the cipher and key.

An analogy of a cryptosystem is a deadbolt lock. A deadbolt lock can be easily identified, and its inner working mechanisms aren't closely guarded state secrets. What makes a deadbolt lock effective is the individual key that controls a specific lock on a specific door. But if the key is weak (imagine only one or two notches on a flat key) or not well protected (left under your doormat), the lock won't protect your belongings as well. Similarly, if an attacker is able to determine what cryptographic algorithm (lock) was used to encrypt a message, it should still be protected, because you're using a strong key that you've kept secret rather than a six-character password that you wrote on a scrap of paper and left under your mouse pad.

Classes of ciphers

Ciphers are cryptographic transformations. The two main classes of ciphers used in symmetric key algorithms are *block* and *stream* (see "Cryptographic Methods," later in this chapter), which describe how the ciphers operate on input data.

The two main classes of ciphers are block and stream.

Block ciphers

Block ciphers operate on a single fixed block (typically, 128 bits) of plaintext to produce the corresponding ciphertext. When a given key is used in a block cipher, the same plaintext block always produces the same ciphertext block. Advantages of block ciphers compared with stream ciphers are

>> **Reusable keys:** Key management is much easier.

>> **Interoperability:** Block ciphers are more widely supported.

Block ciphers are typically implemented in software. Examples of block ciphers include AES, DES, Blowfish, Twofish, and RC5.

A DISPOSABLE CIPHER: THE ONE-TIME PAD

A *one-time pad* (key) is a *keystream* (stream of random or pseudorandom characters) that can be used only once. Considered to be unbreakable because it's random, and is used only once and then destroyed, a one-time pad consists of a pad of the same length as the message to which it's applied. The sender and receiver have an identical pad, which is used by the sender to encrypt the message and by the receiver to decrypt the message. This type of cipher is very effective for short messages but impractical for large messages (several megabytes) due to the computing resources required to create unique keystreams for such messages. One-time pads are typically implemented as stream ciphers.

Remember: A one-time pad is an example of a stream cipher and is considered unbreakable.

Stream ciphers

Stream ciphers operate in real time on a continuous stream of data, typically bit by bit. Stream ciphers generally work faster than block ciphers and require less code to implement. But the keys in a stream cipher are generally used only once (see the nearby sidebar "A disposable cipher: The one–time pad") and then discarded. Key management becomes a serious problem. When a stream cipher is used, the same plaintext bit or byte produces a different ciphertext bit or byte every time it is encrypted. Stream ciphers are typically implemented in hardware.

Examples of stream ciphers include Salsa20 and RC4.

REMEMBER

People often assume that traffic such as a streaming video from a service such as YouTube is encrypted with a stream cipher. Such content consists of individual TCP/IP packets that are encrypted with a block cipher.

Types of ciphers

The two basic types of ciphers are substitution and transposition. Both are involved in the process of transforming plaintext into ciphertext.

REMEMBER

Most modern cryptosystems use both substitution and permutation to achieve encryption.

Substitution ciphers

Substitution ciphers replace bits, characters, or character blocks in plaintext with alternate bits, characters, or character blocks to produce ciphertext. A classic example of a substitution cipher is one that Julius Caesar used: He swapped letters of the message with other letters from the same alphabet. In a simple substitution cipher using the standard English alphabet, a *cryptovariable* (key) is added *modulo 26* to the plaintext message. In modulo 26 addition, the remainder is the final result for any sum equal to or greater than 26. A basic substitution cipher in which the word "boy" is encrypted by adding three characters using modulo 26 math produces the following result:

	b	o	y	PLAINTEXT
	2	15	25	NUMERIC VALUE
+	3	3	3	SUBSTITUTION VALUE
	5	18	2	MODULO 26 RESULT
	e	r	b	CIPHERTEXT

A substitution cipher may be

» **Monoalphabetic:** A single alphabet is used to encrypt the entire plaintext message.

» **Polyalphabetic:** This more complex substitution uses a different alphabet to encrypt each bit, character, or character block of a plaintext message.

A more modern example of a substitution cipher is the S-boxes (substitution boxes) employed in the Data Encryption Standard (DES) algorithm. The S-boxes in that algorithm produce a nonlinear substitution (6 bits in, 4 bits out).

Transposition

Transposition ciphers rearrange bits, characters, or character blocks in plaintext to produce ciphertext. In a simple columnar transposition cipher, a message might be read horizontally but written vertically to produce the ciphertext, as in the following example,

```
THE QUICK BROWN FOX JUMPS OVER THE LAZY DOG
```

written in nine columns as

```
THEQUICKB
ROWNFOXJU
MPSOVERTH
ELAZYDOG
```

CRYPTOGRAPHY ALTERNATIVES

Technology provides valid and interesting alternatives to cryptography when a message needs to be protected during transmission. Following are some useful options.

Steganography: A picture is worth a thousand (hidden) words

Steganography is the art of hiding the very existence of a message. It is related to but different from cryptography. As in cryptography, one purpose of steganography is to protect the contents of a message. But the contents of the message aren't encrypted. Instead, the existence of the message is hidden in some other communications medium. A message may be hidden in a graphic or sound file, in slack space on storage media, in traffic noise over a network, or in a digital image. By using the example of a digital image, the least significant bit (the right-most bit) of each byte in the image file can be used to transmit a hidden message without noticeably altering the image. Because the message itself isn't encrypted, however, if it is discovered, its contents can be easily compromised.

Many popular brands of printers employ hidden (barely visible) watermarks that make it possible to trace a printed document to an individual printer. This technique is known as *printer steganography*.

Digital watermarking: The (ouch) low watermark

Digital watermarking is a technique similar to steganography that can be used to verify the authenticity of an image or data, or to protect the intellectual-property rights of the creator. Watermarking is the visible cousin of steganography; no attempt is made to hide its existence. Watermarks have long been used on currency, letterhead, and stock.

Within the past decade, the use of digital watermarking has become more widespread. To display photo examples on the Internet without risking intellectual-property theft, for example, a copyright notice may be prominently imprinted across the image. As with steganography, nothing is encrypted with digital watermarking; the confidentiality of the material is not protected with a watermark.

and then transposed (encrypted) vertically as

```
TRMEHOPLEWSAQNOZUFVYIOEDCXROKJTGBUH
```

The original letters of the plaintext message are the same; only the order has been changed to achieve encryption.

DES performs permutations through the use of P-boxes (permutation boxes) to spread the influence of a plaintext character over many characters so that they're not easily traced back to the S-boxes used in the substitution cipher.

Other types of ciphers include

>> **Codes:** Codes include words and phrases to communicate a secret message.

>> **Running (or book) ciphers:** The key is page 137 of *The Catcher in the Rye,* for example, and text on that page is added modulo 26 to perform encryption/decryption.

>> **Vernam ciphers:** Also known as one-time pads, these ciphers are keystreams that can be used only once. We discuss these ciphers in the earlier sidebar "A disposable cipher: The one-time pad."

>> **Concealment ciphers:** These ciphers include *steganography,* which we discuss in the nearby sidebar "Cryptography alternatives."

Cryptographic life cycle

The *cryptographic life cycle* is the sequence of events that occurs throughout the use of cryptographic controls in a system. These steps include

>> Development of requirements for a cryptosystem

>> Selection of cryptographic controls

>> Implementation of the cryptosystem

>> Examination of cryptosystem for proper implementation, effective key management, and efficacy of cryptographic algorithms

>> Rotation of cryptographic keys

>> Mitigation of any defects identified

These steps are not altogether different from the selection, implementation, examination, and correction of any other type of security control in a network and computing environment. Like virtually any other components in a network and computing environment, components in a cryptosystem must be examined periodically to ensure that they are still effective and being operated properly.

Cryptographic methods

Cryptographic methods include symmetric, asymmetric, elliptic curves, and quantum.

Symmetric

Symmetric key cryptography — also known as *symmetric algorithm, secret key, single key,* and *private key cryptography* — uses a single key to encrypt and decrypt information. Two parties (for our example, Thomas and Richard) can exchange an encrypted message by using the following procedure:

1. The sender (Thomas) encrypts the plaintext message with a secret key known only to the intended recipient (Richard).

2. The sender transmits the encrypted message to the intended recipient.

3. The recipient decrypts the message with the same secret key to obtain the plaintext message.

For an attacker (Harold) to read the message, he must do one of the following things:

>> Guess the secret key (by using a brute-force attack, for example).

>> Obtain the secret key by using the rubber-hose technique. This technique is another form of brute-force attack. Humans are typically the weakest link, and neither Thomas nor Richard has much tolerance for pain.

>> Get the secret key through social engineering. Thomas and Richard both like money and may be all too willing to help Harold's Nigerian uncle claim his vast fortune.

>> Intercept the secret key during the initial exchange.

The following list includes the main disadvantages of symmetric systems:

>> **Distribution:** Secure distribution of secret keys is required, either through out-of-band methods or asymmetric systems.

>> **Scalability:** A different key is required for each pair of communicating parties.

>> **Limited functionality:** Symmetric systems can't provide authentication or nonrepudiation (discussed later in this chapter).

Symmetric systems also have many advantages:

>> **Speed:** Symmetric systems are much faster than asymmetric systems.

>> **Strength:** Strength is gained when the algorithm uses a large key (128 bit, 192 bit, 256 bit, or larger).

>> **Availability:** Many algorithms available for organizations to select and use.

Symmetric key algorithms include DES, Triple DES (3DES), Advanced Encryption Standard (AES), International Data Encryption Algorithm (IDEA), and Rivest Cipher 5 (RC5).

REMEMBER

Symmetric key systems use a shared secret key.

DATA ENCRYPTION STANDARD

In the early 1970s, NIST solicited vendors to submit encryption algorithm proposals to be evaluated by the National Security Agency in support of a national cryptographic standard. This new encryption standard was used for private-sector and sensitive but unclassified government data. In 1974, IBM submitted a 128-bit algorithm known as Lucifer. After some modifications (the algorithm was shortened to 56 bits, and the S-boxes were changed), the IBM proposal was endorsed by the National Security Agency and formally adopted as the DES. It was published in *Federal Information Processing Standard* (FIPS) PUB 46 in 1977 (updated and revised in 1988 as FIPS PUB 46-1) and *American National Standards Institute* (ANSI) X3.92 in 1981.

REMEMBER

DES is a block cipher that uses a 56-bit key.

The DES algorithm is a symmetric (or private) key cipher consisting of an algorithm and a key. The algorithm is a 64-bit block cipher based on a 56-bit symmetric key. It consists of 56 key bits plus 8 parity bits. Alternatively, you can think of it as being 8 bytes, with each byte containing 7 key bits and 1 parity bit.) During encryption, the original message (plaintext) is divided into 64-bit blocks. Operating on a single block at a time, the algorithm splits each 64-bit plaintext block into two 32-bit blocks. Under control of the 56-bit symmetric key, 16 rounds of transpositions and substitutions are performed on each character to produce the ciphertext output.

TECHNICAL STUFF

A *parity bit* is used to detect errors in a bit pattern. If the bit pattern has 55 key bits (ones and zeros) that add up to an even number, an *odd-parity bit* should be a one, making the total of the bits — including the parity bit — an odd number. For an *even-parity bit*, if the 55 key bits add up to an even number, the parity bit should be a zero, making the total of the bits — including the parity bit — an even number. If an algorithm uses even parity, and the resulting bit pattern (including the parity bit) is an odd number, the transmission has been corrupted.

TECHNICAL STUFF

A *round* is a transformation (permutations and substitutions) that an encryption algorithm performs on a block of plaintext to convert (encrypt) it to ciphertext.

The four distinct modes of operation (the mode of operation defines how the plaintext/ciphertext blocks are processed) in DES are Electronic Code Book, Cipher Block Chaining, Cipher Feedback, and Output Feedback.

TRIPLE DES

The Triple Data Encryption Standard (3DES) effectively extended the life of the DES algorithm. In 3DES implementations, a message is encrypted by using one key, encrypted by using a second key, and then encrypted again by using either the first key or a third key.

ADVANCED ENCRYPTION STANDARD

In May 2002, NIST announced the Rijndael Block Cipher as the new standard to implement the Advanced Encryption Standard (AES), which replaced DES as the U.S. government standard for encrypting sensitive but unclassified data. AES was subsequently approved for encrypting classified U.S. government data up to the top secret level (using 192- or 256-key lengths).

The Rijndael Block Cipher, developed by Dr. Joan Daemen and Dr. Vincent Rijmen, uses variable block and key lengths (128, 192, or 256 bits) and 10 to 14 rounds. It was designed to be simple, resistant to known attacks, and fast. It can be implemented in either hardware or software and has relatively low memory requirements.

REMEMBER

AES is based on the Rijndael Block Cipher.

Until recently, the only known successful attacks against AES were *side-channel attacks,* which don't attack the encryption algorithm directly; instead, they attack the system on which the encryption algorithm is implemented. Side-channel attacks using cache-timing techniques are most common against AES implementations. In 2009, a theoretical related-key attack against AES was published. The attack method is considered to be theoretical because although it reduces the mathematical complexity required to break an AES key, it is still well beyond the computational capability available today.

BLOWFISH AND TWOFISH

The *Blowfish* algorithm operates on 64-bit blocks, employs 16 rounds, and uses variable key lengths of up to 448 bits. The *Twofish* algorithm, a finalist in the AES selection process, is a symmetric block cipher that operates on 128-bit blocks, employing 16 rounds with variable key lengths up to 256 bits. Both Blowfish and Twofish were designed by Bruce Schneier (and others) and are freely available in the public domain. (Neither algorithm has been patented.) To date, no known successful cryptanalytic attacks have been made against either algorithm.

RIVEST CIPHERS

Dr. Ron Rivest, Dr. Adi Shamir, and Dr. Len Adleman invented the RSA (Rivest, Shamir, Adleman) algorithm and founded the company RSA Data Security. The *Rivest ciphers* are a series of symmetric algorithms that include the following:

» **RC2:** A block-mode cipher that encrypts 64-bit blocks of data by using a variable-length key.

» **RC4:** A stream cipher (data is encrypted in real time) that uses a variable-length key (128 bits is standard).

» **RC5:** Similar to RC2 but including a variable-length key (0 to 2,048 bits), variable block size (32, 64, or 128 bits), and a variable number of processing rounds (0 to 255).

» **RC6:** Derived from RC5 and a finalist in the AES selection process. It uses a 128-bit block size and variable-length keys of 128, 192, or 256 bits.

Note: RC1 was never published, and RC3 was broken during development.

IDEA CIPHER

The *International Data Encryption Algorithm (IDEA)* Cipher evolved from the Proposed Encryption Standard and the Improved Proposed Encryption Standard, developed in 1990. IDEA is a block cipher that operates on 64-bit plaintext blocks by using a 128-bit key. IDEA performs eight rounds on 16-bit sub-blocks and can operate in four distinct modes similar to DES. The IDEA Cipher provides stronger encryption than RC4 and 3DES, but because it is patented, it's not widely used today. The patents were set to expire in various countries between 2010 and 2012. It is currently used in some software applications, including Pretty Good Privacy email.

Asymmetric

Asymmetric key cryptography (also known as *asymmetric algorithm cryptography* or *public-key cryptography*) uses two separate keys: one key to encrypt and a different key to decrypt information. These keys are known as *public* and *private key pairs*. When two parties want to exchange an encrypted message by using asymmetric key cryptography, they follow these steps, shown in Figure 5-7:

1. The sender (Thomas) encrypts the plaintext message with the intended recipient's (Richard) public key.

2. This produces a ciphertext message that can be transmitted to the intended recipient (Richard).

3. The recipient (Richard) decrypts the message with his private key, known only to him.

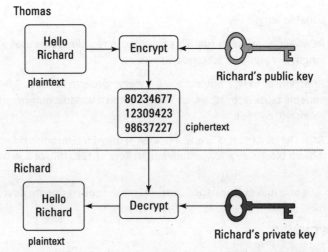

FIGURE 5-7: Sending a message using asymmetric key cryptography.

Image courtesy of authors

Only the private key can decrypt the message; thus, an attacker (Harold) who possesses only the public key can't decrypt the message. Not even the original sender can decrypt the message. This use of an asymmetric key system is known as a *secure message*. A secure message guarantees the confidentiality of the message.

REMEMBER

Asymmetric key systems use a public key and a private key.

Secure message format uses the recipient's private key to protect confidentiality.

REMEMBER

If the sender wants to guarantee the authenticity of a message (or, more correctly, the authenticity of the sender), they can digitally sign the message with this procedure, shown in Figure 5-8:

1. The sender (Thomas) digitally signs the plaintext message with his own private key.

2. The sender transmits the signed message to the intended recipient (Richard).

3. To verify that the message is from the purported sender, the recipient (Richard) applies the sender's (Thomas's) public key (which is known to every Tom, Dick, and Harry).

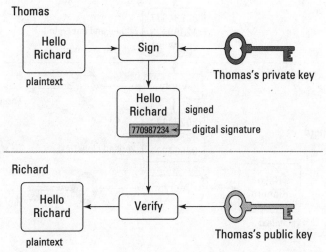

FIGURE 5-8: Verifying message authenticity using asymmetric key cryptography.

An attacker can also verify the authenticity of the message, of course. This use of an asymmetric key system is known as an *open message format* because it guarantees only authenticity, not confidentiality.

REMEMBER

Open message format uses the sender's private key to ensure authenticity.

If the sender wants to guarantee both the confidentiality and authenticity of a message, they can do so by using this procedure, shown in Figure 5-9:

1. The sender (Thomas) encrypts the message first with the intended recipient's (Richard's) public key and then signs with his own private key.

2. The sender transmits the ciphertext message to the intended recipient (Richard).

3. The recipient (Richard) uses the sender's (Thomas's) public key to verify the authenticity of the message and then uses his own private key to decrypt the message's contents.

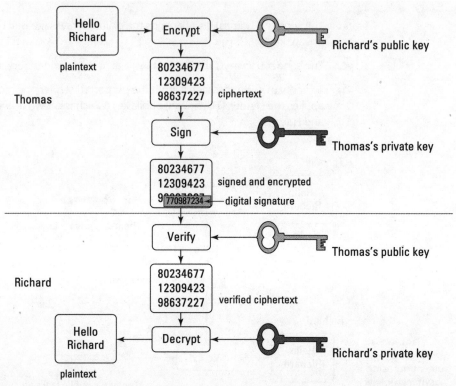

FIGURE 5-9:
Encrypting
and signing a
message using
asymmetric key
cryptography.

Image courtesy of authors

If an attacker intercepts the message, they can apply the sender's public key, but then they have an encrypted message that they can't decrypt without the intended recipient's private key. Thus, both confidentiality and authenticity are assured. This use of an asymmetric key system is known as a *secure and signed message format*.

REMEMBER

A secure and signed message format uses the sender's private key and the recipient's public key to protect confidentiality and ensure authenticity.

A public key and a private key are mathematically related, but theoretically, no one can compute or derive the private key from the public key. This property of asymmetric systems is based on the concept of a one-way function. A *one-way function* is a problem that you can easily compute in one direction but not in the reverse direction. In asymmetric key systems, a *trapdoor* (private key) resolves the reverse operation of the one-way function.

A trapdoor one-way function is like a lock box that is supplied to the user in an opened configuration. Any user may place an item inside the box and then close the lid, which latches the lock closed as it does so. Only the person who has the key can then open the box to obtain the item inside. In this analogy, the lock box itself is the public key, and the key that opens the box is the private key.

Because of their complexity, asymmetric key systems are more commonly used for key management or digital signatures than for encryption of bulk information. Often, a hybrid system is employed, using an asymmetric system to securely distribute the secret keys of a symmetric key system that's used to encrypt the data.

The main disadvantage of asymmetric systems is their lower speed. Because of the types of algorithms that are used to achieve the one-way hash functions, very large keys are required. (A 128-bit symmetric key has strength equivalent to that of a 2,304-bit asymmetric key.) Those large keys in turn require more computational power, causing a significant loss of speed (up to 10,000 times slower than a comparable symmetric key system).

Asymmetric systems also have many significant advantages, including

>> **Extended functionality:** Asymmetric key systems can provide both confidentiality and authentication; symmetric systems can provide only confidentiality.

>> **Scalability:** Because symmetric key systems require secret key exchanges among all the communicating parties, their scalability is limited. Asymmetric key systems, which do not require secret key exchanges, resolve key management issues associated with symmetric key systems and therefore are more scalable.

Asymmetric key algorithms include RSA, Diffie-Hellman Key Exchange, El Gamal, Merkle-Hellman (Trapdoor) Knapsack, and Elliptic Curve, which we talk about in the following sections.

RSA

The RSA algorithm is a key transport algorithm based on the difficulty of factoring a number that's the product of two large prime numbers (typically, 512 bits). Two users (Thomas and Richard) can securely transport symmetric keys by using RSA as follows:

1. Thomas creates a symmetric key, encrypts it with Richard's public key, and transmits it to Richard.

2. Richard decrypts the symmetric key by using his own private key.

REMEMBER

RSA is an asymmetric key algorithm based on factoring prime numbers.

DIFFIE-HELLMAN KEY EXCHANGE

Dr. Whitfield Diffie and Dr. Martin Hellman published a paper titled "New Directions in Cryptography" that detailed a new paradigm for secure key exchange based on discrete logarithms. *Diffie-Hellman* is described as a key agreement algorithm. Two users (Thomas and Richard, who have never met) can exchange symmetric keys by using Diffie-Hellman as follows and as depicted in Figure 5-10:

1. Thomas and Richard obtain each other's public keys.

2. Thomas and Richard ten combine their own private keys with the public key of the other person, producing a symmetric key that only the two users involved in the exchange know.

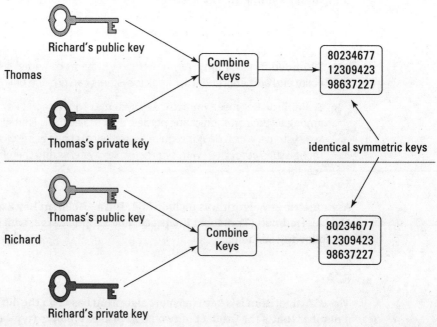

FIGURE 5-10: Diffie-Hellman key exchange is used to generate a symmetric key for two users.

Image courtesy of authors

Diffie-Hellman is vulnerable to man-in-the-middle attacks, in which an attacker intercepts the public keys during the initial exchange and substitutes their own private key to create a session key that can decrypt the session. (You can read more about these attacks in the section "Man-in-the-middle" later in this chapter.) A separate authentication mechanism is necessary to protect against this type of attack, ensuring that the two parties communicating in the session are in fact the legitimate parties.

Diffie-Hellman is an asymmetric key algorithm based on discrete logarithms.

EL GAMAL

El Gamal is an unpatented, asymmetric key algorithm based on the discrete logarithm problem used in Diffie-Hellman (discussed in the preceding section). El Gamal extends the functionality of Diffie-Hellman to include encryption and digital signatures.

MERKLE-HELLMAN (TRAPDOOR) KNAPSACK

The *Merkle-Hellman (Trapdoor) Knapsack*, published in 1978, employs a unique approach to asymmetric cryptography. It's based on the problem of determining what items, in a set of items that have fixed weights, can be combined to obtain a given total weight. Knapsack was broken in 1982.

Knapsack is an asymmetric key algorithm based on fixed weights.

ELLIPTIC CURVE

Elliptic Curves is far more difficult to compute than conventional discrete logarithm problems or factoring prime numbers. (A 160-bit EC key is equivalent to a 1,024-bit RSA key.) The use of smaller keys means that Elliptic Curve is significantly faster than other asymmetric algorithms (and many symmetric algorithms) and can be widely implemented in various hardware applications, including wireless devices and smart cards.

Elliptic Curve is more efficient than other asymmetric key systems, and many symmetric key systems because it can use a smaller key.

QUANTUM COMPUTING

Quantum computing is an emerging computing processor design that uses the properties of quantum states to perform computation. Although quantum computing is still in its infancy, it may someday pose a significant threat to encryption. A quantum computer may eventually be able to break the most advanced encryption in very short periods of time.

Realizing that quantum computing may eventually be used to break cryptosystems, cryptographers are revisiting the designs of their cryptosystems and developing new ways to ensure that they can resist quantum-computing cryptanalysis.

Quantum computing is currently a theoretical threat to cryptosystems.

Public key infrastructure

A *public key infrastructure* (PKI) is an arrangement whereby a designated authority stores encryption keys or certificates (an electronic document that uses the public key of an organization or person to establish identity and a digital signature to establish authenticity) associated with users and systems, thereby enabling secure communications through the integration of digital signatures, digital certificates, and other services necessary to ensure confidentiality, integrity, authentication, nonrepudiation, and access control.

REMEMBER

The four basic components of a PKI are the Certificate Authority, Registration Authority, repository, and archive:

>> **Certificate Authority:** The Certificate Authority (CA) comprises hardware, software, and the personnel administering the PKI. It issues certificates, maintains and publishes status information and Certificate Revocation Lists, and maintains archives.

>> **Registration Authority:** The Registration Authority (RA) also comprises hardware, software, and the personnel administering the PKI. It's responsible for verifying certificate contents for the CA.

>> **Repository:** A *repository* is a system that accepts certificates and Certificate Revocation Lists from a CA and distributes them to authorized parties.

>> **Archive:** An *archive* offers long-term storage of archived information from the CA.

Key management practices

Like physical keys, encryption keys must be safeguarded. Most successful attacks against encryption exploit some vulnerability in key management functions rather than some inherent weakness in the encryption algorithm. Following are the major functions associated with managing encryption keys:

>> **Generation:** Keys must be generated randomly on a secure system, and the generation sequence itself shouldn't provide potential clues regarding the contents of the keyspace. Generated keys shouldn't be displayed in the clear.

>> **Distribution:** Keys must be securely distributed. Distribution is a major vulnerability in symmetric key systems. Using an asymmetric system to distribute secret keys securely is one solution.

>> **Installation:** Key installation is often a manual process. This process should ensure that the key isn't compromised during installation, incorrectly entered, or too difficult to be used readily.

- » **Storage:** Keys must be stored on protected or encrypted storage media, or the application using the keys should include safeguards that prevent extraction of the keys.

- » **Change:** Keys, like passwords, should be changed regularly, relative to the value of the information being protected and the frequency of use. Keys used frequently are more likely to be compromised through interception and statistical analysis.

- » **Control:** Key control addresses the proper use of keys. Different keys have different functions and may be approved for only certain levels of classification.

- » **Disposal:** Keys (and any distribution media) must be properly disposed of, erased, or destroyed so that the key's contents are not disclosed, possibly providing an attacker insight into the key management system.

A *cryptoperiod* is the length of time that an encryption key can be considered valid. Various factors influence the length of the cryptoperiod, including the length of the key and the strength of the encryption algorithm. When an encryption key has reached the end of its cryptoperiod, it should be discarded, and a new key should be generated. This process may require deciphering existing ciphertext and encrypting it the new key.

REMEMBER

The seven key management issues are generation, distribution, installation, storage, change, control, and disposal.

KEY ESCROW AND KEY RECOVERY

Law enforcement has always been concerned about the potential use of encryption for criminal purposes. To counter this threat, NIST published the Escrowed Encryption Standard in Federal Information Processing Standards Publication 185 (1994). The premise of the standard is to divide a secret key into two parts and place those two parts into escrow with two separate trusted organizations. With a court order, law enforcement officials can obtain the two parts, recover the secret key, and decrypt the suspected communications. One implementation is the Clipper Chip proposed by the U.S. government. The Clipper Chip uses the Skipjack Secret Key algorithm for encryption and an 80-bit secret key.

Digital signatures and digital certificates

Message authentication guarantees the authenticity and integrity of a message by ensuring that

>> A message hasn't been altered (maliciously or accidentally) during transmission.

>> A message isn't a replay of a previous message.

>> The message was sent from the origin stated (and is not a forgery).

>> The message is sent to the intended recipient.

Checksums, CRC values, and parity checks are examples of basic message authentication and integrity controls. More-advanced message authentication is performed by using digital signatures and message digests.

REMEMBER

Digital signatures and message digests can be used to provide message authentication.

The Digital Signature Standard (DSS), published by NIST in Federal Information Processing Standard (FIPS) 186-4, specifies three acceptable algorithms in its standard: the RSA Digital Signature Algorithm; the Digital Signature Algorithm (DSA); which is based on a modified El Gamal algorithm; and the Elliptic Curve Digital Signature Algorithm.

A digital signature is a simple way to verify the authenticity (and integrity) of a message. Instead of encrypting a message with the intended receiver's public key, the sender encrypts it with their own private key. The sender's public key properly decrypts the message, authenticating the originator of the message. This process is known as an open message format in asymmetric key systems, as we discuss in the section "Asymmetric" earlier in this chapter.

Nonrepudiation

To *repudiate* is to deny. *Nonrepudiation* means that an action (such as an online transaction, email communication, and so on) or occurrence can't be easily denied. Nonrepudiation is a related function of identification and authentication and accountability. It's difficult for a user to deny sending an email message that was digitally signed with that user's private key, for example. Likewise, it's difficult to deny responsibility for an enterprise-wide outage if the accounting logs positively identify you (from username and strong authentication) as the poor soul who inadvertently issued the write-erase command on the core routers two seconds before everything dropped!

Integrity (hashing)

It's often impractical to encrypt a message with the receiver's public key to protect confidentiality and then encrypt the entire message again by using the sender's private key to protect authenticity and integrity. Instead, a representation of the encrypted message is encrypted with the sender's private key to produce a digital signature. The intended recipient decrypts this representation by using the sender's public key and then independently calculates the expected results of the decrypted representation by using the same known one-way hashing algorithm. If the results are the same, the integrity of the original message is assured. This representation of the entire message is known as a *message digest*.

To *digest* means to reduce or condense something, and a message digest does precisely that. (Conversely, *indigestion* means to expand, like gases. How do you spell *relief?*) A message digest is a condensed representation of a message; think *Reader's Digest*. Ideally, a message digest has the following properties:

>> The original message can't be re-created from the message digest.

>> Finding a message that produces a particular digest shouldn't be computationally feasible.

>> No two messages should produce the same message digest (a situation known as a *collision*).

>> The message digest should be calculated by using the entire contents of the original message; it shouldn't be a representation of a representation.

Message digests are produced by using a one-way hash function. There are several types of one-way hashing algorithms (digest algorithms), including MD5, SHA-2 variants, and HMAC.

A collision results when two messages produce the same digest or when a message produces the same digest as a different message.

A one-way function ensures that the same key can't encrypt and decrypt a message in an asymmetric key system. One key encrypts the message (produces ciphertext), and a second key (the trapdoor) decrypts the message (produces plaintext), effectively reversing the one-way function.

A one-way hashing algorithm produces a hashing value (or message digest) that can't be reversed; that is, it can't be decrypted. In other words, no trapdoor exists for a one-way hashing algorithm. The purpose of a one-way hashing algorithm is to ensure integrity and authentication.

REMEMBER

MD5, SHA-2, SHA-3, and HMAC are examples of commonly used message authentication algorithms.

MD

MD (Message Digest) is a family of one-way hashing algorithms developed by Dr. Ron Rivest that includes MD (obsolete), MD2, MD3 (not widely used), MD4, MD5, and MD6:

>> **MD2:** Developed in 1989 and still widely used today, MD2 takes a variable-size input (message) and produces a fixed-size output (128-bit message digest). MD2 is very slow (originally developed for 8-bit computers) and highly susceptible to collisions.

>> **MD4:** Developed in 1990, MD4 produces a 128-bit digest and is used to compute NT-password hashes for various Microsoft Windows OSes, including NT, XP, and Vista. An MD4 hash is typically represented as a 32-digit hexadecimal number. Several known weaknesses are associated with MD4, and it's also susceptible to collision attacks.

>> **MD5:** Developed in 1991, MD5 is one of the most popular hashing algorithms in use today, commonly used to store passwords and to check the integrity of files. Like MD2 and MD4, MD5 produces a 128-bit digest. Messages are processed in 512-bit blocks, using four rounds of transformation. The resulting hash is typically represented as a 32-digit hexadecimal number. MD5 is also susceptible to collisions and is now considered to be cryptographically broken by the U.S. Department of Homeland Security.

>> **MD6:** Developed in 2008, MD6 uses very large input message blocks (up to 512 bytes) and produces variable-length digests (up to 512 bits). MD6 was originally submitted for consideration as the new SHA-3 standard but was eliminated from further consideration after the first round in July 2009. Unfortunately, the first widespread use of MD6 (albeit unauthorized and illicit) was in the Conficker.B worm in late 2008, shortly after the algorithm was published.

SHA

Like MD, SHA (Secure Hash Algorithm) is another family of one-way hash functions. The SHA family of algorithms is designed by the U.S. National Security Agency and published by NIST. The SHA family of algorithms includes SHA-1, SHA-2, and SHA-3:

>> **SHA-1:** Published in 1995, SHA-1 takes a variable-size input (message) and produces a fixed-size output (160-bit message digest versus MD5's 128-bit

message digest). SHA-1 processes messages in 512-bit blocks and adds padding to a message length, if necessary, to produce a total message length that's a multiple of 512. Note that SHA-1 is no longer considered to be a viable hash algorithm.

>> **SHA-2:** Published in 2001, SHA-2 consists of four hash functions — SHA-224, SHA-256, SHA-384, and SHA-512 — that have digest lengths of 224, 256, 384, and 512 bits, respectively. SHA-2 processes messages in 512-bit blocks for the 224, 256, and 384 variants, and 1,024-bit blocks for SHA-512.

>> **SHA-3:** Published in 2015, SHA-3 includes SHA3-224, SHA3-256, SHA3-384, and SHA3-512, which produce digests of 224, 256, 384, and 512 bits, respectively. SHAKE128 and SHAKE256 are also variants of SHA3.

WARNING

The SHA-1 digest algorithm is now considered to be obsolete. SHA-3 should be used instead. Similarly, the MD-4 digest algorithm is obsolete, and MD-5 should be used.

HMAC

The Hashed Message Authentication Code (or Checksum) (HMAC) further extends the security of the MD5 and SHA-1 algorithms through the concept of a keyed digest. HMAC incorporates a previously shared secret key and the original message into a single message digest. Thus, even if an attacker intercepts a message, modifies its contents, and calculates a new message digest, the result doesn't match the receiver's hash calculation because the modified message's hash doesn't include the secret key.

Understand Methods of Cryptanalytic Attacks

Attackers employ a variety of methods in their attempts to crack a cryptosystem. The following sections provide a brief overview of the most common methods.

CROSS REFERENCE

This section covers Objective 3.7 of the Security Architecture and Engineering domain in the CISSP Exam Outline (May 1, 2021).

WORK FACTOR: FORCE × EFFORT = WORK

Work factor describes the expenditure required — in terms of time, effort, and resources — to break a cryptosystem. Given enough time, effort, and resources, any cryptosystem can be broken. The goal of all cryptosystems, then, is to achieve a work factor that sufficiently protects the encrypted information against a reasonable estimate of available time, effort, and resources. *Reasonable* can be difficult to estimate, however, as technology continues to improve rapidly.

Moore's Law is based on an observation by Gordon Moore, one of the founders of Intel, that processing power seems to double about every 18 months. To compensate for Moore's Law, some really hard encryption algorithms are used. Today, encrypted information is valuable for perhaps only three months with encryption algorithms that (theoretically) would take several hundred millennia to break. Everybody's confident that by tomorrow, such a feat will be mere child's play.

Brute force

In a *brute-force* (or *exhaustion*) attack, the cryptanalyst attempts every possible combination of key patterns, sometimes using rainbow tables and specialized or scalable computing architectures. This type of attack can be very time-intensive (up to several hundred million years) and resource-intensive, depending on the length of the key, the speed of the attacker's computer, and the life span of the attacker.

TECHNICAL STUFF

A *rainbow table* is a precomputed table used to reverse cryptographic hash functions in a specific algorithm. Examples of password-cracking programs that use rainbow tables include Ophcrack and RainbowCrack.

Ciphertext only

In a *ciphertext-only* attack, the cryptanalyst obtains the ciphertext of several messages, all encrypted by using the same encryption algorithm, but they don't have the associated plaintext. The cryptanalyst attempts to decrypt the data by searching for repeating patterns and using statistical analysis. Certain words in the English language, such as *the* and *or*, occur frequently, for example. This type of attack is generally difficult and requires a large sample of ciphertext.

Known plaintext

In a *known-plaintext* attack, the cryptanalyst has obtained the ciphertext and corresponding plaintext of several past messages, which they use to decipher new messages.

Frequency analysis

Frequency analysis is a method of attack in which an attacker examines ciphertext in an attempt to correlate commonly used words such as *the* and *and* to discover the encryption key or the algorithm in use.

Chosen ciphertext

In a *chosen-ciphertext* attack, the cryptanalyst selects a sample of ciphertext (or plaintext) and obtains the corresponding plaintext (or ciphertext). Several types of chosen ciphertext attacks exist, including

» **Chosen plaintext:** The cryptanalyst chooses plaintext to be encrypted, and the corresponding ciphertext is obtained.

» **Adaptive chosen plaintext:** The cryptanalyst chooses plaintext to be encrypted; then, based on the resulting ciphertext, they choose another sample to be encrypted.

» **Chosen ciphertext:** The cryptanalyst chooses ciphertext to be decrypted, and the corresponding plaintext is obtained.

» **Adaptive chosen ciphertext:** The cryptanalyst chooses ciphertext to be decrypted; then, based on the resulting ciphertext, they choose another sample to be decrypted.

Implementation attacks

Implementation attacks attempt to exploit some weakness in the cryptosystem, such as vulnerability in a protocol or algorithm.

Side channel

A *side-channel attack* is an attack in which the attacker is observing one or more characteristics of a system to discover its secrets. In an attack against a cryptosystem, a side-channel attack attempts to learn more about the cryptosystem, usually to obtain an encryption key. Several methods are used in a side-channel attack, including

>> **Remanence:** Examination of the contents of deleted files

>> **Acoustics:** Examination of the sound produced during computation

>> **Power analysis:** Analysis of the consumption of electric power during computation

>> **Electromagnetic:** Examination of the electromagnetic radiation or electric fields during computation

>> **Timing:** Analysis of the time required to perform various computations

TIP

A side-channel attack can allow an attacker to learn about a cryptosystem through observation and inference.

Fault injection

Fault injection refers to techniques used to stress a system to see how it will behave. When applying fault injection to a cryptosystem, an attacker may be attempting to see whether the cryptosystem can be tricked into malfunctioning (for example, revealing plaintext when an unusual key value, such as null, is entered or a buffer overflow attack is executed) or to trick it into revealing secrets about the cryptosystem.

You could consider fault injection to be a form of fuzzing. In most cases, a cryptosystem is just a program running an algorithm, and that program may have flaws if its inputs are not sanitized properly.

This topic is a specific case in the larger field of software security. If this topic floats your boat, you'll want to bookmark this page and head over to Chapter 10.

Timing

A *timing* (or *replay)* attack occurs when a session key is intercepted and used against a later encrypted session between the same two parties. Replay attacks can be countered by incorporating a time stamp in the session key.

Man in the middle

A *man-in-the-middle (MITM)* attack involves an attacker intercepting messages between two parties on a network and potentially modifying the original message.

In this type of attack, an attacker encrypts known plaintext with each possible key on one end, decrypts the corresponding ciphertext with each possible key, and then compares the results in the middle. Although commonly classified as a brute-force attack, this kind of attack may also be considered to be an analytic attack because it involves some differential analysis.

Pass the hash

Pass the hash is an authentication-bypass attack in which an attacker steals password hashes and uses them to authenticate to a system that uses NTLM authentication. To employ a pass the hash attack, the attacker must first obtain a system's password hashes, generally through another attack.

If an attacker is able to obtain password hashes for a system but cannot successfully execute a pass-the-hash attack, the attacker can also use a rainbow table or employ brute-force password cracking techniques to obtain plaintext passwords, which can be used to log in to a target system.

Kerberos exploitation

Kerberos is a cryptosystem used for authentication and access control in distributed environments. Microsoft Active Directory and other environments such as X use Kerberos.

Attackers may choose to attack Active Directory servers, particularly the Key Distribution Service Account, to gain broad access to an environment. A successful attack can give the attacker the ability to forge valid ticket-granting tickets, thus giving them access to virtually all network resources. Such an attack is called a *golden ticket attack*.

A golden ticket attack can be difficult to detect except through inference by observing the behavior of authenticated users. Although prevention through techniques such as effective vulnerability management and access governance is essential, we know that we cannot stop all attacks at initial stages; thus, we also need to detect them through techniques such as user entity and behavior analytics.

Ransomware

If you've read this entire section, you may wonder why ransomware is included in cryptosystem attacks. That's a good question, and here's the answer: Ransomware is not so much an attack on a cryptosystem; the cryptosystem itself is the attack weapon. We thought this section was a good place to mention this topic. (ISC)² also mentions ransomware in section 3.7 of the CBK, so we're sort of obligated to discuss it here anyway.

Ransomware is an attack on a system in which the attacker, after somehow successfully gaining user-level or administrative-level access on a system, encrypts data on the system and displays a message to the user, informing them of the attack and demanding a ransom if the user wants to recover their encrypted data. Variants of ransomware will also upload the plaintext data to the attacker's server before encryption and then threaten to publish the stolen data. Further variants also inform identifiable people that their personal information has been stolen.

The best mitigations against ransomware include

>> **Robust asset management and vulnerability management:** Organizations that keep all systems up to date and patched can dramatically reduce the likelihood of a successful ransomware attack.

>> **Robust antimalware:** Antimalware that employs both signature and heuristics-based techniques can significantly reduce the chances of a successful attack.

>> **Application whitelisting:** Ransomware is generally unable to execute successfully on a target system that uses application whitelisting.

>> **Strict access control:** Ransomware can succeed where users have broad access to network shares. Limitations on write access to network shares can dramatically reduce the potency of a ransomware attack.

>> **Intrusion prevention systems:** Because most ransomware depends on command-and-control traffic to obtain encryption keys, effective intrusion prevention can block ransomware from executing.

>> **Enterprise backup:** Instead of paying ransoms to recover encrypted data (which, according to the Federal Bureau of Investigation, results in recovery only about half the time), organizations can recover data from backups.

Organizations that are concerned about ransomware should perform threat modeling and other forms of risk analysis to determine what measures should be employed to reduce the probability and effect of a ransomware attack.

Apply Security Principles to Site and Facility Design

Securely designed and built software running on securely designed and built systems must be operated in securely designed and built facilities. Otherwise, an adversary with unrestricted access to a system and its installed software will inevitably succeed in compromising your security efforts. Astute organizations involve security professionals during the design, planning, and construction of new or renovated locations and facilities. Proper site- and facility-requirements planning during the early stages of construction helps ensure that a new building or data center is adequate, safe, and secure, \which can help an organization avoid costly situations later.

CROSS REFERENCE This section covers Objective 3.8 of the Security Architecture and Engineering domain in the CISSP Exam Outline (May 1, 2021).

The principles of Crime Prevention through Environmental Design (CPTED), published in 1971, have been widely adopted by security practitioners in the design of public and private buildings, offices, communities, and campuses. CPTED focuses on designing facilities by using techniques such as unobstructed areas, creative lighting, and functional landscaping, which naturally deter crime through positive psychological effects. By making it difficult for a criminal to hide, gain access to a facility, escape a location, or otherwise perpetrate an illegal and/or violent act, such techniques may cause a would-be criminal to decide against attacking a target or victim and create an environment that's perceived as being safer for people who use the area regularly. CPTED consists of three basic strategies:

>> **Natural access control:** Uses security zones (defensible space) to limit or restrict movement and differentiate between public, semiprivate, and private areas that require different levels of protection. Natural access control can be accomplished by limiting points of entry into a building and using structures such as sidewalks and lighting to guide visitors to main entrances and reception areas. Target hardening complements natural access controls by using mechanical and/or operational controls, such as window and door locks, alarms, security guards, guard docs, picture identification requirements, and visitor sign-in/sign-out procedures.

>> **Natural surveillance:** Reduces criminal threats by making intruder activity more observable and easier to detect. Natural surveillance can be accomplished by maximizing visibility and activity in strategic areas, such as by placing windows to overlook streets and parking areas, landscaping to

eliminate hidden areas and create clear lines of sight, installing open railings on stairways to improve visibility, and using numerous low-intensity lighting fixtures to eliminate shadows and reduce security-camera glare or blind spots (particularly at night).

>> **Territorial reinforcement:** Creates a sense of pride and ownership, which causes intruders to stand out and encourages people to report suspicious activity instead of ignoring it. Territorial reinforcement is accomplished through maintenance activities (picking up litter, cleaning up graffiti, repairing broken windows, and replacing light bulbs), assigning people to be responsible for an area or space, placing amenities (such as benches and water fountains) in common areas, and displaying prominent signage where appropriate. It can also include scheduled activities, such as corporate-sponsored beautification projects and company picnics.

Location, location, location! Although, to a certain degree, this bit of conventional business wisdom may be less important to profitability in the age of e-commerce, it's still a critical factor in physical security. Important factors in considering a location include

>> **Climatology and natural disasters:** Although an organization is unlikely to choose a geographic location solely based on the likelihood of hurricanes or earthquakes, these factors must be considered when designing a safe and secure facility. Related factors may include flood plains, the location of evacuation routes, and the adequacy of civil and emergency preparedness.

>> **Local hazards:** Are high-risk conditions or activities nearby, such as hazardous-materials storage, railway freight lines, or flight paths for the local airport? Is the area heavily industrialized (so that air and noise pollution, including vibration, might affect your systems)?

>> **Crime rate:** Consider whether a location being considered is in or near a high-crime area and whether features or conditions could invite criminal elements.

>> **Visibility:** Will your employees and facilities be targeted for crime, terrorism, vandalism, or social unrest? Is the site near another high-visibility organization that may attract undesired attention? Is your facility located near a government or military target? Keeping a low profile is generally best because you avoid unwanted and unneeded attention; avoid external building markings when possible.

>> **Accessibility:** Consider local traffic patterns, convenience to airports, proximity to emergency services (police, fire, and medical facilities), and availability of adequate housing. Will on-call employees have to drive for an hour to respond when your organization needs them?

>> **Utilities:** It is important to understand where the facility is located in the power grid, and whether electrical power is stable and clean. Also determine whether fiber optic cable is already in place to support current and future telecommunications requirements. Finally, determine whether electric utility and telecommunications facilities feature geographic diversity, in which multiple feeds originating from different locations enter the building.

>> **Joint tenants:** Will you have full access to all necessary environmental controls? Can (and should) physical security costs and responsibilities be shared between joint tenants? Are other tenants potential high-visibility targets? Do other tenants take security as seriously as your organization does?

TIP

If your organization already occupies a facility, and no preoccupation security assessment was performed, consider performing an assessment to identify and document any possible risks. Although it's unlikely that your organization is going to relocate because of the presence of one or more risks, knowing about them is better than not.

Design Site and Facility Security Controls

The CISSP candidate must understand the various threats to physical security; the elements of site- and facility-requirements planning and design; and various physical security controls, including access controls, technical controls, environmental and life safety controls, and administrative controls. In addition, you must know how to support the implementation and operation of these controls, as covered in this section.

CROSS REFERENCE

This section covers Objective 3.9 of the Security Architecture and Engineering domain in the CISSP Exam Outline (May 1, 2021).

Many physical and technical controls should be considered during the initial design of a secure facility to reduce costs and improve the overall effectiveness of these controls. Building design considerations include

>> **Exterior walls:** Ideally, exterior walls should be able to withstand high winds (tornadoes and hurricanes/typhoons) and reduce electronic emanations that can be detected and used to re-create high-value data (such as government or military data). If possible, the use of exterior windows should be avoided throughout the building, particularly on lower levels. Metal bars over windows or reinforced windows on lower levels may be necessary. Any windows should be fixed (meaning that you can't open them), shatterproof, and sufficiently opaque to conceal inside activities.

>> **Interior walls:** Interior walls adjacent to secure or restricted areas must extend from the floor to the ceiling (through raised flooring and drop ceilings) and must comply with applicable building and fire codes. Walls adjacent to storage areas (such as closets containing janitorial supplies, paper, media, or other flammable materials) must meet minimum fire ratings, which are typically higher than for other interior walls. Ideally, bulletproof walls should protect the most sensitive areas.

>> **Security zones:** Access controls to sensitive areas within work facilities should be enacted so that only authorized personnel may access them. This consideration will influence floor plans so that interior security zones are protected.

>> **Floors:** Flooring (both slab and raised) must be capable of bearing loads in accordance with local building codes (typically, 150 pounds per square foot). Additionally, raised flooring must have a nonconductive surface and be grounded properly to reduce safety risks.

>> **Ceilings:** Weight-bearing and fire ratings must be considered. Drop ceilings may temporarily conceal intruders and small water leaks; conversely, stained drop-ceiling tiles can reveal leaks while temporarily impeding water damage.

>> **Doors:** Doors and locks must be sufficiently strong and well designed to resist forcible entry, and they need a fire rating equivalent to that of adjacent walls. Emergency exits must remain unlocked from the inside and should also be clearly marked, as well as monitored or alarmed. Electronic lock mechanisms and other access control devices should fail open (unlock) in the event of an emergency to permit people to exit the building. Many doors swing out to facilitate emergency exiting; thus, door hinges are located on the outside of the room or building. These hinges must be secured to prevent an intruder from easily lifting hinge pins and removing the door. Magnetic locks should be inspected to identify signs of tampering.

>> **Lighting:** Exterior lighting for all physical spaces and buildings in the security perimeter (including entrances and parking areas) should be sufficient to provide safety for personnel, as well as to discourage prowlers and casual intruders.

>> **Wiring:** All wiring, conduits, and cable runs must comply with building and fire codes, and must be properly protected. Plenum cabling must be used below raised floors and above drop ceilings, because PVC-clad cabling releases toxic chemicals when it burns.

TECHNICAL STUFF

A *plenum* is the vacant area above a drop ceiling or below a raised floor. A fire in these areas can spread very rapidly, carrying smoke and noxious fumes to other areas of a burning building. For this reason, non-PVC-coated cabling, known as *plenum cabling,* must be used in these areas in most jurisdictions.

>> **Electricity and heating/cooling/air conditioning(HVAC):** Electrical load and heating, cooling, and air conditioning must be planned carefully to ensure that sufficient power is available in the right locations and that proper climate ranges (temperature and humidity) are maintained.

>> **Pipes:** Locations of shutoff valves for water, steam, or gas pipes should be identified and marked appropriately. Drains should have positive flow, carrying drainage away from the building.

>> **Fire detection and prevention:** Inert gas fire suppression should be used in all rooms where IT equipment resides. Wet and dry pipe should not be used in those locations if local fire codes permit. Advanced smoke detectors can provide earlier warning of precombustion activities.

>> **Lightning strikes:** Approximately 10,000 fires are started every year by lightning strikes in the United States alone, despite the fact that only 20 percent of all lightning ever reaches the ground. Lightning can heat the air in immediate contact with the stroke to 54,000° Fahrenheit (F), which translates to 30,000° Celsius (C), and lightning can discharge 100,000 amperes of electrical current. Now *that's* an inrush!

>> **Magnetic fields:** Monitors and magnetic-based storage media such as backup tape and hard disk drives can be permanently damaged or erased by magnetic fields.

>> **Sabotage/terrorism/war/theft/vandalism:** Both internal and external threats must be considered. A heightened security posture is also prudent during certain other disruptive situations, including labor disputes, corporate downsizing, hostile terminations, bad publicity, demonstrations/protests, and civil unrest.

>> **Equipment failure:** Equipment failures are inevitable. Maintenance and support agreements, ready spare parts, and redundant systems can mitigate the effects.

>> **Loss of communications and utilities:** This category includes voice and data communications, water, and electricity. Loss of communications and utilities may happen because of any of the factors discussed in the preceding items, as well as human error.

>> **Vibration and movement:** Causes may include earthquakes, landslides, and explosions. Equipment may also be damaged by sudden or severe vibrations, falling objects, or equipment racks tipping over. More seriously, vibrations or movement may weaken structural integrity, causing a building to collapse or otherwise be unusable.

>> **Severe weather:** This category includes hurricanes, tornadoes, high winds, severe thunderstorms and lightning, rain, snow, sleet, and ice. Such forces of nature may cause fires, water damage and flooding, structural damage, loss of communications and utilities, and hazards to personnel.

>> **Personnel loss:** This category includes illness, injury, death, transfer, labor disputes, resignations, and terminations. The negative effects of a personnel loss can be mitigated through good security practices, such as documented procedures, job rotations, cross-training, and redundant functions.

TIP

Although much of the information in this section may seem to be common sense, the CISSP exam asks very specific and detailed questions about physical security, and many candidates lack practical experience in fighting fires, so don't underestimate the importance of physical security — in real life and on the CISSP exam!

Wiring closets, server rooms, and more

Wiring closets, intermediate distribution facilities (IDFs), server rooms, data centers, and media and evidence storage facilities contain high-value equipment and/or media that is critical to ongoing business operations or support of investigations. Physical security controls often found in these locations include

>> **Strong access controls:** Typically, this category includes the use of key cards, as well as a PIN pad or biometric. Only those personnel who have a need to know should be authorized to access high-security areas.

>> **Fire suppression:** Often, you'll find inert gas fire suppression instead of water sprinklers, because water can damage computing equipment in case of discharge.

>> **Video surveillance:** Cameras are fixed at entrances to wiring closets, distribution frames, and data center entrances, as well as in the interiors of those facilities, to observe the goings-on of both authorized personnel and intruders.

- **Visitor log:** All visitors, who generally require a continuous escort, are required to sign a visitor log.

- **Asset check-in/check-out log:** All personnel are required to log the introduction and removal of any equipment and media.

In many jurisdictions, visible notices are required when video surveillance is employed. These notices can serve as deterrent controls.

Restricted and work area security

High–security work areas often employ physical security controls above and beyond those used in ordinary work areas. In addition to key card access control systems and video surveillance, additional physical security controls may include

- **Multifactor key card entry:** In addition to using key cards, employees may be required to use a PIN pad or biometric device to access restricted areas.

- **Mantraps:** A set of interlocked double doors or turnstiles can be used to prevent tailgating.

- **Security guards:** Guards are present at ingress and egress points, and roam within the facility to be on the alert for unauthorized personnel or unauthorized activities.

- **Guard dogs:** Guard dogs provide additional deterrence against unauthorized entry and also assist in the capture of unauthorized personnel in a facility.

- **Security walls and fences:** Restricted facilities may employ one or more security walls and fences to keep unauthorized personnel away from facilities. General height requirements for fencing are listed in Table 5-5.

- **Security lighting:** Restricted facilities may have additional lighting to expose and deter any would-be intruders.

- **Security gates, crash gates, and bollards:** These controls limit the movement of vehicles near a facility to reduce vehicle-borne threats.

- **Sally ports:** These controlled entries provide better control for admitting authorized personnel into a work facility while making it more difficult for unauthorized people to enter.

TABLE 5-5

General Fencing Height Requirements

Height	General Effect
3–4 feet (1 meter)	Deters casual trespassers
6–7 feet (2 meters)	Too high to climb easily
8 feet (2.4 meters) plus three-strand barbed wire	Deters determined intruders

Work-area security also makes us think of various safety issues, all of which are important to the security professional, although one or more of the following may be managed by facilities or other personnel:

>> **First aid:** From simple kits for simple matters to those for more complicated situations such as broken bones, first aid kits ensure treatment, comfort, and even life safety for personnel on the job.

>> **Emergency food and water:** Supplies of emergency food and water can be vital for the survival of workers occupying a building when disasters of various types occur.

>> **Automated external defibrillators:** These portable devices can save a person's life. They're kept on hand in many workplaces, including stores and commercial aircraft.

>> **Evacuation signage and drills:** This aspect of security involves exit signs, mustering stations, and personnel trained in ensuring the safe exit of all personnel in the event of an emergency.

>> **Active assailant response:** Simple changes in building design and training for personnel can greatly influence the survivability of an active-assailant situation, in which one or more people are attempting to harm others in a workplace or school.

Utilities and heating, ventilation, and air conditioning

Environmental and life safety controls such as utilities and heating, ventilation, and air conditioning (HVAC) are necessary for maintaining a safe and acceptable operating environment for computers, equipment, and personnel.

HVAC systems maintain the proper environment for computers and personnel. HVAC-requirements planning involves making complex calculations based on numerous factors, including the average BTUs (British Thermal Units) produced by the estimated computers and personnel occupying a given area, the size of the room, insulation characteristics, and ventilation systems.

The ideal temperature range for computer equipment is between 50°F and 80°F (10°C and 27°C). At temperatures as low as 100°F (38°C), magnetic storage media can be damaged.

REMEMBER

The ideal temperature range for computer equipment is between 50°F and 80°F (10°C and 27°C).

The ideal humidity range for computer equipment is between 40 and 60 percent. Higher humidity causes condensation and corrosion. Lower humidity increases the potential for static electricity.

Doors and side panels on computer equipment racks should be kept closed (and locked, as a form of physical access control) to ensure proper airflow for cooling and ventilation. When possible, empty spaces in equipment racks (such as a half-filled rack or gaps between installed equipment) should be covered with blanking panels to reduce hot and cold air mixing between the hot side or hot aisle (typically, the power-supply side of the equipment) and the cold side or cold aisle (typically, the front of the equipment). Such mixing of hot and cold air can reduce the efficiency of cooling systems.

Heating and cooling systems should be maintained properly, and air filters should be cleaned regularly to reduce dust contamination and fire hazards.

Most gas-discharge fire suppression systems automatically shut down HVAC systems before discharging, but a separate emergency power-off switch should be installed near exits to facilitate manual shutdown in an emergency.

Ideally, HVAC equipment should be dedicated, controlled, and monitored. If the systems aren't dedicated or independently controlled, proper liaison with the building manager is necessary to ensure that everyone knows who to call when problems occur. Monitoring systems should alert the appropriate personnel when operating thresholds are exceeded.

Environmental issues

Water damage (and damage from liquids in general) can be caused by many things, including pipe breakage, firefighting, leaking roofs, spilled drinks, flooding, and tsunamis. Wet computers and other electrical equipment pose a potentially lethal hazard.

Both preventive and detective controls are used to ensure that water in unwanted places does not disrupt business operations or destroy expensive assets. Common features include

>> **Water diversion:** Barriers of various types help prevent water from entering sensitive areas.

>> **Water detection alarms:** Sensors that detect the presence of water can alert personnel of the matter and provide valuable time before damage occurs.

Contaminants in the air, unless filtered out by HVAC systems, can be irritating or harmful to personnel and to equipment. A build-up of carbon dioxide and carbon monoxide can also be injurious and even cause death. Air quality sensors can be used to detect and particulates, contaminants, CO_2, and CO and alert facilities personnel.

Fire prevention, detection, and suppression

Threats from a fire can be potentially devastating and lethal. Proper precautions, preparation, and training not only help limit the spread of fire and damage, but also (and more important) save lives.

Saving human lives is the first priority in any life-threatening situation.

Other hazards associated with fires include smoke, explosions, building collapse, release of toxic materials or vapors, and water damage.

For a fire to burn, it requires three elements: heat, oxygen, and fuel. These three elements are sometimes referred to as the *fire triangle*, which is depicted in Figure 5-11. Fire suppression and extinguishing systems fight fires by removing one of these three elements or by temporarily breaking the chemical reaction among these three elements, separating the fire triangle. Fires are classified according to fuel type, as listed in Table 5-6.

You must be able to describe Class A, B, and C fires and their primary extinguishing methods. Class D and K (or F) are not as common as computer fires (unless your server room happens to be located directly above the deep-fat fryers of a local bar and hot-wings restaurant).

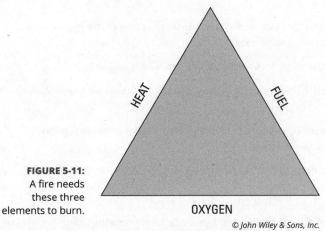

FIGURE 5-11:
A fire needs these three elements to burn.

OXYGEN

© *John Wiley & Sons, Inc.*

TABLE 5-6   **Fire Classes and Suppression/Extinguishing Methods**

Class	Description (Fuel)	Extinguishing Method
A	Common combustibles, such as paper, wood, furniture, and clothing	Water or soda acid
B	Burnable fuels, such as gasoline or oil	CO_2 or soda acid
C	Electrical fires, such as computers or electronics	CO_2 (**Note:** The most important step in fighting a fire in this class is turning off the electricity first.)
D	Special fires, such as combustible metals	May require total immersion or other special techniques
K (or F)	Cooking oils or fats	Water mist or fire blankets

Fire detection and suppression systems are some of the most essential life safety controls for protecting facilities, equipment, and (most important) human lives. The three main types of fire detection systems are

>> **Heat-sensing:** These devices sense either temperatures exceeding a predetermined level (*fixed-temperature detectors*) or rapidly rising temperatures (*rate-of-rise detectors*). Fixed-temperature detectors are more common and exhibit a lower false-alarm rate than rate-of-rise detectors.

>> **Flame-sensing:** These devices sense the flicker (pulsing) of flames or the infrared energy of a flame. These systems are relatively expensive but provide extremely rapid response time.

>> **Smoke-sensing:** These devices detect smoke, which is one of the byproducts of fire. The four types of smoke detectors are

- *Photoelectric:* These detectors sense variations in light intensity.

- *Beam:* Similar to the photoelectric type, these detectors sense when smoke interrupts beams of light.

- *Ionization:* These detectors sense disturbances in the normal ionization current of radioactive materials.

- *Aspirating:* These detectors draw air into a sampling chamber and sense minute amounts of smoke.

REMEMBER

The three main types of fire detection systems are heat-sensing, flame-sensing, and smoke-sensing.

The two primary types of fire suppression systems are

>> **Water sprinkler:** Water extinguishes fire by removing the heat element from the fire triangle, and it's most effective against Class A fires. Water is the primary fire-extinguishing agent for all business environments. Although water can potentially damage equipment, it's one of the most effective, inexpensive, readily available, and least harmful (to humans) extinguishing agents available. The four variations of water sprinkler systems are

- *Wet-pipe (or closed-head):* Most commonly used and considered the most reliable. Pipes are always charged with water and ready for activation. Typically, a fusible link in the nozzle melts or ruptures, opening a gate valve that releases the water flow. Disadvantages include flooding because of nozzle or pipe failure and because of frozen pipes in cold weather.

- *Dry-pipe:* No standing water in the pipes. At activation, a clapper valve opens, air is blown out of the pipe, and water flows. This type of system is less efficient than the wet-pipe system but reduces the risk of accidental flooding; the time delay provides an opportunity to shut down computer systems (or remove power) if conditions permit.

- *Deluge:* Operates similarly to a dry-pipe system but is designed to deliver large volumes of water quickly. Deluge systems typically are not used for information processing areas.

- *Preaction:* Combines wet- and dry-pipe systems. Pipes are initially dry. When a heat sensor is triggered, the pipes are charged with water, and an alarm is activated. Water isn't actually discharged until a fusible link melts (as in wet-pipe systems). This system is recommended for information processing areas because it reduces the risk of accidental discharge by permitting manual intervention.

The four main types of water sprinkler systems are wet-pipe, dry-pipe, deluge, and preaction.

» **Gas discharge:** Gas discharge systems may be portable (such as a CO_2 extinguisher) or fixed (beneath a raised floor). These systems are typically classified according to the extinguishing agent that's employed. These agents include

- *Carbon dioxide (CO_2):* CO_2 is a colorless, odorless gas that extinguishes fire by removing the oxygen element from the fire triangle. CO_2 is most effective against Class B and C fires. Because this gas removes oxygen, it is potentially lethal and therefore is best suited for use in unmanned areas or on a delay (including manual override) in staffed areas.

  CO_2 is also used in portable fire extinguishers, which should be located near all exits and within 50 feet (15 meters) of any electrical equipment. All portable fire extinguishers (CO_2, water, and soda acid) should be clearly marked (listing the extinguisher type and the fire classes it can be used for) and periodically inspected. Additionally, all personnel should receive training in the proper use of fire extinguishers.

- *Soda acid:* Includes a variety of chemical compounds that extinguish fires by removing the fuel element (suppressing the flammable components of the fuel) of the fire triangle. Soda acid is most effective against Class A and B fires. It is not used for Class C fires because of the highly corrosive nature of many of the chemicals used.

- *Inert gas-discharge:* Gas-discharge systems suppress fire by separating the elements of the fire triangle; they are most effective against Class B and C fires. Inert gases don't damage computer equipment, don't leave liquid or solid residue, mix thoroughly with the air, and spread extremely quickly. But in concentrations higher than 10 percent, the gases are harmful if inhaled, and some types degrade into toxic chemicals (hydrogen fluoride, hydrogen bromide, and bromine) when used on fires that burn at temperatures above 900°F (482°C).

  Halon used to be the gas of choice in gas-discharge fire suppression systems. But because of Halon's ozone-depleting characteristics, the Montreal Protocol of 1987 prohibited the further production and installation of Halon systems (beginning in 1994) and encouraged the replacement of existing systems. Acceptable replacements include FM-200 (most effective), CEA-410 or CEA-308, NAF-S-III, FE-13, Argon or Argonite, and Inergen.

Halon is an ozone-depleting substance. Acceptable replacements include FM-200, CEA-410 or CEA-308, NAF-S-III, FE-13, Argon or Argonite, and Inergen.

Power

General considerations for electrical power include having one or more dedicated feeders from one or more utility substations or power grids, as well as ensuring that adequate physical access controls are implemented for electrical distribution panels and circuit breakers. An emergency power-off switch should be installed near major systems and exit doors to shut down power in case of fire or electrical shock. Additionally, a backup power source should be established, such as a diesel or natural-gas power generator, along with an uninterruptible power supply (UPS). Backup power should be provided for critical facilities and systems, including emergency lighting, fire detection and suppression, mainframes and servers (and certain workstations), HVAC, physical access control systems, and telecommunications equipment.

WARNING

Although natural gas can be a cleaner alternative than diesel for backup power, in terms of air and noise pollution, it's generally not used for emergency life systems (such as emergency lighting and fire protection systems) because the fuel source (natural gas) can't be locally stored, so the system relies instead on an external fuel source that must be supplied by pipelines.

Protective controls for electrostatic discharge include the following:

>> Maintain proper humidity levels (40 to 60 percent).

>> Ensure proper grounding.

>> Use antistatic flooring, antistatic carpeting, and floor mats.

>> Forbid crepe-soled shoes. (Okay, we're *mostly* kidding on this one.)

Protective controls for electrical noise include the following:

>> Install power-line conditioners.

>> Ensure proper grounding.

>> Use shielded cabling.

A UPS is perhaps the most important protection against electrical anomalies because it provides clean power to sensitive systems and a temporary power source during electrical outages (blackouts, brownouts, and sags). This power supply must be sufficient to shut down the protected systems properly.

REMEMBER

A UPS shouldn't be used as a backup power source. A UPS — even a building UPS — is designed to provide temporary power, typically for 5 to 30 minutes, to give a backup generator time to start or to allow a controlled, proper shutdown of protected systems.

Sensitive equipment can be damaged or affected by various electrical hazards and anomalies, including

>> **Electrostatic discharge (ESD):** The ideal humidity range for computer equipment is 40 to 60 percent. Higher humidity causes condensation and corrosion. Lower humidity increases the potential for static electricity. A static charge of as little as 40 volts can damage sensitive circuits, and 2,000 volts can cause a system shutdown. The minimum discharge that can be felt by humans is 3,000 volts, and electrostatic discharges of more than 25,000 volts are possible. If you can feel it, it's a problem for your equipment!

REMEMBER

The ideal humidity range for computer equipment is 40 to 60 percent. Also, remember that it's not the volts that kill; it's the amps!

>> **Electrical noise:** This category includes electromagnetic interference (EMI) and radio frequency interference (RFI). Electromagnetic interference is generated by the different charges among the three electrical wires (hot, neutral, and ground) and can be *common-mode noise* (caused by hot and ground) or *traverse-mode noise* (caused by a difference in power between the hot and neutral wires). Radio frequency interference is caused by electrical components, such as fluorescent lighting and electric cables. A *transient* is a momentary line-noise disturbance.

>> **Electrical anomalies:** These anomalies include the ones listed in Table 5-7.

TABLE 5-7

Electrical Anomalies

Electrical Event	Definition
Blackout	Total loss of power
Fault	Momentary loss of power
Brownout	Prolonged drop in voltage
Sag	Short drop in voltage
Inrush	Initial power rush
Spike	Momentary rush of power
Surge	Prolonged rush of power
Voltage drop	Decrease in electric voltage

TIP

You may want to come up with some meaningless mnemonic for the list in Table 5-7, such as *Bob Frequently Buys Shoes In Shoe Stores Verbosely.* You need to know these terms for the CISSP exam.

WARNING

Surge protectors and surge suppressors provide only minimal protection for sensitive computer systems, and they're commonly (and dangerously) used to overload an electrical outlet or as a daisy-chained extension cord. The protective circuitry in most of these units costs less than $1 (compare the cost of a low-end surge protector with that of a 6-foot extension cord), and you get what you pay for. These glorified extension cords provide only minimal spike protection. True, a surge protector provides more protection than nothing at all, but don't be lured into complacency by these units; check them regularly for proper use and operation, and don't accept them as being viable alternatives to a UPS.

Chapter 6

Communication and Network Security

The Communication and Network Security domain requires a thorough understanding of network fundamentals, secure network design, concepts of network operation, networking technologies, and network management techniques. This domain represents 14 percent of the CISSP certification exam.

TIP A thorough understanding of networking protocols, services, practices, and architecture will definitely help you pass the exam. If your network experience is light, we recommend that you pick up a copy of *Networking All-In-One For Dummies* (John Wiley & Sons, Inc.). Also consider earning a networking certification, such as CompTIA Network+ or Cisco Certified Network Associate (CCNA), before taking the CISSP exam. These materials are extremely helpful in preparing for this portion of the CISSP exam if your background is light in network technology.

Assess and Implement Secure Design Principles in Network Architectures

A solid understanding of networking concepts and fundamentals is essential for creating a secure network architecture. This understanding requires knowledge of network topologies; IP addressing; various networking protocols (including multilayer and converged protocols); wireless networks; communication security; and new and evolving networking trends, such as software-defined networks, microsegmentation, and cloud computing.

CROSS REFERENCE

This section covers Objective 4.1 of the Communication and Network Security domain in the CISSP Exam Outline (May 1, 2021).

Data networks are commonly classified as local area networks and wide area networks. Although these classifications are basic, you should understand the fundamental distinctions between these two types of networks.

A *local area network* (LAN) is a data network that operates across a relatively small geographic area, such as a single building or floor. A LAN connects workstations, servers, printers, and other devices so that network resources, such as files and email, can be shared. Key characteristics of LANs include the following:

>> Can connect networked resources over a small geographic area, such as a floor, a building, or a group of buildings.

>> Are relatively inexpensive to set up and maintain, typically consisting of readily available equipment such as servers, desktop PCs, printers, switches, hubs, bridges, repeaters, wireless access points, and various security devices such as firewalls and intrusion prevention systems.

>> Can be wired, wireless, or a combination of wired and wireless.

>> Perform at relatively high speeds — typically 10 megabits per second (Mbps), 100 Mbps, 1000 Mbps (also referred to as 1 gigabit per second [1 Gbps]), 10 Gbps, and 40 Gbps for wired networks, and 11 Mbps, 54 Mbps, or 600 Mbps for wireless networks. We cover LAN speeds in the section "Physical Layer (Layer 1)" later in this chapter.

WARNING

Be careful when referring to data capacity (and their abbreviations) and data storage. *100 Mbps* is100 megabits per second, and *100 MB* is 100 megabytes. The distinction is subtle (a little *b* versus a big *B* and *bits* rather than *bytes)*, but the difference is significant: A byte is equal to 8 bits. Data speeds are typically referred to in bits per second; data storage is typically referred to in bytes.

REMEMBER

A LAN is a data network that operates across a relatively small geographic area, such as a building or group of buildings.

A *wide area network* (WAN) connects multiple LANs and other WANs by using telecommunications devices and facilities to form an internetwork. Key characteristics of WANs include the following:

» Connects multiple LANs over large geographic areas, such as a small city (such as a metropolitan area network), a region or country, a global corporate network, the entire planet (such as the Internet), or beyond (such as the International Space Station via satellite).

» Can be relatively expensive to set up and maintain, typically consisting of equipment such as routers, channel service unit/data service unit devices, firewalls, virtual private networks (VPNs), gateways, and various other security devices.

» Perform at relatively low speeds by using various technologies, such as digital subscriber line (DSL; 128 Kbps to 16 Mbps, for example), T-1 (1.544 Mbps), DS-3 (approximately 45 Mbps), OC-12 (approximately 622 Mbps), and OC-255 (approximately 13 Gbps). We cover WAN speeds in the section "Data Link Layer (Layer 2)" later in this chapter.

A WAN is a data network that operates across a relatively large geographic area and includes portions supplied by telecommunications carriers.

REMEMBER

OSI and TCP/IP models

The OSI and TCP/IP models define standard protocols for network communication and interoperability by using a layered approach. This approach divides complex networking issues into simpler functional components that aid in the understanding, design, and development of networking solutions. It provides the following specific advantages:

» Clarifies the general functions of a communications process instead of focusing on specific issues

» Reduces complex networking processes into simpler sublayers and components

» Promotes interoperability by defining standard interfaces

» Aids development by allowing vendors to change individual features at a single layer instead of rebuilding the entire protocol stack

» Facilitates easier (and more logical) troubleshooting

The OSI model isn't just a theoretical model to be pondered by intellectuals; it really is helpful for explaining complex networking topics. For this reason, much of the information you need to know for the Communication and Network Security domain is presented in this chapter in the context of the OSI model. You don't necessarily need to know which protocols and equipment belong to which layers of the OSI model for the CISSP exam or in your day-to-day work. We've presented it this way to logically organize the information you need to know for this domain.

TIP

The OSI Reference Model

In 1984, the International Organization for Standardization (ISO) adopted the Open Systems Interconnection (OSI) Reference Model (or simply the OSI model) to facilitate interoperability between network devices independent of the manufacturer.

The OSI model consists of seven distinct layers that describe how data is communicated between systems and applications on a computer network, as shown in Figure 6-1. These layers include

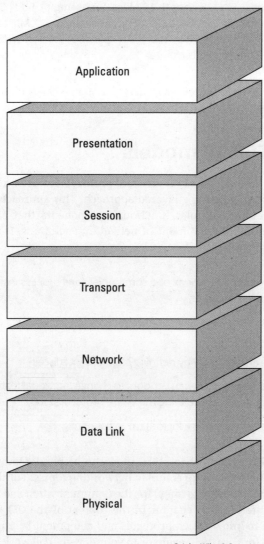

FIGURE 6-1:
The seven layers
of the OSI model.

» Application (Layer 7)

» Presentation (Layer 6)

» Session (Layer 5)

» Transport (Layer 4)

» Network (Layer 3)

» Data Link (Layer 2)

» Physical (Layer 1)

TIP

Try creating a mnemonic to recall the layers of the OSI model, such as *All People Seem To Need Delicious Pizza* and, in reverse, *Please Do Not Throw Sausage Pizza Away.* (Now we're getting hungry.)

In the OSI model, data is passed from the highest layer (Application, Layer 7) downward through each layer to the lowest layer (Physical, Layer 1) and then transmitted across the network medium to the destination node, where it's passed upward from the lowest layer to the highest layer. Each layer communicates only with the layer immediately above and below it (adjacent layers). This communication is achieved through a process known as *data encapsulation*, which wraps protocol information from the layer immediately above in the data section of the layer immediately below. Figure 6-2 illustrates this process.

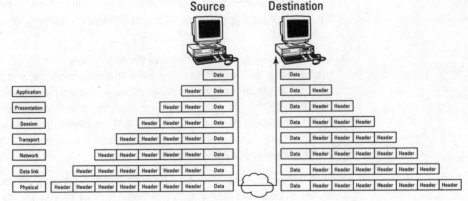

FIGURE 6-2:
Data
encapsulation in
the OSI model.

© *John Wiley & Sons, Inc.*

Application Layer (Layer 7)

The Application Layer (Layer 7) is the highest layer of the OSI model. It supports the components that deal with the communication aspects of an application that requires network access, and it provides an interface to the user. So both the Application Layer and the end user interact directly with the application.

The Application Layer is responsible for the following:

>> Identifying and establishing availability of communication partners

>> Determining resource availability

>> Synchronizing communication

REMEMBER

The Application Layer is responsible for identifying and establishing availability of communication partners, determining resource availability, and synchronizing communication.

Don't confuse the Application Layer with software applications such as Microsoft Word and Excel. Applications that function at the Application Layer include

>> **File Transfer Protocol (FTP) and Secure File Transfer Protocol (SFTP):** A program used to copy files from one system to another over a network. FTP operates on TCP ports 20 (the data port) and 21 (the control port), while SFTP only requires TCP port 22 by default.

>> **Hypertext Transfer Protocol (HTTP):** The language of the World Wide Web, used by web servers and browsers for non-sensitive content. HTTP operates on TCP port 80.

>> **Hypertext Transfer Protocol Secure (HTTPS):** The language of commercial transactions on the World Wide Web. HTTPS is actually the HTTP protocol used in combination with SSL/TLS (discussed in the section "Transport Layer (Layer 4)"). HTTPS operates on TCP port 443 but occasionally on other ports such as 8443.

>> **Internet Message Access Protocol (IMAP):** A store-and-forward electronic mail protocol that allows an email client to access, manage, and synchronize email on a remote mail server. IMAP provides more functionality and security than POP3, such as requiring users to explicitly delete emails from the server. The most current version is IMAPv4 (or IMAP4), which operates on TCP and UDP port 143. Email clients that use IMAP can be secured by using TLS or SSL encryption over TCP/UDP port 993.

>> **Post Office Protocol Version 3 (POP3):** An email retrieval protocol that allows an email client to access email on a remote mail server by using TCP port 110. Inherently insecure, POP3 allows users to authenticate over the Internet by using plaintext passwords. Email clients that use POP3 can be secured by using TLS or SSL encryption over TCP/UDP port 995.

>> **Privacy Enhanced Mail (PEM):** PEM is an Internet Engineering Task Force standard (IETF) for providing email confidentiality and authentication. PEM is not widely used.

>> **Secure Multipurpose Internet Mail Extensions (S/MIME):** S/MIME is a secure method of sending email incorporated into several popular browsers and email applications.

>> **Simple Mail Transfer Protocol (SMTP):** Used to send and receive email across the Internet. This protocol has several well-known vulnerabilities that make it inherently insecure. SMTP operates on TCP/UDP port 25. SMTP over SSL/TLS (SMTPS) uses TCP/UDP port 465.

>> **Simple Network Management Protocol (SNMP):** Used to collect network information by polling stations and sending traps (alerts) to a management station. SNMP has many well-known vulnerabilities, including default cleartext community strings (passwords). SNMP operates on TCP/UDP ports 161 (agent) and 162 (manager). Secure SNMP uses TCP/UDP ports 10161 (agent) and 10162 (manager).

>> **Telnet:** Provides terminal emulation for remote access to system resources. Telnet operates on TCP/UDP port 23. Because Telnet transmits passwords in cleartext, it is no longer considered safe; SSH (discussed in the section "Session Layer (Layer 5)" later in this chapter) is preferred.

>> **Trivial File Transfer Protocol (TFTP):** A lean, mean version of FTP without directory-browsing capabilities or user authentication. Generally considered to be less secure than FTP, TFTP operates on UDP port 69.

CONTENT DISTRIBUTION NETWORKS

Content distribution networks (CDNs) are large distributed networks of servers that cache web content, such as static web pages, downloadable objects, on-demand and streaming music and video, and web applications for subscriber organizations, and then serve that content to Internet users over the most optimal network path available. CDNs that filter content based on HTTP header data operate at Layer 7. CDNs that filter traffic based on IP addresses are more specifically categorized as load balancers and operate at Layer 4.

CDNs offload much of the performance demand on Internet-facing systems for subscriber organizations, and many offer optional security services, such as mitigation of distributed denial-of-service attacks.

CDNs operate data centers throughout a large geographic region or worldwide and must ensure the security of their data center systems and networks for their customers. Service-level agreements and applicable regulatory compliance must be addressed when evaluating CDN providers.

Some CDN providers include optional web application firewall capabilities that protect web servers from Application Layer attacks.

Presentation Layer (Layer 6)

The Presentation Layer (Layer 6) provides coding and conversion functions that are applied to data being presented to the Application Layer (Layer 7). These functions ensure that data sent from the Application Layer of one system are compatible with the Application Layer of the receiving system.

REMEMBER

The Presentation Layer is responsible for coding and conversion functions.

Tasks associated with this layer include

>> **Data representation:** Use of common data representation formats (standard image, sound, and video formats) enables application data to be exchanged between different types of computer systems.

>> **Character conversion:** Information is exchanged between different systems by using common character conversion schemes.

>> **Data compression:** Common data compression schemes enable compressed data to be decompressed properly at the destination.

>> **Data encryption:** Common data encryption schemes enable encrypted data to be decrypted properly at the destination.

Examples of Presentation Layer protocols include

>> **American Standard Code for Information Interchange (ASCII):** A character-encoding scheme based on the English alphabet, consisting of 128 characters.

>> **Extended Binary-Coded Decimal Interchange Code (EBCDIC):** An 8-bit character-encoding scheme largely used on mainframe and mid-range computers.

>> **Graphics Interchange Format (GIF):** A widely used bitmap image format that allows up to 256 colors and is suitable for images or logos (but not photographs).

>> **Joint Photographic Experts Group (JPEG):** A photographic compression method widely used to store and transmit photographs.

>> **Motion Picture Experts Group (MPEG):** An audio and video compression method widely used to store and transmit audio and video files.

Session Layer (Layer 5)

The Session Layer (Layer 5) establishes, coordinates, and terminates communication sessions (service requests and service responses) between networked systems.

REMEMBER

The Session Layer is responsible for establishing, coordinating, and terminating communication sessions between systems.

A communication session is divided into three distinct phases:

» **Connection establishment:** Initial contact between communicating systems is made, and the end devices agree on communications parameters and protocols to be used, including the mode of operation:

- *Simplex mode:* In simplex mode, a one-way communications path is established with a transmitter at one end of the connection and a receiver at the other end. An analogy is AM radio; a radio station broadcasts music, and the radio receiver can only receive the broadcast.

- *Half-duplex mode:* In half-duplex mode, both communicating devices are capable of transmitting and receiving messages, but they can't do so at the same time. An analogy is a two-way radio; a button must be pressed to transmit and then released to receive a signal.

- *Full-duplex mode:* In full-duplex mode, both communicating devices are capable of transmitting and receiving simultaneously. An analogy is a telephone; it can transmit and receive signals (but not necessarily communicate) at the same time.

» **Data transfer:** Information is exchanged between end devices.

» **Connection release:** When data transfer is complete, end devices systematically end the session.

Examples of Session Layer protocols include

» **Network Basic Input/Output System (NetBIOS):** A Microsoft protocol that allows applications to communicate over a LAN. When NetBIOS is combined with protocols such as TCP/IP, known as NetBIOS over TCP/IP (or NBT), applications can communicate over large networks.

» **Network File System (NFS):** Developed by Sun Microsystems to facilitate transparent user access to remote file-system resources on a Unix-based TCP/IP network.

» **Remote Procedure Call (RPC):** A client-server network redirection tool. Procedures are created on clients and performed on servers.

» **Secure Remote Procedure Call (S-RPC):** S-RPC is a secure client-server protocol that's defined at multiple upper layers of the OSI model. RPC is used to request services from another computer on the network. S-RPC provides public and private keys to clients and servers by using Diffie-Hellman. After S-RPC operations initially authenticate, they're transparent to the end user.

>> **Secure Shell (SSH and SSH-2):** SSH provides a secure alternative to Telnet (discussed in the section "Application Layer (Layer 7)" later in this chapter) for remote access. SSH establishes an encrypted tunnel between the client and the server, and can authenticate the client to the server. SSH can be used to protect the confidentiality and integrity of network communications. SSH-2 establishes an encrypted tunnel between the SSH client and SSH server, and can authenticate the client to the server. SSH version 1 is also widely used but has inherent vulnerabilities that are easily exploited.

REMEMBER

SSH-2 (or SSH) is an Internet security application that provides secure remote access.

>> **Session Initiation Protocol (SIP):** An open signaling protocol standard for establishing, managing and terminating real-time communications — such as voice, video, and text — over large IP-based networks.

Transport Layer (Layer 4)

The Transport Layer (Layer 4) provides transparent, reliable data transport and end-to-end transmission control. The Transport Layer hides the details of the lower layer functions from the upper layers.

Specific Transport Layer functions include

>> **Flow control:** Manages data transmission between devices, ensuring that the transmitting device doesn't send more data than the receiving device can process

>> **Multiplexing:** Enables data from multiple applications to be transmitted over a single physical link

>> **Virtual circuit management:** Establishes, maintains, and terminates virtual circuits

>> **Error checking and recovery:** Implements various mechanisms for detecting transmission errors and taking action to resolve any errors that occur, such as requesting that data be retransmitted

REMEMBER

The Transport Layer is responsible for providing transparent data transport and end-to-end transmission control.

Several important protocols defined at the Transport Layer include

>> **Transmission Control Protocol (TCP):** A full-duplex (capable of simultaneous transmission and reception), connection-oriented protocol that provides reliable delivery of packets across a network. A connection-oriented protocol

requires a direct connection between two communicating devices before any data transfer occurs. In TCP, this connection is accomplished via a three-way handshake. The receiving device acknowledges packets, and packets are retransmitted if an error occurs. The following characteristics and features are associated with TCP:

- *Connection-oriented:* TCP establishes and manages a direct virtual connection to the remote device.

- *Reliable:* TCP guarantees delivery by acknowledging received packets and requesting retransmission of missing or corrupted packets.

- *Slow:* Because of the additional overhead associated with initial handshaking, acknowledging packets, and error correction, TCP is generally slower than connectionless protocols, such as User Datagram Protocol (UDP).

TECHNICAL STUFF

>> A *three-way handshake* is the method used to establish a TCP connection (see Figure 6-3). A PC attempting to establish a connection with a server initiates the connection by sending a TCP SYN (Synchronize) packet, which is the first part of the handshake. In the second part of the handshake, the server replies to the PC with a SYN ACK packet (Synchronize Acknowledgement). Finally, the PC completes the handshake by sending an ACK or SYN-ACK-ACK packet, acknowledging the server's acknowledgement, and the data communications commence.

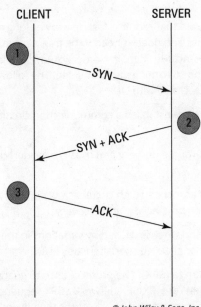

FIGURE 6-3: The TCP three-way handshake.

© *John Wiley & Sons, Inc.*

A *socket* is a logical endpoint on a system or device used to communicate over a network to another system or device (or even on the same device). A socket usually is expressed as an IP address and port number, such as 192.168.100.2:25.

» **User Datagram Protocol (UDP):** A connectionless protocol that provides fast best-effort delivery of datagrams across a network. A connectionless protocol doesn't guarantee delivery of transmitted packets (datagrams) and is thus considered unreliable. It doesn't

- Attempt to establish a connection with the destination network prior to transmitting data

- Acknowledge received datagrams

- Perform re-sequencing

- Perform error checking or recovery

TECHNICAL
STUFF

» A *datagram* is a self-contained unit of data that is capable of being routed between a source and a destination. Similar to a packet, which is used in the Internet Protocol (IP), datagrams are commonly used in UDP and other protocols.

TECHNICAL
STUFF

The term Protocol Data Unit (PDU) is used to describe the unit of data used at a particular layer of a protocol. In OSI, the Layer 1 PDU is a *bit*, Layer 2's PDU is a *frame*, Layer 3's is a *packet*, and Layer 4's is a *segment* or *datagram*, and Layer 7's is *message*.

UDP is ideally suited for data that requires fast delivery as long as that data isn't sensitive to packet loss and doesn't need to be fragmented. Examples of applications that use UDP include Domain Name System, Simple Network Management Protocol, and streaming audio or video. The following characteristics and features are associated with UDP:

- *Connectionless:* Doesn't pre-establish a communication circuit with the destination network

- *Best effort:* Doesn't guarantee delivery and thus is considered to be unreliable

- *Fast:* Has no overhead associated with circuit establishment, acknowledgement, sequencing, or error-checking and recovery

TECHNICAL
STUFF

Jitter in streaming audio and video is caused by variations in the delay of received packets, which is a negative characteristic of UDP.

» **Sequenced Packet Exchange (SPX):** The protocol used to guarantee data delivery in older Novell NetWare IPX/SPX networks. SPX sequences transmitted packets, reassembles received packets, confirms that all packets are received, and requests retransmission of packets that aren't received. SPX is to IPX as TCP is to IP, though the order is stated as IPX/SPX rather than SPX/IPX (as in TCP/IP): SPX and TCP are Layer 4 protocols, and IPX and IP are Layer 3 protocols. Just think of the SPX-IPS and TCP-IP relationships as yang and yin, rather than yin and yang!

Several examples of connection-oriented and connectionless-oriented protocols are identified in Table 6-1.

» **Secure Sockets Layer/Transport Layer Security (SSL/TLS):** The SSL/TLS protocol provides session-based encryption and authentication for secure communication between clients and servers on the Internet. SSL/TLS provides server authentication with optional client authentication.

» **Stream Control Transmission Protocol (SCTP):** A message-oriented protocol (similar to UDP) that ensures reliable, in-sequence transport with congestion control (similar to TCP). SCTP also provides multihoming and redundant paths for resiliency and reliability.

TABLE 6-1 **Connection-Oriented and Connectionless-Oriented Protocols**

Protocol	Layer	Type
TCP (Transmission Control Protocol)	4 (Transport)	Connection-oriented
UDP (User Datagram Protocol)	4 (Transport)	Connectionless-oriented
IP (Internet Protocol)	3 (Network)	Connectionless-oriented
ICMP (Internet Control Message Protocol)	3 (Network)	Connectionless-oriented
IPX (Internetwork Packet Exchange)	3 (Network)	Connectionless-oriented
SPX (Sequenced Packet Exchange)	4 (Transport)	Connection-oriented

Network Layer (Layer 3)

The Network Layer (Layer 3) provides routing and related functions that enable data to be transported between systems on the same network or on interconnected networks (or *internetworks*). Routing protocols — such as the Routing Information Protocol (RIP), Open Shortest Path First (OSPF), and Border Gateway Protocol (BGP) — are defined at this layer. Logical addressing of devices on the network is accomplished at this layer by using routed protocols, including IP and Internetwork Packet Exchange (IPX).

REMEMBER

The Network Layer is responsible primarily for routing. Routing protocols move routed protocol messages across a network. Routing protocols include RIP, OSPF, IS-IS, IGRP, and BGP. Routed protocols include IP and IPX.

ROUTING PROTOCOLS

Routing protocols are defined at the Network Layer and specify how routers communicate with one another on a WAN. Routing protocols are classified as static or dynamic.

A *static* routing protocol requires an administrator to create and update routes manually on the router. If the route is down, the network is down. The router can't reroute traffic dynamically to an alternative destination (unless a different route is specified manually). Also, if a given route is congested, but an alternative route is available and relatively fast, the router with static routes can't route data dynamically over the faster route. Static routing is practical only on very small networks or for very limited, special-case routing scenarios (such as a destination that's reachable only via a single router). Despite the limitations of static routing, it has a few advantages, such as low bandwidth requirements (routing information isn't broadcast across the network) and some built-in security (users can get only to destinations that are specified in the routing table).

A *dynamic* routing protocol can discover routes and determine the best route to a given destination at any given time. The routing table is periodically updated with current routing information. Dynamic routing protocols are further classified as link-state and distance-vector (for intradomain routing) and path-vector (for interdomain routing) protocols.

A *distance-vector* protocol makes routing decisions based on two factors: the distance (hop count or other metric) and vector (the egress router interface). It periodically informs its peers and/or neighbors of topology changes. *Convergence* — the time it takes for all routers in a network to update their routing tables with the most current information, such as link status changes — can be a significant problem for distance-vector protocols. Without convergence, some routers in a network may be unaware of topology changes, causing the router to send traffic to an invalid destination. During convergence, routing information is exchanged between routers, and the network slows considerably.

Routing Information Protocol (RIP) is a distance-vector routing protocol that uses hop count as its routing metric. To prevent routing loops, in which packets effectively get stuck bouncing between various router nodes, RIP implements a hop limit of 15, which significantly limits the size of networks that RIP can support. After a data packet crosses 15 router nodes (hops) between a source and a destination, the destination is considered to be unreachable. In addition to hop limits, RIP employs three mechanisms to prevent routing loops:

>> **Split horizon:** Prevents a router from advertising a route back out through the same interface from which the route was learned.

>> **Route poisoning:** Sets the hop count on a bad route to 16, effectively advertising the route as unreachable if it takes more than 15 hops to reach.

>> **Holddown timers:** Cause a router to start a timer when the router first receives information that a destination is unreachable. Subsequent updates about that destination will not be accepted until the timer expires. This mechanism also helps prevent problems associated with *flapping,* which occurs when a route (or interface) changes state (up, down, up, down) repeatedly over a short period.

RIP uses UDP port 520 as its transport protocol and port; thus, it is a connectionless-oriented protocol. Other disadvantages of RIP include slow convergence and insufficient security. (RIPv1 has no authentication, and RIPv2 transmits passwords in cleartext.) RIP is a legacy protocol, but it's still in widespread use on networks today despite its limitations, because of its simplicity.

TECHNICAL
STUFF

Hop count generally refers to the number of router nodes that a packet must pass through to reach its destination.

A *link-state* protocol requires every router to calculate and maintain a complete map, or routing table, of the entire network. Routers that use a link-state protocol periodically transmit updates that contain information about adjacent connections (called *link states*) to all other routers in the network. Link-state protocols are computation-intensive but can calculate the most efficient route to a destination, taking into account numerous factors such as link speed, delay, load, reliability, and cost (an arbitrarily assigned weight or metric). Convergence occurs very rapidly (within seconds) with link-state protocols; distance-vector protocols usually take longer (several minutes or even hours in very large networks). Two examples of link-state routing protocols are

>> **Open Shortest Path First (OSPF):** OSPF is a link-state routing protocol widely used in large enterprise networks. It's considered to be an interior gateway protocol because it performs routing within a single autonomous system. OSPF is encapsulated directly into IP datagrams, as opposed to using a Transport Layer protocol such as TCP or UDP. OSPF networks are divided into areas identified by 32-bit area identifiers. *Area identifiers* can (but don't have to) correspond to network IP addresses and can duplicate IP addresses without conflicts. Special OSPF areas include the *backbone area* (also known as *area 0*), *stub area,* and *not-so-stubby area.*

>> **Intermediate System to Intermediate System (IS-IS):** IS-IS is a link-state routing protocol used to route datagrams through a packet-switched network. This interior gateway protocol is used for routing within an autonomous system and is, used extensively in large service-provider backbone networks.

TECHNICAL STUFF

An *autonomous system (AS)* is a group of contiguous IP address ranges under the control of a single Internet entity. Individual autonomous systems are assigned a 16-bit or 32-bit number (ASN) that uniquely identifies the network on the Internet, assigned by the Internet Assigned Numbers Authority (IANA).

A *path-vector* protocol is similar in concept to a distance-vector protocol, but without the scalability issues associated with limited hop counts. Border Gateway Protocol (BGP) is an example of a path-vector protocol. BGP is a path-vector routing protocol used between separate autonomous systems. It's considered to be an exterior gateway protocol because it performs routing between separate autonomous systems. It's the core protocol used by Internet service providers, network service providers, and very large private IP networks. When BGP runs between autonomous systems, it's called external BGP (eBGP). When BGP runs within an autonomous system (such as on a private IP network), it's called internal BGP (iBGP).

ROUTED PROTOCOLS

Routed protocols are Network Layer protocols, such as Internetwork Packet Exchange (IPX) and IP, which address packets with routing information and allow those packets to be transported across networks via routing protocols (discussed in the preceding section).

Internetwork Packet Exchange (IPX) is a connectionless protocol used primarily in older Novell NetWare networks for routing packets across the network. It's part of the Internetwork Packet Exchange/Sequenced Packet Exchange (IPX/SPX) protocol suite, which is analogous to the TCP/IP suite.

IP contains addressing information that enables packets to be routed. IP is part of the Transmission Control Protocol/Internet Protocol (TCP/IP) suite, which is the language of the Internet. IP has two primary responsibilities:

>> Connectionless, best-effort (no guarantee of) delivery of datagrams

>> Fragmentation and reassembly of datagrams

IP Version 4 (IPv4), which is currently the most commonly used version, uses a 32-bit logical IP address that's divided into four 8-bit sections (*octets*) and consists of two main parts: the network number and the host number. The first four bits in an octet are known as the high-order bits, and the last four bits in an octet are known as the low-order bits. The first bit in the octet is referred to as the most significant bit, and the last bit in the octet is referred to as the least significant bit. Each bit position represents its value (see Table 6-2) if the bit is on (1); otherwise, its value is zero (off or 0).

TABLE 6-2

Bit Position Values in an IPv4 Address

High-Order Bits			Low-Order Bits				
Most significant bit							Least significant bit
128	64	32	16	8	4	2	1

Each octet contains an 8-bit number with a value of 0 to 255. Table 6-3 shows a partial list of octet values in binary notation.

TABLE 6-3

Binary Notation of Octet Values

Decimal	Binary	Decimal	Binary	Decimal	Binary
255	1111 1111	200	1100 1000	9	0000 1001
254	1111 1110	180	1011 0100	8	0000 1000
253	1111 1101	160	1010 0000	7	0000 0111
252	1111 1100	140	1000 1100	6	0000 0110
251	1111 1011	120	0111 1000	5	0000 0101
250	1111 1010	100	0110 0100	4	0000 0100
249	1111 1001	80	0101 0000	3	0000 0011
248	1111 1000	60	0011 1100	2	0000 0010
247	1111 0111	40	0010 1000	1	0000 0001
246	1111 0110	20	0001 0100	0	0000 0000

IPv4 addressing supports five address classes, indicated by the high-order (left-most) bits in the IP address, as listed in Table 6-4.

TABLE 6-4

IP Address Classes

Class	Purpose	High-Order Bits	Address Range	Maximum Number of Hosts
A	Large networks	0	1 to 126	16,777,214 (224-2)
B	Medium networks	10	128 to 191	65,534 (216-2)
C	Small networks	110	192 to 223	254 (28-2)
D	Multicast	1110	224 to 239	N/A
E	Experimental	1111	240 to 254	N/A

The address range 127.0.0.1 to 127.255.255.255 is a loopback network used for testing and troubleshooting. Packets sent to a 127 address are immediately routed back to the source device. The most commonly used loopback (or localhost) address for devices is 127.0.0.1 (sometimes called *home*), although any address in the 127 network range can be used for this purpose.

Several IPv4 address ranges are also reserved for use in private networks, including

>> 10.0.0.0–10.255.255.255 (Class A)

>> 172.16.0.0–172.31.255.255 (Class B)

>> 192.168.0.0–192.168.255.255 (Class C)

These addresses aren't routable on the Internet and thus are often implemented behind firewalls and gateways by using Network Address Translation (NAT) to conserve IP addresses, mask the network architecture, and enhance security. NAT translates private, non-routable addresses on internal network devices to registered IP addresses when communication across the Internet is required. The widespread use of NAT and private network addresses somewhat delayed the inevitable depletion of IPv4 addresses, which is limited to approximately 4.3 billion due to its 32-bit format (2^{32} = 4,294,967,296 possible addresses). But the thing about inevitability is that it's . . . well, inevitable. Factors such as the proliferation of mobile devices worldwide, always-on Internet connections, inefficient use of assigned IPv4 addresses, and the spectacular miscalculation of IBM's Thomas Watson — who in 1943 predicted that there would be a worldwide market for "maybe five computers" (he was no Nostradamus) — have led to the depletion of IPv4 addresses.

Technically, we're not completely out of IPv4 addresses. Each regional Internet registry has reserved a very small pool of IPv4 addresses to facilitate the transition to IPv6.

In 1998, the IETF formally defined IP Version 6 (IPv6) as the replacement for IPv4. IPv6 uses a 128-bit hexadecimal IP address (versus 32 bits for IPv4) and incorporates additional functionality to provide security, multimedia support, plug-and-play compatibility, and backward compatibility with IPv4. The main reason for developing IPv6 was to provide infinitely more network addresses than are available with IPv4 addresses. Okay, it's not infinite, but it is ginormous — 2^{128} or approximately 3.4×10^{38} (that's 340 hundred undecillion) unique addresses!

IPv6 addresses consist of 32 hexadecimal numbers grouped into eight blocks (sometimes referred to as *hextels*) of four hexadecimal digits, separated by a colon.

REMEMBER

A hexadecimal digit is represented by 4 bits (see Table 6-5), so each hextel is 16 bits (four 4-bit hexadecimal digits), and eight 16-bit hextels equals 128 bits. An IPv6 address is further divided into two 64-bit segments. The first (also referred to as the top or upper) 64 bits represent the network part of the address, and the last (also referred to as the bottom or lower) 64 bits represent the node or interface part of the address. The network part is further subdivided into a 48-bit global network address and a 16-bit subnet. The node or interface part of the address is based on the IEEE Extended Unique Identifier (EUI-64) physical or media access control (MAC) address (discussed in the "Data Link Layer (Layer 2)" section later in this chapter) of the node or interface.

TABLE 6-5 ## Decimal, Hexadecimal, and Binary Notation

Decimal	Hexadecimal	Binary
0	0	0000
1	1	0001
2	2	0010
3	3	0011
4	4	0100
5	5	0101
6	6	0110
7	7	0111
8	8	1000
9	9	1001
10	A	1010
11	B	1011
12	C	1100
13	D	1101
14	E	1110
15	F	1111

The basic format for an IPv6 address is

xxxx:xxxx:xxxx:xxxx:xxxx:xxxx:xxxx:xxxx

where x represents a hexadecimal digit (0–f).

Following is an example of an IPv6 address:

2001:0db8:0000:0000:0008:0800:200c:417a

There are several rules the IETF has defined to shorten an IPv6 address:

>> **Leading zeroes in an individual hextel can be omitted, but there must be at least one hexadecimal digit in each hextel (except as noted in the next rule).** Applying this rule to the previous example yields the following result: 2001:db8:0:0:8:800:200c:417a.

>> **Two colons (::) can be used to represent one or more groups of 16 bits of zeros, as well as leading or trailing zeroes in an address, and can only appear once in an IPv6 address.** Applying this rule to the previous example yields the following result: 2001:db8::8:800:200c:417a.

>> **In mixed IPv4 and IPv6 environments, the form x:x:x:x:x;x:d.d.d.d can be used, in which x represents the six high-order 16-bit hextels of the address and d represents the four low-order 8-bit octets (in standard IPv4 notation) of the address.** 0db8:0:0:0:0:FFFF:129.144.52.38 is a valid IPv6 address. Applying the previous two rules to this example yields the following result: db8::ffff:129.144.52.38.

Letters in hexadecimal notation are not case-sensitive (*A* is the same as *a*, *B* is the same as *b*, and so on), so either form can be used in IPv6 addresses, although IETF recommends using lowercase letters.

Security features in IPv6 include network-layer security via Internet Protocol Security and requirements defined in Request For Comments 7112 to prevent fragmentation exploits in IPv6 headers.

Although you don't need to know all the intricate details of IPv6 addressing for the CISSP exam, as its use becomes more commonplace — particularly in Internet of Things (IoT) devices — you need to be familiar with the security enhancements in IPv6 and be able to recognize a valid IPv6 address.

IMPLICATIONS OF MULTILAYER PROTOCOLS

Multilayer protocols are groups of protocols that are purpose-built for some type of specialized communications need. Multilayer protocols have their own schemes for encapsulation, like TCP/IP itself.

One good example of a multilayer protocol is Distributed Network Protocol (DNP3), which is used in industrial control systems (ICS) and supervisory control and data acquisition (SCADA) networks. DNP3 has a Data Frame Layer, Transport Layer, and Application Layer.

DNP3's original design lacks security features, such as authentication and encryption. Recent updates to the standard have introduced security protocols. Without security features, relatively simple attacks (such as eavesdropping, spoofing, and perhaps denial of service) can be carried out easily on specialized multiprotocol networks.

CONVERGED PROTOCOLS

Converged protocols refers to an implementation of two or more protocols for a specific communications purpose. Examples of converged protocols include

>> Multiprotocol Label Switching (MPLS)

>> Fibre Channel over Ethernet (FCoE)

>> Voice over Internet Protocol (VoIP)

>> Session Initiation Protocol (SIP)

>> Internet Small Computer System Interface (iSCSI)

SOFTWARE-DEFINED NETWORKS

Software-defined networks (SDN) represent the ability to create, configure, manage, secure, and monitor network elements rapidly and efficiently. SDN uses an open standards architecture that enables intelligent network functions, such as routing, switching, and load balancing (the overlay function), to be performed on virtual software that is installed on commodity network hardware (the physical underlay), similar to server virtualization. In SDN, network elements and network architectures are virtual, which enables organizations to build and modify their networks and network elements quickly.

As with other virtualization technologies, correct management of SDN requires policy, process, and discipline to avoid network sprawl (the phenomenon in which undisciplined administrators bypass change control processes and unilaterally create virtual network elements).

Related to SDN, software-defined WAN is discussed later in this chapter.

IPSEC

Internet Protocol Security (IPsec) is an IETF open standard for VPNs that operates at the Network Layer (Layer 3) of the OSI model. It's the most popular and robust VPN protocol in use today. IPsec ensures confidentiality, integrity, and authenticity by using Layer 3 encryption and authentication to provide an end-to-end solution. IPsec operates in two modes:

>> **Transport:** Only the data is encrypted.

>> **Tunnel:** The entire packet is encrypted.

The two modes of IPsec are Transport mode (on the LAN) and Tunnel mode (on the WAN).

The two main protocols used in IPsec are

>> **Authentication Header (AH):** Provides integrity, authentication, and nonrepudiation

>> **Encapsulating Security Payload (ESP):** Provides confidentiality (encryption) and limited authentication

Each pair of hosts communicating in an IPsec session must establish a security association (SA), which is a one-way connection between two communicating parties; thus, two associations are required for each pair of communicating hosts. Additionally, each association supports only a single protocol (AH or ESP). Therefore, using both an AH and an ESP between two communicating hosts requires four security associations. A security association has three parameters that uniquely identify it in an IPsec session:

>> **Security Parameter Index (SPI):** The SPI is a 32-bit string used by the receiving station to differentiate between SAs terminating on that station. The SPI is located within the AH or ESP header.

>> **Destination IP address:** The destination address could be the end station or an intermediate gateway or firewall, but it must be a unicast address.

>> **Security Protocol ID:** The Security Protocol ID must be an AH or ESP association.

REMEMBER

In IPsec, a security association is a one-way connection. You need a minimum of two security associations for two-way communications.

Key management is provided in IPsec by using the Internet Key Exchange (IKE), which is a combination of three complementary protocols: the Internet Security Association and Key Management Protocol (ISAKMP), the Secure Key Exchange Mechanism, and the Oakley Key Exchange Protocol. Internet Key Exchange operates in three modes: Main, Aggressive, and Quick.

OTHER NETWORK LAYER PROTOCOLS

Other protocols defined at the Network Layer include Internet Control Message Protocol (ICMP) and Simple Key Management for Internet Protocols (SKIP).

ICMP is used for network control and diagnostics. Commonly used ICMP commands include `ping` and `traceroute`. Although ICMP is very helpful in troubleshooting routing and connectivity issues in a network, it is also commonly used by attackers for network reconnaissance, device discovery, and denial-of-service (DoS) attacks (such as an ICMP flood).

SKIP is a Network Layer key management protocol used to share encryption keys. An advantage of SKIP is that it doesn't require a communication session to be established before it sends encrypted keys or packets. The protocol is bandwidth-intensive, however, because of the size of additional header information in encrypted packets.

NETWORKING EQUIPMENT AT THE NETWORK LAYER

The primary networking equipment defined at Layer 3 is routers and gateways.

Routers are intelligent devices that link dissimilar networks and use logical or physical addresses to forward data packets only to the destination network (or along the network path). Routers employ various routing algorithms (such as RIP, OSPF, and BGP) to determine the best path to a destination, based on variables that include bandwidth, cost, delay, and distance.

Gateways are created with software running on a computer (workstation or server) or router. Gateways link dissimilar programs and protocols by examining the entire Layer 7 data packet so as to translate incompatibilities. A gateway can be used, for example, to link an IP network to an IPX network or a Microsoft Exchange mail server to a Lotus Notes server (a mail gateway).

Data Link Layer (Layer 2)

The Data Link Layer ensures that messages are delivered to the proper device across a physical network. This layer also defines the networking protocol (such as Ethernet, USB, Wi-Fi, or token ring) used to send and receive data between individual devices. The Data Link Layer formats messages from layers above into frames for transmission, handles point-to-point synchronization and error control, and can perform link encryption.

The IEEE 802 standards and protocols further divide the Data Link Layer into two sublayers: Logical Link Control (LLC) and Media Access Control (MAC), as shown in Figure 6-4.

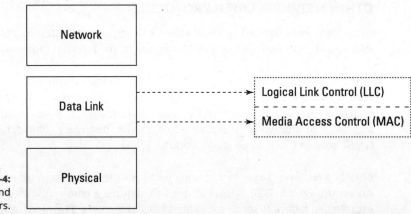

FIGURE 6-4:
The LLC and MAC sublayers.

REMEMBER

The Data Link Layer is responsible for ensuring that messages are delivered to the proper device across a physical network.

The LLC sublayer operates between the Network Layer above and the MAC sublayer below. The LLC sublayer performs the following three functions:

» Provides an interface for the MAC sublayer by using Source Service Access Points and Destination Service Access Points.

» Manages the control, sequencing, and acknowledgement of frames being passed up to the Network Layer or down to the Physical Layer.

» Bears responsibility for timing and flow control. *Flow control* monitors the flow of data between devices to ensure that a receiving device, which may not necessarily be operating at the same speed as the transmitting device, isn't overwhelmed and dropping packets.

REMEMBER

LLC and MAC are sublayers of the Data Link Layer.

The MAC sublayer operates between the LLC sublayer above and the Physical Layer below. It's responsible primarily for framing and has the following three functions:

>> **Performs error control:** Error control uses a cyclic redundancy check (CRC), a simple mathematical calculation or checksum used to create a message profile. The CRC is recalculated by the receiving device. If the calculated CRC doesn't match the received CRC, the packet is dropped, and a request to resend is transmitted back to the device that sent it.

>> **Identifies hardware device (or MAC) addresses:** A *MAC address* (also known as a *hardware address* or *physical address*) is a 48-bit address that's encoded on each device by its manufacturer. The first 24 bits identify the manufacturer or vendor. The second 24 bits uniquely identify the device.

>> **Controls media access:** The three basic types of media access are

- *Contention:* In contention-based networks such as Ethernet, individual devices must vie for control of the physical network medium. This type of access is ideally suited to networks characterized by small bursts of traffic. Ethernet networks use a contention-based method, known as Carrier Sense Multiple Access with Collision Detection (CSMA/CD), in which all stations listen for traffic on the physical network medium. If the line is clear, any station can transmit data. If another station attempts to transmit data at the same time, a collision occurs, the traffic is dropped, and both stations must wait a random period of time before attempting to retransmit. Another method, used in Wi-Fi networks, is known as Carrier Sense Multiple Access with Collision Avoidance (CSMA/CA). Unlike CSMA/CD, CSMA/CA prevents collisions from occurring in the first place. This is accomplished by listening for traffic on the network. If the network is clear (or idle), a station waits a period of time (known as the inter-frame gap, or IFG) and then sends a frame. The sender station then sets a timer and waits for acknowledgement from the receiver that the frame was successfully received. If the acknowledgement is not received, the sender waits for a back-off time period and then re-transmits the frame.

- *Token passing:* In token-passing networks such as token ring and Fiber Distributed Data Interface (FDDI), individual devices must wait for a special frame, known as a *token,* before they transmit data across the physical network medium. This type of network is considered to be deterministic (transmission delay can be reliably calculated, and collisions don't occur) and is ideally suited for networks that have large, bandwidth-consuming applications that are delay-sensitive. Token ring, FDDI, and ARCnet networks use various token-passing methods for media access control.

- *Polling:* In polling networks, individual devices (secondary hosts) are polled by a primary host to see whether they have data to be transmitted. Secondary hosts can't transmit until permission is granted by the primary host. Polling is typically used in mainframe environments and on wireless networks.

LAN PROTOCOLS AND TRANSMISSION METHODS

Common LAN protocols are defined at the Data Link and Physical layers, and include the following:

>> **Ethernet:** The Ethernet protocol transports data to the physical LAN medium by using CSMA/CD (discussed in the preceding section). It is designed for networks characterized by sporadic, sometimes heavy traffic requirements. Ethernet is by far the most common LAN protocol used today, most often implemented with twisted-pair cabling (discussed in the section "Cable and connector types"). Ethernet operates at speeds up to 10 Mbps, Fast Ethernet operates at speeds up to 100 Mbps (over Cat 5 twisted-pair or fiber-optic cabling), and Gigabit Ethernet operates at speeds up to 40 Gbps (over Cat 5e, Cat 6, or Cat 7 twisted-pair or fiber-optic cabling).

>> **ARCNET:** The Attached Resource Computer NETwork (ARCNET) protocol is one of the earliest LAN technologies developed. It transports data to the physical LAN medium by using the token-passing media access method that we discuss in the preceding section. It's implemented in a star topology by using coaxial cable. ARCNET provides slow but predictable network performance.

>> **Token ring:** The token ring protocol transports data to the physical LAN medium by using the token-passing media access method that we discuss in the preceding section. In a token ring network, all nodes are attached to a Multistation Access Unit (MAU) in a logical ring (but physical star) topology. One node on the token ring network is designated as the active monitor and ensures that no more than one token is on the network at any given time. (Variations permit more than one token on the network.) If the token is lost, the active monitor is responsible for ensuring that a replacement token is generated. Token ring networks operate at speeds of 4 and 16 Mbps — pretty slow by today's standards. Token ring networks are rarely seen nowadays.

>> **Fiber Distributed Data Interface (FDDI):** The FDDI protocol transports data to the physical LAN medium by using the token-passing media access method that we discuss in the preceding section. It's implemented as a dual counter-rotating ring over fiber-optic cabling at speeds up to 100 Mbps. All stations on a FDDI network are connected to both rings. During normal operation, only one ring is active. In the event of a network break or fault, the ring wraps back through the nearest node onto the second ring.

>> **Address Resolution Protocol (ARP):** ARP maps Network Layer IP addresses to MAC addresses. ARP discovers physical addresses of attached devices by broadcasting ARP query messages on the network segment. Then IP-address-to-MAC-address translations are maintained in a dynamic table that's cached on the system.

>> **Reverse Address Resolution Protocol (RARP):** RARP maps MAC addresses to IP addresses. This process is necessary when a system, such as a diskless machine, needs to discover its IP address. The system broadcasts a RARP message that provides the system's MAC address and requests to be informed of its IP address. A RARP server replies with the requested information.

REMEMBER

Both ARP and RARP are Layer 2 protocols. ARP maps an IP address to a MAC address and is used to identify a device's hardware address when only the IP address is known. RARP maps a MAC address to an IP address and is used to identify a device's IP address when only the MAC address is known.

LAN data transmissions are classified as

>> **Unicast:** Packets are sent from the source to a single destination device by using a specific destination IP address.

>> **Multicast:** Packets are copied and sent from the source to multiple destination devices by using a special multicast IP address that the destination stations have been specifically configured to use.

>> **Anycast:** Packets are copied and sent from the source to a single destination IP address that is shared by several devices in multiple locations. Anycast is commonly used in CDNs (discussed earlier in this chapter).

>> **Broadcast:** Packets are copied and sent from the source to every device on a destination network by using a broadcast IP address.

REMEMBER

LAN data transmissions are classified as unicast, multicast, or broadcast.

WIRELESS NETWORKS

Important wireless networks to be familiar with for the CISSP exam include Li-Fi, Wi-Fi, Near-Field Communication (NFC), ZigBee, and satellite.

Similar to fiber-optic networks, which use light rather than electrical signals to transmit data over wired networks, *Li-Fi* uses light in the visible, ultraviolet, and infrared spectrums rather than radio frequency to transmit data up to 100 Gbps over wireless networks.

WLAN (wireless LAN) technologies, commonly known as *Wi-Fi*, function at the lower layers of the OSI Reference Model. WLAN protocols define how frames are transmitted over the air. See Table 6-6 for a description of the most common IEEE 802.11 WLAN standards.

TABLE 6-6

Wireless LAN Standards

Type	Speed	Description
802.11a	54 Mbps	Operates at 5 GHz (less interference than at 2.4 GHz)
802.11b	11 Mbps	Operates at 2.4 GHz (first widely used protocol)
802.11g	54 Mbps	Operates at 2.4 GHz (backward-compatible with 802.11b)
802.11n	600 Mbps	Operates at 5 GHz or 2.4 GHz
802.11ac	1 Gbps	Operates at 5 GHz
802.11ad (WiGig)	6.7 Gbps	Operates at 60 GHz (range limited to 30 ft)
802.11ah (WiFi HaLow)	347 Mbps	Operates at 900 MHz
802.11ax (WiFi 6)	10 Gbps	Operates at 5 GHz or 2.4 GHz (backwards-compatible with 802.11a/b/g/n/ac)

ZigBee is a low-cost, low-power wireless mesh network protocol based on the IEEE 802.15.4 standard. ZigBee is commonly used in industrial environments and smart home products. Various ZigBee specifications include ZigBee Pro, ZigBee Radio Frequency for Consumer Electronics (RF4CE), and ZigBee IP.

Near-Field Communication (NFC) is a set of communication protocols used over short distances (up to 4 cm), often used with smartphones for access control and contactless payments.

SATELLITE NETWORKS

Satellite broadband technologies are commonly used in remote areas where wired, wireless, or cellular services may not be readily available, as well as to provide backup connectivity. Other applications include providing connectivity for logistics (ships, planes, and trains) and communication backbones for IoT devices (such as fleet management and remote maintenance).

CELLULAR NETWORKS

Smartphones, IoT devices, and other mobile devices use cellular networks as well as Wi-Fi networks to communicate. The Third Generation Partnership Project defines the various generations and develops the protocols for mobile telecommunications (that is, cellular networks). The International Telecommunication Union (ITU) is a regulatory organization of the United Nations that is responsible for the global use of mobile telecommunication. Relevant cellular network generations include

» **3G:** The first 3G networks provided minimum information transfer rates of 200 kilobits per second (Kbps). The ITU has never formally defined a standard for 3G data rates, so downlink data speeds vary widely from 384 Kbps for Wideband Code Division Multiple Access (W-CDMA) to 168 Mbps for Advanced Evolved High Speed Packet Access (HSPA+). 3G cellular networks are still commonly used in many countries and are increasingly being used as offload networks for 4G and 5G, as well as IoT applications.

» **4G:** 4G Long-Term Evolution (LTE) mobile networks are currently the most commonly deployed cellular networks used for mobile technology. LTE networks are built on packet-switched all-IP core networks. LTE Advanced and LTE Advanced Pro have theoretical peak data rates of 1 Gbps and 3 Gbps, respectively.

» **5G:** 5G networks deliver significantly higher speeds and lower latency than previous generations but also introduce many other important innovations to support a wide variety of use cases, including massive Machine-Type Communications (mMTC), Ultra Reliable and Low Latency Communication (URLLC), and enhanced Mobile Broadband (eMBB).

WAN TECHNOLOGIES AND PROTOCOLS

WAN technologies function at the lower three layers of the OSI Reference Model (the Physical, Data Link, and Network layers), primarily at the Data Link Layer. WAN protocols define how frames are carried across a single data link between two devices. These protocols include

» **Point-to-point links:** These links provide a single, pre-established WAN communications path from the customer's network across a carrier network (such as a Public Switched Telephone Network (PSTN)) to a remote network. These point-to-point links include

- *Layer 2 Forwarding Protocol (L2F):* A tunneling (data encapsulation) protocol developed by Cisco and used to implement VPNs, specifically Point-to-Point Protocol (discussed later in this section) traffic. L2F provides encapsulation but doesn't provide encryption or confidentiality.

- *Layer 2 Tunneling Protocol (L2TP):* A tunneling protocol used to implement VPNs. L2TP is derived from L2F (described in the preceding item) and PPTP (described in this list) and uses UDP port 1701 (see the section "Network Layer [Layer 3]" earlier in this chapter) to create a tunneling session. L2TP is commonly implemented along with an encryption protocol, such as IPsec, because it doesn't encrypt traffic or provide confidentiality by itself. We discuss L2TP and IPsec in more detail in the section "Remote access" later in this chapter.

- *Point-to-Point Protocol (PPP):* The successor to SLIP (see the discussion later in this section), PPP provides router-to-router and host-to-network connections over synchronous and asynchronous circuits. It's a more robust protocol than SLIP and provides additional built-in security mechanisms. PPP is far more common than SLIP in modern networking environments.

- *Point-to-Point Tunneling Protocol (PPTP):* A tunneling protocol developed by Microsoft and commonly used to implement VPNs, specifically PPP traffic. PPTP doesn't provide encryption or confidentiality, instead relying on other protocols, such as PAP, CHAP, and EAP, for security. We discuss PPTP, PAP, CHAP, and EAP in more detail in the section "Remote access" later in this chapter.

- *Serial Line IP (SLIP):* The predecessor of PPP, SLIP was originally developed to support TCP/IP networking over low-speed asynchronous serial lines (such as dial-up modems) for Berkeley Unix computers. SLIP is rarely seen today except in computer museums.

» **Circuit-switched networks:** In a circuit-switched network, a dedicated physical circuit path is established, maintained, and terminated between the sender and receiver across a carrier network for each communications session (the *call*). This network type is used extensively in telephone company networks and functions similarly to a regular telephone call. Examples include

- *Digital Subscriber Line (xDSL):* xDSL uses existing analog phone lines to deliver high-bandwidth connectivity to remote customers.

- *Integrated Services Digital Network (ISDN):* ISDN is a communications protocol that operates over analog phone lines that have been converted to use digital signaling. ISDN lines are capable of transmitting both voice and data traffic. ISDN defines a B-channel for data, voice, and other services, and a D-channel for control and signaling information.

With the introduction and widespread adoption of DSL and DOCSIS, ISDN has largely fallen out of favor in the United States and is no longer available in many areas.

» Circuit-switched networks are ideally suited for always-on connections that experience constant traffic.

REMEMBER

>> **Packet-switched networks:** In a packet-switched network, devices share bandwidth (by using statistical multiplexing) on communications links to transport packets between a sender and receiver across a carrier network. This type of network is more resilient to error and congestion than circuit-switched networks. We compare packet-switched and circuit-switched networks in Table 6-7.

Examples of packet-switched networks include

- *Asynchronous Transfer Mode (ATM):* A very high-speed, low-delay technology that uses switching and multiplexing techniques to rapidly relay fixed-length (53-byte) cells that contain voice, video, or data. Cell processing occurs in hardware that reduces transit delays. ATM is ideally suited for fiber-optic networks that carry bursty (uneven) traffic.

- *Data Over Cable Service Interface Specification (DOCSIS):* A communications protocol for transmitting high-speed data over an existing cable TV system.

- *Frame Relay:* A packet-switched standard protocol that handles multiple virtual circuits by using High-level Data Link Control encapsulation (which we discuss later in this section) between connected devices. Frame Relay uses a simplified framing approach that has no error correction and Data Link Connection Identifiers to achieve high speeds across the WAN. Frame Relay can be used on Switched Virtual Circuits (SVCs) or Permanent Virtual Circuits (PVCs). An SVC is a temporary connection that's dynamically created (in the circuit establishment phase) to transmit data (which happens during the data transfer phase) and then disconnected (in the circuit termination phase). PVCs are permanently established connections. Because the connection is permanent, a PVC doesn't require the bandwidth overhead associated with circuit establishment and termination. PVCs are generally more expensive than SVCs.

- *Multi-Protocol Label Switching (MPLS):* A packet-switched, high-speed, highly scalable, highly versatile technology used to create fully meshed VPNs. It can carry IP packets as well as ATM, SONET (Synchronous Optical Networking), or Ethernet frames. MPLS is specified at both Layer 2 and Layer 3. Label Edge Routers (LERs) in an MPLS network push or encapsulate a packet (or frame) with an MPLS label. The label information is used to switch the payload through the MPLS cloud at very high speeds. Label Switch Routers within the MPLS cloud make routing decisions based solely on the label information without actually examining the payload. At the egress point, an LER *pops* (decapsulates) the packet, removing the MPLS label when the packet exits the MPLS network. One advantage of an MPLS network is that a customer loses visibility into the cloud. If you're a glass-half-full type, one advantage of an MPLS network is that an attacker loses visibility into the cloud.

- *Synchronous Optical Network (SONET) and Synchronous Digital Hierarchy (SDH):* A high-availability, high-speed, multiplexed, low-latency technology used on fiber-optic networks. SONET was originally designed for the public telephone network and is widely used throughout the United States and Canada, particularly within the energy industry. SDH was developed after SONET and is used throughout the rest of the world. Data rates for SONET and SDH are defined at optical carrier (OC) levels (see Table 6-8).

- *Switched Multimegabit Data Service (SMDS):* A high-speed, packet-switched, connectionless-oriented, datagram-based technology available over public switched networks. Typically, companies that exchange large amounts of data bursts with other remote networks use SMDS.

- *X.25:* The first packet-switching network, X.25 is an International Telecommunication Union – Telecommunications (ITU-T) standard that defines how point-to-point connections between data terminal equipment (DTE) and data carrier equipment (DCE) are established and maintained. X.25 specifies the Link Access Procedure, Balanced (LAPB) protocol at the Data Link Layer and the Packet Level Protocol (PLP; also known as X.25 Level 3) at the Network Layer. X.25 is more common outside the United States but largely has been superseded by MPLS and Frame Relay.

REMEMBER

Packet-switched networks are ideally suited for on-demand connections that have bursty traffic.

>> **Other WAN protocols:** Two other important WAN protocols defined at the Data Link Layer include

- *High-level Data Link Control (HDLC):* A bit-oriented, synchronous protocol that was created by the ISO to support point-to-point and multipoint configurations. Derived from SDLC, it specifies a data encapsulation method for synchronous serial links and is the default for serial links on Cisco routers. Unfortunately, various vendor implementations of the HDLC protocol are incompatible.

- *Synchronous Data Link Control (SDLC):* A bit-oriented, full-duplex serial protocol that was developed by IBM to facilitate communications between mainframes and remote offices. It defines and implements a polling method of media access, in which the primary (front end) polls the secondaries (remote stations) to determine whether communication is required.

WAN protocols and technologies are implemented over telecommunications circuits. Refer to Table 6-8 for a description of common telecommunications circuits and speeds.

TABLE 6-7

Circuit Switching versus Packet Switching

Circuit Switching	Packet Switching
Ideal for always-on connections, constant traffic, and voice communications	Ideal for bursty traffic and data communications
Connection-oriented	Connectionless-oriented
Fixed delays	Variable delays

TABLE 6-8

Common Telecommunications Circuits

Type	Speed	Description
DS0	64 Kbps	Digital Signal Level 0, framing specification used in transmitting digital signals over a single channel at 64 Kbps on a T1 facility
DS1	1.544 Mbps or 2.048 Mbps	Digital Signal Level 1, framing specification used in transmitting digital signals at 1.544 Mbps on a T1 facility (U.S.) or at 2.048 Mbps on an E1 facility (EU)
DS3	44.736 Mbps	Digital Signal Level 3, framing specification used in transmitting digital signals at 44.736 Mbps on a T3 facility
T1	1.544 Mbps	Digital WAN carrier facility; transmits DS1-formatted data at 1.544 Mbps (24 DS0 user channels at 64 Kbps each)
T3	44.736 Mbps	Digital WAN carrier facility; transmits DS3-formatted data at 44.736 Mbps (672 DS0 user channels at 64 Kbps each)
E1	2.048 Mbps	Wide-area digital transmission scheme used primarily in Europe that carries data at a rate of 2.048 Mbps
E3	34.368 Mbps	Wide-area digital transmission scheme used primarily in Europe that carries data at a rate of 34.368 Mbps (16 E1 signals)
OC-1	51.84 Mbps	SONET (Synchronous Optical Networking) Optical Carrier WAN specification
OC-3	155.52 Mbps	SONET
OC-12	622.08 Mbps	SONET
OC-48	2.488 Gbps	SONET
OC-192	9.9 Gbps	SONET
OC-768	39 Gbps	SONET

ASYNCHRONOUS AND SYNCHRONOUS COMMUNICATIONS

Asynchronous communication transmits data in a serial stream that has control data (start and stop bits) embedded in the stream to indicate the beginning and end of characters. Asynchronous devices must communicate at the same speed, which is controlled by the slower of the two communicating devices. Because no internal clocking is used, parity bits are used to reduce transmission errors.

Synchronous communications use an internal clocking signal to transmit large blocks of data, known as *frames*. Synchronous communication is characterized by very high-speed transmission rates.

NETWORKING EQUIPMENT AT THE DATA LINK LAYER

Networking devices that operate at the Data Link Layer include bridges, switches, DTEs/DCEs, and wireless equipment:

» A *bridge* is a semi-intelligent repeater used to connect two or more (similar or dissimilar) network segments. A bridge maintains an Address Resolution Protocol (ARP) cache that contains the MAC addresses of individual devices on connected network segments. When a bridge receives a data signal, it checks its ARP cache to determine whether the destination MAC address is on the local network segment. If the data signal turns out to be local, it isn't forwarded to a different network. If the MAC address isn't local, however, the bridge forwards (and amplifies) the data signal to all other connected network segments. A serious networking problem associated with bridges is a broadcast storm, in which broadcast traffic is automatically forwarded by a bridge, effectively flooding a network. Network bridges have been superseded by switches (discussed next).

» *Data Terminal Equipment* (DTE) is a general term used to classify devices at the user end of a user-to-network interface (such as computer workstations). A DTE connects to Data Carrier Equipment (DCE; also known as Data Circuit-Terminating Equipment), which consists of devices at the network end of a user-to-network interface. The DCE provides the physical connection to the network, forwards network traffic, and provides a clocking signal to synchronize transmissions between the DCE and the DTE. Examples of DCEs include NICs (Network Interface Cards), modems, and CSUs/DSUs (Channel Service Units/Data Service Units).

» *Wireless Access Points* (APs) are transceivers that connect wireless clients to the wired network. Access points are base stations for the wireless network.

They're essentially hubs (or routers) operating in half-duplex mode — they can only receive or transmit at a given time; they can't do both at the same time (unless they have multiple antennas). Wireless access points use antennas to transmit and receive data. The four basic types of wireless antennas are

- *Omnidirectional:* The most common type of wireless antenna, omnidirectional antennas are essentially short poles that transmit and receive wireless signals with equal strength in all directions around a horizontal axis. Omnidirectional antennas often employ a dipole design.

- *Parabolic:* Also known as *dish antennas,* parabolic antennas are directional dish antennas made of meshed wire grid or solid metal. Parabolic antennas are used to extend wireless ranges over great distances.

- *Sectorized:* Similar in shape to omnidirectional antennas, sectorized antennas have reflectors that direct transmitted signals in a specific direction (usually, a 60- to 120-degree pattern) to provide additional range and decrease interference in a specific direction.

- *Yagi:* Similar in appearance to a small aerial TV antenna, yagi antennas are used for long distances in point-to-point or point-to-multipoint wireless applications.

Client devices in a Wi-Fi network include desktop and laptop PCs, as well as mobile devices and other endpoints (such as smartphones, medical devices, bar-code scanners, and many so-called smart devices such as thermostats and other home automation devices). Wireless network interface cards (or wireless cards) come in a variety of forms, such as PCI adapters, PC cards, and USB adapters or are built into wireless-enabled devices, such as laptop PCs, tablets, and smartphones.

Access points and the wireless cards that connect to them must use the same WLAN 802.11 standard or be backward-compatible. See the section "Wireless networks" earlier in this chapter for a list of the 802.11 specifications.

Access points (APs) can operate in one of four modes:

- *Root* (also *infrastructure*): The default configuration for most APs. The AP is directly connected to the wired network, and wireless clients access the wired network via the wireless access point.

- *Repeater* (also *stand-alone*): The AP doesn't connect directly to the wired network but provides an upstream link to another AP, effectively extending the range of the WLAN.

- *Bridge:* A rare configuration that isn't supported in most APs. Bridge mode is used to connect two separate wired network segments via a wireless access point.

- *Mesh:* Multiple APs work together to create the appearance of a single Wi-Fi network for larger homes and workspaces.

TIP

Ad hoc is a type of WLAN architecture that doesn't have any APs. The wireless devices communicate directly in a peer-to-peer network, such as between two notebook computers.

Physical Layer (Layer 1)

The Physical Layer sends and receives bits across the network medium (cabling or wireless links) from one device to another.

It specifies the electrical, mechanical, and functional requirements of the network, including network topology, cabling and connectors, and interface types, as well as the process for converting bits to electrical (or light) signals that can be transmitted across the physical medium. Various network topologies, made from copper or fiber-optic wires and cables, hubs, and other physical materials, comprise the Physical Layer.

NETWORK TOPOLOGIES

Four basic network topologies are defined at the Physical Layer. Although there are many variations of the basic types — such as FDDI, star-bus (or tree), and star-ring — we stick to the basics here:

>> **Star:** Each individual node on the network is connected directly to a switch, hub, or concentrator. All data communications must pass through the switch (or hub), which can become a bottleneck or single point of failure. A star topology is ideal for practically any size environment and is the most common basic topology in use today. A star topology is also easy to install and maintain, and network faults are easily isolated without affecting the rest of the network.

>> **Mesh:** All systems are interconnected to provide multiple paths to all other resources. In most networks, a partial mesh is implemented for only the most critical network components, such as routers, switches, and servers (by using multiple network interface cards or server clustering) to eliminate single points of failure.

>> **Ring:** A closed loop connects end devices in a continuous ring. Functionally, this topology is achieved by connecting individual devices to a Multistation Access Unit (MSAU or MAU). Physically, this setup gives the ring topology the appearance of a star topology. Ring topologies are common in token ring and FDDI networks. In a ring topology, all communication travels in a single direction around the ring.

>> **Bus:** In a bus (or linear bus) topology, all devices are connected to a single cable (the backbone) that's terminated on both ends. Bus networks were commonly used for very small networks because they were inexpensive and

easy to install. In large environments, they were impractical because the media had physical limitations (namely, the length of the cabling), the backbone was a single point of failure (a break anywhere on the network affects the entire network), and tracing a fault in a large network could be extremely difficult. Bus networks are extremely rare today and are no longer the least expensive or easiest to install network option.

CABLE AND CONNECTOR TYPES

Cables carry the electrical or light signals that represent data between devices on a network. Data signaling is described by several characteristics, including type (see the sidebar "Analog and digital signaling" in this chapter), control mechanism (see the sidebar "Asynchronous and synchronous communications," in this chapter), and classification (baseband or broadband). Baseband signaling uses a single channel for transmission of digital signals and is common in LANs that use twisted–pair cabling. Broadband signaling uses many channels over a range of frequencies for transmission of analog signals, including voice, video, and data. The four basic cable types used in networks include

>> **Coaxial cable:** Coaxial (abbreviated as coax and pronounced *KOH-axe*) cable consists of a single, solid-copper-wire core, surrounded by a plastic or Teflon insulator, braided-metal shielding, and (sometimes) a metal foil wrap, all covered with a plastic sheath. This construction makes the cable very durable and resistant to Electromagnetic Interference (EMI) and Radio Frequency Interference (RFI) signals. Coax cable is commonly used to connect cable or satellite television receivers (the cable that goes from the black box to the wall). Note that coax cable used for television signals is not compatible with coax cable used for computer networking due to different capacitance. Coax cable comes in two flavors, thick and thin:

   • *Thick:* Also known as *RG8* or *RG11* or *thicknet,* thick cable uses a screw-type connector, known as an Attachment Unit Interface (AUI).

   • *Thin:* Also known as *RG58* or *thinnet*, thin cable is typically connected to network devices by a Bayonet Neill-Concelman (BNC) connector.

>> **Twinaxial cable:** Twinaxial (also known as twinax) cable is very similar to coax cable, but it consists of two solid copper-wire cores rather than a single core. Twinax is used to achieve high data transmission speeds (such as 10 Gb Ethernet over very short distances such as 10 meters) at a relatively low cost. Typical applications for twinax cabling include SANs and top-of-rack network switches that connect critical servers to a high-speed core. Other advantages of twinax cabling include lower transceiver latency (delay in transmitter/receiver devices), low power consumption (compared with 10 GbE twisted-pair cables), and low bit error ratios.

Bit error ratio (BER) is the ratio of incorrectly received bits to total received bits over a specified period of time.

» **Twisted-pair cable:** Twisted-pair cable is the most popular LAN cable in use today. It's lightweight, flexible, inexpensive, and easy to install. One easily recognized example of twisted-pair cable is common telephone wire.

Twisted-pair cable consists of four copper-wire pairs that are twisted together to improve the transmission quality of the cable by reducing crosstalk and attenuation. The tighter the twisted pairs, the better the transmission speed and quality.

Crosstalk occurs when a signal transmitted over one channel or circuit negatively affects the signal transmitted over another channel or circuit. An (ancient) example of crosstalk occurred over analog phone lines when you could hear parts of other conversations over the phone. *Attenuation* is the gradual loss of intensity of a wave (such as electrical or light) while it travels over or through a medium.

Currently, 10 categories of twisted-pair cabling exist, but only Cat 5/5e, Cat 6/6a, and Cat 7/7a cable are typically used for networking today (see Table 6-9).

Twisted-pair cable can be unshielded (UTP) or shielded (STP). UTP cabling is more common because it's easier to work with and less expensive than STP. STP is used when noise is a problem or when security is a major concern and is popular in IBM rings. Noise is produced by external sources and can distort or otherwise impair the quality of a signal. Examples of noise include RFI and EMI from sources such as electrical motors, radio signals, fluorescent lights, microwave ovens, and electronic equipment. Shielded cabling also reduces electromagnetic emissions that may be intercepted by an attacker.

TEMPEST is a (previously classified) U.S. military term that refers to the study of electromagnetic emissions from computers and related equipment.

Twisted-pair cable is terminated with an RJ-type terminator. The three common types of RJ-type connectors are RJ-11, RJ-45, and RJ-49. Although these connectors are similar in appearance (particularly RJ-45 and RJ-49), only RJ-45 connectors are used for LANs. RJ-11 connectors are used for analog phone lines, and RJ-49 connectors are commonly used for ISDN lines and WAN interfaces.

» **Fiber-optic cable:** Fiber-optic cable is typically used in backbone networks and high-availability networks (such as FDDI). Fiber-optic cable carries data as light signals, rather than as electrical signals. The cable consists of a glass or plastic core or bundle, a glass insulator (commonly known as *cladding*), Kevlar fiber strands for strength, and a polyvinyl chloride (PVC) or Teflon outer sheath. Advantages of fiber-optic cable include high speeds, long distances, and resistance to interception and interference. Fiber-optic cable is terminated with an SC-type (square), ST-type (straight tip), or LC-type (Lucent, little, or local connector) connector. See Table 6-10 for a comparison of the various cable types and their characteristics.

TABLE 6-9   ## Common Twisted-Pair Cable Categories

Category	Use and Speed	Example
5 (not a TIA/EIA standard)	Data (up to 100 Mbps)	Fast Ethernet
5e	Data (up to 1000 Mbps at 100 MHz)	Gigabit Ethernet
6	Data (up to 1000 Mbps at 250 MHz)	Gigabit Ethernet
6a	Data (up to 10 Gbps at 500 MHz)	10 Gigabit Ethernet
7	Data (up to 10 Gbps at 600 MHz up to 100 meters)	10 Gigabit Ethernet
7a	Data (up to 100 Gbps at 1000 MHz up to 15 meters)	40 Gigabit Ethernet

TABLE 6-10   ## Cable Types and Characteristics

Cable Type	Ethernet Designation	Maximum Length	EMI/RFI Resistance
RG58 (thinnet)	10Base-2	185 m	Good
RG8/11 (thicknet)	10Base-5	500 m	Better
UTP	10Base-T 100Base-TX 1000Base-T 10GbE	100 m	Poor
STP	10Base-T 100Base-TX 1000Base-T 10GbE	100 m	Fair to good
Fiber-optic	100Base-F	2,000 m	Best (EFI and RFI have no effect on fiber-optic cable)

TECHNICAL STUFF

Ethernet designations, such as 10Base-T and 100Base-TX, refer to the capacity of the cable and the signaling type (baseband). The last part of the designation is less strictly defined. It may refer to the approximate maximum length (as in 10Base-2 and 10Base-5), the type of connector (as in 10Base-T, 100Base-TX, and 100Base-F), or the type and speed of the connector (as in 1000Base-T/GbE).

ANALOG AND DIGITAL SIGNALING

Analog signaling conveys information through a continuous signal by using variations of wave amplitude, frequency, and phase.

Digital signaling conveys information in pulses through the presence or absence (on-off) of electrical signals.

INTERFACE TYPES

The interface between the Data Terminal Equipment (DTE) and Data Communications Equipment (DCE), which we discuss in the following section, is specified at the Physical Layer.

REMEMBER

Network topologies, cable and connector types, and interfaces are defined at the Physical Layer of the OSI model.

Common interface standards include

>> **EIA/TIA-232-F:** This standard (formerly known as RS-232) supports circuits at signal speeds of up to 115,200 bits per second.

>> **V.24. ITU-T:** This standard is essentially the same as the EIA/TIA-232 standard.

>> **V.35. ITU-T:** This standard describes a synchronous communications protocol between network access devices and a packet network that supports speeds of up to 48 Kbps.

>> **X.21bis. ITU-T:** This standard (formerly known as CCITT) defines the communications protocol between DCE and DTE in an X.25 network. It's essentially the same as the EIA/TIA-232 standard.

>> **High-Speed Serial Interface (HSSI):** This network standard was developed to address the need for high-speed (up to 52 Mbps) serial connections over WAN links.

NETWORKING EQUIPMENT

Networking devices that operate at the Physical Layer include network interface cards (NICs), network media (cabling, connectors, and interfaces, all of which we discuss in the section "Cable and connector types" earlier in this chapter), repeaters, and hubs.

NICs are used to connect a computer to the network. NICs may be integrated on a computer motherboard or installed as an adapter card, such as an ISA, PCI, or PC card. Similar to a NIC, a WAN interface card (WIC) contains a built-in CSU/DSU and is used to connect a router to a digital circuit. Variations of WICs include high-speed WAN interface cards and voice WAN interface cards.

A *repeater* is a nonintelligent device that simply amplifies a signal to compensate for attenuation (signal loss) so that one can extend the length of the cable segment.

A *hub* or concentrator is used to connect multiple LAN devices, such as servers and workstations. The two basic types of hubs are

>> **Passive:** Data enters one port and exits all other ports without any signal amplification or regeneration.

>> **Active:** This hub combines the features of a passive hub and repeater, and is also known as a multiport repeater.

A *switch* is used to connect multiple LAN devices. Unlike a hub, a switch doesn't send outgoing packets to all devices on the network; it sends packets only to actual destination devices.

The TCP/IP Model

The Transmission Control Protocol/Internet Protocol (TCP/IP) Model is similar to the OSI Reference Model. It was originally developed by the U.S. Department of Defense and actually preceded the OSI model. The TCP/IP Model is not as widely used as a learning and troubleshooting tool as the OSI model. The most notable difference between the models is that the TCP/IP model consists of only four layers rather than seven (see Figure 6-5):

>> **Application Layer:** Consists of network applications and processes, and loosely corresponds to the upper layers of the OSI model (Application, Presentation, and Session layers)

>> **Transport Layer:** Provides end-to-end delivery and corresponds to the OSI Transport Layer

>> **Internet Layer:** Defines the IP datagram and routing, and corresponds to the OSI Network Layer

>> **Network Access (or Link) Layer:** Contains routines for accessing physical networks, and corresponds to the OSI Data Link and Physical layers

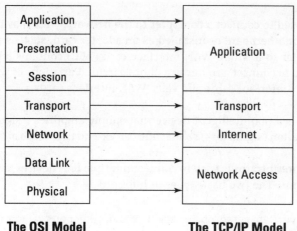

FIGURE 6-5:
Comparing the
OSI model and
the TCP/IP Model.

The OSI Model   **The TCP/IP Model**

© *John Wiley & Sons, Inc.*

Secure Network Components

Network equipment such as routers, switches, wireless access points, and other network components must be securely operated and maintained. The CISSP candidate must understand general security principles and unique security considerations associated with different types of network equipment.

CROSS REFERENCE

This section covers Objective 4.2 of the Communication and Network Security domain in the CISSP Exam Outline (May 1, 2021).

TIP

The CISSP exam doesn't test your knowledge of specific security products, such as how to configure a Cisco router or Checkpoint firewall. Instead, you need to understand security functions, fundamentals, and concepts specifically related to different types of network equipment in general.

Operation of hardware

Network equipment such as routers and switches (discussed earlier in this chapter), as well as firewalls, intrusion detection systems, wireless access points, and other components (discussed in the following sections) must be securely deployed, operated, and maintained. Aspects of proper operation of hardware include

>> **Training:** Personnel who deploy and manage hardware devices should receive proper training on the management of those devices.

- **Procedures:** Routine actions taken on hardware devices should be formally documented so that personnel will perform them consistently.

- **Standards:** The organization should establish standards for the secure and consistent configuration of hardware devices to ensure that multiple devices will be similarly (if not identically) configured and that such configuration will not compromise the security of the organization's environment.

- **Monitoring:** The organization should monitor its hardware devices so that appropriate personnel are informed of security incidents, malfunctions, and other notable events.

- **Managed change:** Configuration changes, software updates, and security patching on hardware devices should be made through the organization's change-management processes.

- **Support:** Organizations should ensure that the hardware it uses and the software that runs on that hardware are supported by their manufacturers. Having current support ensures that security and reliability fixes are still produced and can be applied. Application of these fixes (and other changes) should be controlled by a change-management process, which is discussed in Chapter 9.

- **Power and environment:** Information processing hardware is fairly finicky about electric power, temperature, and humidity. Backup power and power conditioners ensure that power is clean and continuous, which leads to improved uptime and long life. These topics are discussed in detail in Chapter 5.

Transmission media

Network transmission media includes wired (such as copper and fiber) and wireless. Wired transmission media is defined at the Physical Layer of the OSI model (discussed previously in this chapter). Wireless transmission media is defined at the Data Link Layer of the OSI model (discussed previously in this chapter). Additionally, the CISSP candidate must understand Wi-Fi security techniques and protocols.

Protecting wired networks

Aside from the use of encryption to render any intercepted communications unreadable by unauthorized parties, it's important to protect communication media from eavesdropping and sabotage. Techniques available to protect wired network media include

>> **Conduit:** Running communications cabling through conduit is a great way to make wiring more difficult to access.

>> **Physical access control:** Where communications cabling passes through rooms and corridors, to the greatest extent possible, physical access controls should be used so that only authorized personnel are permitted to get near any cabling. Controls may include key-card access systems, locking cabinets, and video surveillance.

Protecting Wi-Fi networks

Security on wireless networks, as with all security, is best implemented by using a defense-in-depth approach. Security techniques and protocols include broadcast of SSIDs, authentication, and encryption using Wi-Fi Protected Access (WPA).

TECHNICAL STUFF

An SSID is a name (up to 32 characters) that uniquely identifies a wireless network. A wireless client must know the SSID to connect to the WLAN. Most APs broadcast their SSID (or the SSID can be easily sniffed), however, so the security provided by an SSID is largely inconsequential.

WPA2 and WPA3 provide significant security enhancements over Wired Equivalent Privacy (WEP) and WPA. WPA2 supports various EAP extensions (see the section "Remote access" later in this chapter) to enhance WLAN security. These extensions include EAP-TLS (Transport Layer Security), EAP-TTLS (Tunneled Transport Layer Security), and Protected EAP (PEAPv0 and v1). WPA2 uses the AES-based algorithm Counter Mode with Cipher Block Chaining Message Authentication Code Protocol (CCMP), which replaces TKIP and WEP to produce a WLAN protocol that is far more secure. WPA3 requires either AES-128 or CCMP-128 as the minimum encryption algorithm in WPA3-Personal Mode.

REMEMBER

WEP and WPA are weak protocols that have long been deprecated and should never be used in a Wi-Fi network.

Network access control devices

Network access control devices include firewalls (as well as proxies, web application firewalls, next-generation firewalls, and unified threat management), intrusion detection systems (and intrusion prevention systems), web content filters, data loss prevention, and cloud access security brokers.

Firewalls and firewall types

A *firewall* controls traffic flow between a trusted network (such as a home network or corporate LAN) and an untrusted or public network (such as the Internet), known as *north-south traffic*. A firewall can comprise hardware, software, or a combination of both hardware and software.

The three basic classifications of firewalls are packet-filtering, circuit-level gateway, and application-level gateway. Web application firewalls (WAFs), next-generation firewalls (NGFWs), and unified threat management (UTM) platforms are specialized firewall types.

REMEMBER

Three basic types of firewalls are packet-filtering, circuit-level gateway, and application-level gateway.

PACKET-FILTERING

A packet-filtering firewall (or screening router), one of the most basic (and inexpensive) types of firewalls, is ideally suited to a low-risk environment. A packet-filtering firewall permits or denies traffic based solely on the TCP, UDP, ICMP, and IP headers of the individual packets. It examines the traffic direction (inbound or outbound), the source and destination IP addresses, and the source and destination TCP or UDP port numbers. This information is compared with predefined rules that have been configured in an access control list (ACL) to determine whether each packet should be permitted or denied. A packet-filtering firewall typically operates at the Network Layer or Transport Layer of the OSI model. Some advantages of a packet-filtering firewall are

>> It's inexpensive (can be implemented as a router ACL, which is free [the ACL, not the router!]).

>> It's fast and flexible.

>> It's transparent to users.

Disadvantages of packet-filtering firewalls are

>> Access decisions are based only on address and port information rather than more sophisticated information such as the packet's content, context, or application.

>> It has no protection from IP or DNS address spoofing (forged addresses).

>> It doesn't support strong user authentication.

>> Configuring and maintaining ACLs can be difficult.

>> Logging information may be limited.

A more advanced variation of the packet-filtering firewall is the dynamic packet-filtering firewall. This type of firewall supports dynamic modification of the firewall rule base by using context-based access control (CBAC) or reflexive ACLs, both of which create dynamic access list rules for individual sessions as they are established. An ACL might be created automatically to allow a user working from the corporate network (inside the firewall) to connect to an FTP server outside the firewall to upload and download files between her PC and the FTP server. When the file transfer is complete, the ACL is deleted from the firewall automatically.

CIRCUIT-LEVEL GATEWAY

A circuit-level gateway controls access by maintaining state information about established connections. When a permitted connection is established between two hosts, a tunnel (or virtual circuit) is created for the session, allowing packets to flow freely between the two hosts without the need for further inspection of individual packets. This type of firewall operates at the Session Layer (Layer 5) of the OSI model.

Advantages of this type of firewall include

>> Speed (because after a connection is established, individual packets aren't analyzed)
>> Support for many protocols
>> Easy maintenance

Disadvantages of this type of firewall include

>> Dependence on the trustworthiness of the communicating users or hosts. (After a connection is established, individual packets aren't analyzed.)
>> Limited logging information about individual data packets is available after the initial connection is established.

A stateful inspection firewall is a type of circuit-level gateway that captures data packets at the Network Layer and then queues and analyzes these packets at the upper layers of the OSI model.

APPLICATION-LEVEL GATEWAY

An application-level (or Application Layer) gateway operates at the Application Layer of the OSI model, processing data packets for specific IP applications. This type of firewall is generally considered to be the most secure and is commonly implemented as a proxy server. In a proxy server, no direct communication between two hosts is permitted. Instead, data packets are intercepted by the proxy server, which analyzes the packet's contents and — if permitted by the firewall rules — sends a copy of the original packet to the intended host.

Advantages of this type of firewall include

>> Data packets aren't transmitted directly to communicating hosts — a tactic that masks the internal network's design and prevents direct access to services on internal hosts.

>> It can be used to implement strong user authentication in applications.

Disadvantages of this type of firewall include

>> It reduces network performance because packets must be passed up to the Application Layer of the OSI model to be analyzed.

>> It must be tailored to specific applications. (Such customization can be difficult to maintain or update for new or changing protocols.)

WEB APPLICATION FIREWALL

A WAF is used to protect a web server (or group of web servers) from various types of web application attacks such as script injection and buffer overflow attacks. A WAF examines HTTP traffic at the Application Layer before it reaches the web server and employs rules to determine whether the traffic is considered routine and friendly or hostile.

NEXT-GENERATION FIREWALLS AND UNIFIED THREAT MANAGEMENT DEVICES

Next-generation firewalls (often termed *next-gen firewalls* or NGFWs) and unified threat management devices (often called UTMs) are similar terms describing firewalls with multiple functions, including combinations of the following security devices:

>> Firewall (of course!)

>> IDS/IPS (discussed later in this chapter)

>> VPN (discussed earlier in this chapter)

>> Web content filtering (discussed later in this chapter)

>> Cloud access security broker (CASB) (discussed later in this chapter)

>> Web application firewall (WAF) (discussed earlier in this chapter)

>> Data loss prevention (DLP) (discussed later in this chapter)

REMEMBER

The main advantage of next-gen firewalls and UTM is greater simplicity. Rather than having to manage many separate security systems, you can perform all of these security functions are performed within a single device.

Firewall architectures

The basic firewall types that we discuss in the preceding sections may be implemented by using one of the firewall architectures described in the following sections. The four basic types of firewall architectures are screening router, dual-homed gateway, screened-host gateway, and screened subnet.

SCREENING ROUTER

A *screening router* is the most basic type of firewall architecture employed. An external router is placed between the untrusted and trusted networks, and a security policy is implemented by using ACLs. Although a router functions as a choke point between a trusted network and an untrusted network, an attacker — after gaining access to a host on the trusted network — may be able to compromise the entire network.

Advantages of a screening router architecture include

>> It's completely transparent.

>> It's relatively simple to use and inexpensive.

Disadvantages of the screening router architecture include

>> It may have difficulty handling certain traffic.

>> It has limited or no logging.

>> It doesn't employ user authentication.

>> It makes masking the internal network structure difficult.

>> It has a single point of failure.

>> It doesn't truly implement a firewall choke-point strategy because it isn't truly a firewall or a choke point, but a router that passes traffic between two networks (private and public).

Still, using a screening router architecture is better than using nothing.

DUAL-HOMED GATEWAYS

Another common firewall architecture is the *dual-homed gateway,* a system that has two network interfaces and sits between an untrusted network and a trusted network. *Bastion host* is a general term often used to refer to proxies, gateways, firewalls, or any server that provides applications or services directly to an untrusted network. Because it's often the target of attackers, a bastion host is sometimes referred to as a *sacrificial lamb.* This term is misleading, however, because a bastion host is typically a hardened system that employs robust security mechanisms.

A dual-homed gateway is often connected to the untrusted network via an external screening router. The dual-homed gateway functions as a proxy server for the trusted network and may be configured to require user authentication. A dual-homed gateway offers a more fail-safe operation than a screening router does because by default, data normally isn't forwarded across the two interfaces.

Advantages of the dual-homed gateway architecture include

>> It operates in fail-safe mode. If it fails, it allows no access.

>> Internal network structure is masked.

Disadvantages of the dual-homed gateway architecture include

>> It may inconvenience users by requiring them to authenticate to a proxy server.

>> It may introduce latency in the network.

>> Proxies may not be available for some services.

>> It may cause slower network performance.

>> It increases cost (adding a device to the architecture).

SCREENED-HOST GATEWAYS

A *screened-host gateway* architecture employs an external screening router and an internal bastion host. The screening router is configured so that the bastion host is the only host accessible from the untrusted network (such as the Internet). The bastion host provides any required web services to the untrusted network, such as HTTP and FTP, as permitted by the security policy. Connections to the Internet from the trusted network are routed via an application proxy on the bastion host or directly through the screening router.

Here are some of the advantages of the screened-host gateway:

>> It provides distributed security between two devices rather than relying on a single device to perform all security functions.

>> It has transparent outbound access.

>> It has restricted inbound access.

Here are some disadvantages of the screened-host gateway:

>> It's considered to be less secure because the screening router can bypass the bastion host for certain trusted services.

>> Masking the internal network structure is difficult.

>> It can have multiple single points of failure (the router or bastion host).

>> It increases cost (adding a device to the architecture).

SCREENED SUBNET

Screened subnet is perhaps the most secure of the current firewall architectures. The screened subnet employs an external screening router, a dual-homed (or multi-homed) host, and a second internal screening router. This firewall type implements the concept of a network demilitarized zone (DMZ). Publicly available services are placed on bastion hosts in the DMZ.

Advantages of the screened-subnet architecture include

>> It's transparent to end users.

>> It's flexible.

>> Internal network structure can be masked.

>> It provides defense in depth instead of relying on a single device to provide security for the entire network.

Disadvantages of a screened-subnet architecture include

>> It's more expensive.

>> It's more difficult to configure and maintain.

>> It can be more difficult to troubleshoot.

MICROSEGMENTATION

Microsegmentation refers to techniques used to isolate groups of systems or individual systems in a network to further protect them from attack. Micro-segmentation can be implemented with OS firewalls, or with routers, switches, or firewalls configured to limit east-west traffic between systems.

Intrusion detection and prevention systems

Intrusion detection is defined as real-time monitoring and analysis of network activity and data for potential vulnerabilities and attacks in progress. One major limitation of current intrusion detection system (IDS) technologies is the requirement to filter false alarms to prevent the operator (the system or security administrator) from being overwhelmed with data. IDSes are classified in many ways, including active and passive, network-based and host-based, and knowledge-based and behavior-based.

IDSes and intrusion prevention systems (IPS) are sometimes referred to as intrusion detection and prevention systems (IDPS).

TIP

ACTIVE AND PASSIVE IDS

An active IDS is configured to block suspected attacks in progress automatically, without requiring any intervention by an operator. IPS has the advantage of providing real-time corrective action in response to an attack, but it has many disadvantages as well. An IPS must be placed inline along a network boundary; thus, the IPS itself is susceptible to attack. Also, if false alarms and legitimate traffic haven't been properly identified and filtered, authorized users and applications may be improperly denied access. Finally, the IPS itself may be used to effect a denial-of-service attack, which involves intentionally flooding the system with alarms that cause it to block connections until no connection or bandwidth is available.

A *passive* IDS is configured to monitor and analyze network traffic activity and alert an operator to potential vulnerabilities and attacks. It can't perform any protective or corrective functions on its own. The major advantages are that these systems can be easily and rapidly deployed and aren't normally susceptible to attack themselves. Passive IDS is usually connected to a network segment via a tap (physical or virtual) or switched port analyzer (SPAN) port.

IDS is a passive system that monitors, analyzes, and alerts. IPS is an active system that monitors, analyzes, and acts (such as alert, block, or drop).

REMEMBER

NETWORK-BASED AND HOST-BASED IDs

A *network-based IDS* (NIDS) usually consists of a network appliance (or sensor) that includes a NIC operating in Promiscuous mode (meaning that it listens to, or "sniffs," all traffic on the network, not just traffic addressed to a specific host) and a separate management interface. The IDS is placed along a network segment or boundary, and it monitors all traffic on that segment.

A *host-based IDS* (HIDS) requires small programs (or agents) to be installed on the individual systems that are to be monitored. The agents monitor the operating system and write data to log files and/or trigger alarms. A host-based IDS can monitor only the individual host systems on which the agents are installed; it doesn't monitor the entire network.

KNOWLEDGE-BASED AND BEHAVIOR-BASED IDS

A *knowledge-based* (or *signature-based*) IDS references a database of previous attack profiles and known system vulnerabilities to identify active intrusion attempts. Knowledge-based IDSes are more common than behavior-based IDSes. Advantages of knowledge-based systems include

» They have lower false-alarm rates than behavior-based IDSes.

» Alarms are more standardized and more easily understood than behavior-based IDS alarms.

Disadvantages of knowledge-based systems include

» The signature database must be continually updated and maintained.

» New, unique, or original attacks may not be detected or may be improperly classified.

A *behavior-based* (or *statistical anomaly-based*) IDS references a baseline or learned pattern of normal system activity to identify active intrusion attempts. Deviations from this baseline or pattern cause an alarm to be triggered. Advantages of behavior-based systems include

» They adapt dynamically to new, unique, or original attacks.

» They are less dependent on identifying specific operating system vulnerabilities than knowledge-based IDSes are.

Disadvantages of behavior-based systems include

>> They have higher false alarm rates than knowledge-based IDSs.

>> They cannot adapt to legitimate usage patterns that may change often and therefore aren't static enough to implement an effective behavior-based IDS.

Web content filters

A web content filter typically is an inline device that monitors and controls internal users' access to websites. Web content filters can be configured to block access to both specific websites and categories of websites (blocking access to sites that discuss polka music, for example).

Organizations that use web content filters to block access to categories of web sites are often trying to keep employees from accessing sites that are not related to work. The use of web content filters also helps enforce policies and protect the organization from potential liability. Blocking access to pornographic and hate-related websites, for example, to enforce sexual harassment and racial discrimination/safe working environment policies, and can help demonstrate due diligence.

Web content filters typically employ large databases of websites that constantly evaluated and updated by the vendor of the content filtering software. These databases often contain errors in classification, which require policies and procedures for employees to request access to legitimate websites or access to blocked websites for legitimate work purposes. These processes can be frustrating for employees, particularly if it takes more than a few minutes for the security team to respond to the request. An alternative policy that many organizations use is "Trust but verify." Websites are not blocked, but users are warned before navigating to a potentially suspicious, dangerous, offensive, or otherwise inappropriate website; also, each user must positively acknowledge that they understand the risk and that they are visiting the site for a legitimate purpose. The website visit is logged and reported. Typically, appropriate security or human resources personnel will follow up with the employee if necessary.

Tech-savvy users often use various proxy software programs in an attempt to circumvent web content filters. Proxy software is a significant risk to enterprise security and should be explicitly forbidden by policy. Next-generation firewalls and certain advanced web content filters are capable of detecting proxy software in the enterprise.

Data loss prevention

Data loss prevention (DLP) refers to a class of security products that are designed to detect and (optionally) prevent the exfiltration of sensitive data over an organization's network connections. DLP systems work by performing pattern matching (such as XXX–XX–XXXX representing a Social Security Number, or XXXX XXXX XXXX XXXX representing a credit card number) against data transmitted over the network. Depending on the type of DLP system and its configuration, the system can either generate an alert describing the suspected data exfiltration or block the transmission.

Another class of DLP products scans file servers and database management systems in search of sensitive data. The idea is that people sometimes extract sensitive data from sanctioned repositories and then make copies of that data for storage in less-secure locations.

Cloud access security brokers

Cloud access security brokers (CASB) monitor and control access to cloud-based applications and services. If an organization uses Box.com for unstructured file storage, for example, CASB can be configured to block access to alternative storage services such as Dropbox and Skydrive.

Organizations generally use CASB to limit the exfiltration of sensitive information and steer personnel to officially sanctioned applications. You can think of them as being security policy enforcement points.

Endpoint security

Endpoints, including desktop and laptop computers, smartphones, tablets, and other mobile equipment (such as medical devices, bar-code scanners, and other so-called smart devices), have become very attractive targets for cybercriminals. Endpoints are particularly vulnerable to attack for many reasons, including

>> **Number and variety:** The sheer number and variety of endpoints on the network creates numerous opportunities for an attacker to exploit vulnerabilities in different operating systems and applications. Keeping all endpoints patched properly and in a timely manner is also a challenge.

>> **Users:** Endpoints are operated by users with varying computer skill levels and awareness of security and privacy issues. Users are susceptible to social engineering, and many of them willingly circumvent security measures on endpoints for the sake of convenience (such as rooting a smartphone to install free or unauthorized apps).

>> **Privilege:** In some organizations, user accounts have the role of local administrator. This role means that any action carried out by a user is executed at the highest level of privilege on the system, which can make malware attacks on endpoints far more potent.

>> **Prioritization:** Endpoints are often treated as being lower-value assets in the network. For this reason, security efforts typically focus on the data center and higher-value assets such as servers and databases.

At its most basic level, endpoint security consists of antimalware (or antivirus) software. *Signature-based* software is the most common type of antivirus software used on endpoints. Signature-based antivirus software scans an endpoint's hard drive and memory in real time and at scheduled times. If a known malware signature is detected, the software performs an action, such as the following:

>> **Quarantine:** Isolates the infected file on the endpoint so that it can't infect other files

>> **Delete:** Removes the infected file

>> **Alert:** Notifies the user (and/or security administrator) that malware has been detected

Signature-based antivirus software must be kept up to date to be effective, and it can detect only known threats. The endpoint is vulnerable to any new zero-day malware threats until a signature is created by the software vendor and uploaded to the endpoint.

Application whitelisting is another common antimalware approach used for endpoint protection. This approach requires a positive control model on the endpoint; only applications that have been explicitly authorized can be run on the endpoint. Trends such as Bring Your Own Device (BYOD) that allow users to use their personal devices for work-related purposes make application whitelisting approaches difficult to implement in the enterprise. Another limitation of application whitelisting is that an application (such as Microsoft Word or Adobe Acrobat) that has already been whitelisted can be run on an endpoint, even if that application is exploited (perhaps with a malicious Word document or Adobe PDF file).

Behavior-based (also known as *heuristics-based* or *anomaly-based)* endpoint protection attempts to create a baseline of normal activity on the endpoint. Any unusual activity (as determined by the baseline) is detected and stopped. Unfortunately, behavior-based software is prone to high false positives and typically requires significant computing resources.

Container-based endpoint protection isolates any vulnerable processes running on an endpoint by creating virtual barriers around individual processes. If a malicious process is detected, the software kills the process before the malicious process can infect any legitimate processes on the endpoint. Container-based approaches typically require significant computing resources and extensive knowledge of any applications running on the endpoint.

In addition to antimalware prevention, endpoint protection should include

- » **Access controls:** Access controls should be enabled and enforced on all endpoints, including smartphones and tablets (such as PINs, passwords, passphrases, swipe patterns, and biometrics such as fingerprint and facial recognition).

- » **Automatic lockout:** Endpoints should be configured to lock automatically after a few minutes of nonactivity so that others will not be able to use them.

- » **Encryption:** Drive encryption should be enabled to protect data on the endpoint device.

- » **Firewalls:** An OS-based or third-party firewall should be installed and configured on each endpoint.

- » **Patch management:** Applications and the endpoint OS must be kept patched and up to date.

- » **Host-based intrusion prevention systems (HIPS):** Some organizations deploy HIPS on endpoints to provide additional protection.

- » **Network control:.** Network controls include next-generation firewalls (that can identify and authenticate endpoints and users), VPNs, IPSes, and network segmentation.

- » **Administrative control:** Endpoints should be configured so that any firmware settings and boot control is password-protected. Endpoint operating systems should be configured so that users are not local administrators.

- » **Physical security:** Endpoints should be protected from unauthorized access and theft. Endpoints should not be left unattended; they should be locked so that they cannot be accessed by unauthorized personnel and physically tethered or locked so that they cannot be stolen.

Implement Secure Communication Channels According to Design

The CISSP exam requires knowledge of secure design principles and implementation of various communication technologies, including voice, multimedia collaboration, remote access, data communications, virtualized networks, and third-party connectivity.

CROSS REFERENCE

This section covers Objective 4.3 of the Communication and Network Security domain in the CISSP Exam Outline (May 1, 2021).

Voice

Private Branch Exchange (PBX) switches, Plain Old Telephone Systems (POTS), and Voice over Internet Protocol (VoIP) switches are some of the most overlooked and costly aspects of a corporate telecommunications infrastructure. Many employees don't think twice about using a company telephone system for extended personal use, including long-distance calls. Personal use of company-supplied mobile phones is another area of widespread abuse. Perhaps the simplest and most effective countermeasure against internal abuse is publishing and enforcing a corporate telephone-use policy. Regular auditing of telephone records is also effective for deterring and detecting telephone abuse. Similarly, as both voice communications and the global workforce have become increasingly mobile, organizations need to define and implement appropriate BYOD, Choose Your Own Device, or corporate-owned, personally-enabled mobile device policies.

Cloud communication has become a viable alternative to PBX and on-premises VoIP systems for many organizations, from small and midsize businesses to large enterprises. Many cloud communication providers offer the same advanced features and functionality as on-premises PBX and VoIP systems, with all the business and technical benefits of the cloud.

Similarly, over-the-top services such as Jabber, Vonage, Vimeo, and Zoom are increasingly common in business communications.

Finally, mobile operators have introduced innovations such as Voice over Long-Term Evolution (VoLTE), Voice over Wi-Fi (VoWiFi), and Wi-Fi calling, providing improved voice communications capabilities.

Types of attacks on voice communications systems include

>> **Identify fraud,** such as caller ID spoofing, eavesdropping, and vishing.

>> **Toll fraud,** such as number harvesting, call hijacking, spam over Internet telephony (SPIT), spam over instant messaging (SPIM), and voice over misconfigured Internet telephones (VOMIT).

>> **Eavesdropping,** whereby an attacker uses techniques to intercept and monitor communications messages.

>> **Denial of service,** such as distributed denial of service and telephony denial of service attacks.

Multimedia collaboration

Multimedia collaboration includes remote meeting software, certain VoIP applications, and instant messaging, among others.

Remote meeting (such as Cisco WebEx, Microsoft Teams, and Zoom) software has become immensely popular and enables rich collaboration over the Internet. Potential security issues associated with remote meeting software include downloading and installing potentially vulnerable add-on components or other required software. Other security issues arise from the inherent capabilities of remote meeting software, such as remote desktop control, file sharing, sound, and video. An unauthorized user who connects to an endpoint via remote meeting software could potentially have access to all of these capabilities.

Instant messaging (IM) applications enable simple, convenient communications within an organization and can boost productivity significantly. IM has long been a favorite attack vector for cybercriminals, however. Users need to be aware that IM is no more secure than any other communication method. Communications can be intercepted (IMs are rarely encrypted), and malware can be spread via these messages.

Remote access

Remote access to corporate networks has become more ubiquitous over the past decade. Such trends such as telecommuting and mobile computing blur the distinction between work and personal lives for many people. Safely enabling ubiquitous access to corporate network resources from any device requires extensive knowledge of various remote access security methods, protocols, and technologies.

Remote access security methods

Remote access security methods include restricted allowed addresses, geolocation, caller ID, callback, and multifactor authentication:

>> **Restricted address:** The restricted address method blocks access to the network based on allowed IP addresses, essentially performing rudimentary node authentication but not user authentication.

>> **Geolocation:** This method blocks access based on the geographic location of the user. This method is a useful countermeasure in the case of theft of remote access credentials.

>> **Caller ID:** The caller ID method restricts access to the network based on allowed phone numbers, thus performing a slightly more secure form of node authentication because phone numbers are more difficult to spoof than IP addresses. This method can be difficult to administer for road warriors who routinely travel to different cities, however.

>> **Callback:** The callback method restricts access to the network by requiring a remote user to first authenticate to the remote access service (RAS) server. The RAS server disconnects and calls the user back at a preconfigured phone number. As with caller ID, this method can be difficult to administer for road warriors.

>> **Multifactor authentication:** Requiring users to authenticate with a user ID and password, plus an additional factor such as a one-time passcode (which may be sent to a mobile device via SMS text message), token, or biometric, reduces the risk of compromised login credentials.

REMEMBER

One limitation of callback is that it can be easily defeated by call forwarding.

Remote access security

Remote access security technologies include RAS servers that use various authentication protocols associated with PPP, RADIUS, and TACACS:

>> **RAS:** Remote access service (RAS) servers use PPP to encapsulate IP packets and establish dial-in connections over serial and ISDN links. PPP incorporates the following three authentication protocols:

- *PAP:* Password Authentication Protocol (PAP) uses a two-way handshake to authenticate a peer to a server when a link is initially established. PAP transmits passwords in cleartext and provides no protection from replay or brute-force attacks.

- *CHAP:* Challenge Handshake Protocol (CHAP) uses a three-way handshake to authenticate both a peer and a server when a link is established and, optionally, at regular intervals throughout the session. CHAP requires both the peer and the server to be preconfigured with a shared secret that must be stored in cleartext. The peer uses the secret to calculate the response to a server challenge by using an MD5 one-way hash function. MS-CHAP, a Microsoft enhancement to CHAP, allows the shared secret to be stored in encrypted form.

- *EAP:* Extensible Authentication Protocol (EAP) adds flexibility to PPP authentication by implementing various authentication mechanisms, including MD5-challenge, S/Key, generic token card, and digital certificates. EAP is implemented on many wireless networks.

» **RADIUS:** The Remote Authentication Dial-In User Service (RADIUS) protocol is an open-source, UDP-based (usually ports 1812 and 1813, and sometimes ports 1645 and 1646), client-server protocol, which provides authentication and accountability. A user provides username/password information to a RADIUS client by using PAP or CHAP.

The RADIUS client encrypts the password and sends the username and encrypted password to the RADIUS server for authentication.

Note: Passwords exchanged between the RADIUS client and the RADIUS server are encrypted, but passwords exchanged between the PC client and the RADIUS client aren't necessarily encrypted, such as when PAP authentication is used. If the PC client happens to also be the RADIUS client, however, all password exchanges are encrypted, regardless of the authentication protocol being used.

RADIUS is an authentication, authorization, and accounting (AAA) protocol that manages access in an AAA transaction.

TECHNICAL STUFF

» **Diameter:** The Diameter protocol is the next-generation RADIUS protocol. Diameter overcomes several RADIUS shortcomings. It uses TCP rather than UDP, supports IPsec or TLS, and has a larger address space than RADIUS.

» **TACACS:** The Terminal Access Controller Access Control System (TACACS) is a UDP-based access control protocol (originally developed for the MILNET), which provides AAA. The original TACACS protocol has been significantly enhanced, primarily by Cisco, as XTACACS (no longer used) and TACACS+ (the most common implementation of TACACS). TACACS+ is TCP-based (port 49) and supports practically any authentication mechanism (PAP, CHAP, MS-CHAP, EAP, token cards, Kerberos, and so on). The basic operation of TACACS+ is similar to RADIUS, including the caveat about encrypted passwords between client and server. The major advantages of TACACS+ are its wide support of various authentication mechanisms and granular control of authorization parameters.

A VPN creates a secure tunnel over a public network, such as the Internet. Encrypting the data as it's transmitted across the VPN creates a secure tunnel. The two ends of a VPN are commonly implemented by using one of the following methods:

>> Client-to-VPN-concentrator (or device)

>> Client-to-firewall

>> Firewall-to-firewall

>> Router-to-router

REMEMBER

Common VPN protocol standards include PPTP, L2F, L2TP, IPsec, and SSL.

POINT-TO-POINT TUNNELING PROTOCOL

PPTP was developed by Microsoft to enable PPP to be tunneled through a public network. PPTP uses native PPP authentication and encryption services (such as PAP, CHAP, and EAP). PPTP is commonly used for dial–up connections. PPTP operates at the Data Link Layer (Layer 2) of the OSI model and is designed for individual client–server connections.

LAYER 2 FORWARDING PROTOCOL

L2F was developed by Cisco and provides similar functionality to PPTP. As its name implies, L2F operates at the Data Link Layer of the OSI model and permits tunneling of Layer 2 WAN protocols such as HDLC and SLIP.

LAYER 2 TUNNELING PROTOCOL

L2TP is an IETF standard that combines Microsoft (and others') PPTP and Cisco L2F protocols. Like PPTP and L2F, L2TP operates at the Data Link Layer of the OSI model to create secure VPN connections for individual client–server connections. L2TP addresses the following end–user requirements:

>> **Transparency:** Requires no additional software.

>> **Robust authentication:** Supports PPP authentication protocols, Remote Authentication Dial-In User Service (RADIUS), Terminal Access Controller Access Control System (TACACS), smart cards, and one-time passwords.

>> **Local addressing:** The VPN entities, rather than the ISP, assign IP addresses.

>> **Authorization:** Authorization is managed by the VPN server-side, similar to direct dial-up connections.

>> **Accounting:** Both the ISP and the user perform AAA accounting.

SECURE SOCKETS LAYER/TRANSPORT LAYER SECURITY

The SSL protocol provides session-based encryption and authentication for secure communication between clients and servers on the Internet. SSL operates at the Transport Layer (Layer 4) of the OSI model. SSL has been deprecated and superseded by Transport Layer Security (TLS). As of August 2018, TLS 1.3 is the most current version of TLS. The terms SSL and TLS are often used interchangeably. However, when referring to most "SSL" implementations today (for example, SSL VPNs), TLS is typically the protocol that is actually used. SSL VPNs (using TLS) have rapidly gained widespread popularity and acceptance in recent years because of their ease of use and low cost. An SSL VPN requires no special client hardware or software (other than a web browser) and little or no client configuration. SSL VPNs provide secure access to web-enabled applications and thus are somewhat more granular in control, as a user is granted access to a specific application rather than to the entire private network. This granularity can also be a limitation of SSL VPNs; not all applications will work over an SSL VPN, and many convenient network functions (file and print sharing) may not be available over an SSL VPN.

Data communications

Network data communications are secured by means of several technologies and protocols.

Virtual LANs (VLANs) are used to logically segment a network, for example by department or resource. VLANs are configured on network switches and restrict VLAN access to devices that are connected to ports that are configured on the switch as VLAN members.

The SSL/TLS protocol (discussed in the preceding section) is commonly used to encrypt network communications.

Virtualized networks

Virtualization technology emulates physical computing resources, such as desktop computers and servers, processors, memory, storage, networking, and individual applications. The core component of virtualization technology is the hypervisor, which runs between a hardware kernel and an OS, and enables multiple guest virtual machines (VMs) to run on a single physical host machine.

Two common types of hypervisors are Type 1 (native or bare metal) hypervisors, which run directly on host hardware, and Type 2 (hosted) hypervisors, which run within an operating system environment.

In addition to virtualized servers, virtualization technology is used for

- **Containerization:** In the same manner in which a hypervisor can facilitate the use of multiple operating system instances, containerization facilitates the use of multiple application instances within a single operating system. Each application executes in a container, which is isolated from other containers. Containerization is useful in environments where applications are designed to be run by themselves in a running operating system.

- **Desktop and application virtualization:** Desktop virtualization is increasingly popular for remote desktop applications used in conjunction with VPN software. Application virtualization allows various use cases such as legacy applications that can't run on newer operating systems, multiple versions of the same application running on a desktop, and multiple versions of software components (such as Java) running on a desktop. An implementation of desktop virtualization is known as a virtual desktop infrastructure, or VDI.

- **Storage virtualization:** Storage virtualization enables storage administrators to manage enterprise storage space that uses commodity or standard off-the-shelf compute and storage hardware components, with storage management functions performed in the virtual software.

- **Network virtualization:** Network virtualization abstracts network functions (such as routing, switching, and traffic management) from the underlying hardware. Popular network virtualization technologies and capabilities include software-defined networks (SDN), software-defined wide area networks (SD-WAN), microsegmentation, virtual extensible LAN, and network functions virtualization.

Security in virtualized environments begins with the hypervisor. A compromised hypervisor can potentially give an attacker access to and control of an entire virtualized environment.

Operational security issues associated with virtualized environments include

- **VM and container sprawl:** Virtualization technology enables organizations to deploy VMs and containers in minutes rather than days or weeks, causing VMs and containers to proliferate in many data centers.

- **Guest operating systems:** All the various OSes and OS versions that exist in a virtualized environment need to be patched regularly and kept updated.

- **Dormant VMs:** VMs that are no longer needed are often turned off rather than deprovisioned. If a dormant VM is turned on later, it will be missing critical security patches and may be vulnerable to attack.

>> **Network visibility:** Most organizations begin their virtualization journey with virtualized servers, often installing multiple NICs in a single physical server; all network traffic flowing to and from the VMs on that server runs over the NICs. Without network virtualization, network administrators have limited visibility into this traffic for troubleshooting and security-monitoring purposes.

Third-party connectivity

Organizations frequently provide access for various third parties such as partners, contractors, and vendors. Third-party connectivity is an attack vector that is frequently exploited by threat actors, due to various vulnerabilities in these types of connections, such as

>> Weak VPN implementations or unprotected out-of-band management (that is, dialup) channels

>> Outdated and/or unpatched operating systems and software, particularly in maintenance systems and operational technology

>> Default admin credentials that have not been changed

>> Weak credentials that are infrequently changed, poorly managed throughout the account life cycle (not disabled when no longer needed, for example), and/or shared by numerous technicians

Examples of data breaches via third-party connectivity include

>> **Morgan Stanley (2021):** Attackers exploited a vulnerability in software used by a third-party vendor that provides account maintenance services to Morgan Stanley's corporate customers.

>> **Solar Winds (2020):** Attackers compromised the Solar Winds company's software publishing infrastructure to implant malware in Solar Winds software used by its customers.

>> **Quest Diagnostics (2018):** Although Quest Diagnostics' systems and networks were not directly compromised, sensitive data shared with a third-party payment and collections partner was breached.

>> **Target (2013):** Up to 100 million payment card records were stolen when attackers used stolen credentials used for remote maintenance by a heating, ventilation, and air conditioning vendor to breach the Target network.

» **Managing identification and authentication of people, devices, and services**

» **Federating identity with a third-party service**

» **Implementing and managing authorization mechanisms**

» **Managing the identity and access provisioning life cycle**

» **Implementing authentication systems**

Chapter **7**

Identity and Access Management

dentity and access management (IAM) is often the first — and sometimes the *only* — line of defense between adversaries and sensitive information. In fact, in the modern cloud era, with ubiquitous mobile computing and anywhere, anytime access to applications and data, many security practitioners now refer to identity as "the new perimeter." Security professionals must have a thorough understanding of the concepts and technologies involved. This domain represents 13 percent of the CISSP certification exam.

IAM is a collection of processes and technologies that are used to control access to critical assets. Together with other critical controls, IAM is part of the core of information security; when it's implemented correctly, unauthorized people are not permitted to access critical assets. Breaches and other abuses of information and assets are less likely to occur.

Control Physical and Logical Access to Assets

The purpose of IAM systems and processes is the management of access to information, systems, devices, and facilities. A variety of controls are used for this purpose in several contexts that are discussed in this section. Chapter 3 contains a discussion of the types and categories of controls.

CROSS REFERENCE

This section covers Objective 5.1 of the Identity and Access Management (IAM) domain in the CISSP Exam Outline (May 1, 2021).

Information

Controlling access to information assets is achieved primarily through logical controls that determine which people or systems (known as *subjects*) are permitted to access which files, directories, databases, tables, records, or fields (known as *objects*). The mechanisms used to control access to information include

>> **File- and directory-level permissions:** These permissions are typically managed at the operating-system level or within a file sharing system (such as a file server, SharePoint, or Box).

>> **Database table, view, field, and row permissions:** Usually managed within a database management system or a third-party tool, permissions can be granted at various levels.

Systems and devices

Controlling access to systems and devices is achieved mainly through mechanisms built into those systems, including

>> **Port-level access control:** At the network level, a system can be configured to accept incoming connection requests based upon their origin (such as IP address, IP network, or geographic region), as well as the port number.

>> **Console login:** A physical or logical console controls access to the system, generally based on the proven identity of the subject who wants to connect.

>> **Remote console login:** A system can be accessed via a remote console connection, which has the general appearance of a local, physical console but is accessed via a network. Again, access permission is based on the proven identity of the subject who wants to connect.

>> **Application programming interfaces (APIs):** A system or application can be accessed programmatically through an API, typically used by an application that needs to access data or functions.

Systems and devices include far more than servers and routers. Many kinds of business and consumer products are marketed as "smart" devices and equipped with Ethernet, Wi-Fi, and Bluetooth connectivity. When pondering systems and devices, be sure to include the vast array of things that are connected to networks, including the following:

>> **Industrial control systems:** This category includes systems that perform remote monitoring and control of utility infrastructure, including electric power and distribution, water supplies, and sewage treatment. Don't forget automated manufacturing, 3D printing, environmental systems, and voting machines.

>> **Medical devices:** These devices include equipment used in hospitals, such as patient monitoring and IV pumps, as well as things on or in our bodies, including insulin pumps and pacemakers.

>> **Wearables:** This category consists of watches, fitness devices, video glasses, and the like.

>> **Transportation:** This category includes automobiles, self-driving cars, drones, and satellites, as well as GPS navigation, autopilots, air traffic control, and more.

DEVICE SECURITY AND LIFE SAFETY

A staggering variety of smart devices is available today — smart automobiles, smart television sets, and smart appliances, among others. These devices include wearable and life safety products, such as those that monitor vital signs (heart rate, respiration, and so on), insulin pumps, IV pumps, patient monitoring, robotic surgery, pacemakers, autonomous vehicles, and aircraft navigation and control.

Security experts have observed that many of these new products have security capabilities that range from well-designed to poorly designed to outright absent. But IAM has never been so important: Exceptionally good authentication and authorization are needed for all these new types of devices to prevent unauthorized access to them. The consequences of doing IAM wrong can literally cost someone their life.

Facilities

The purpose of controlling access to facilities is to ensure the safety of personnel who work in those facilities, as well as to protect information systems and other assets located there. Controlling access to facilities is accomplished by different means, including

>> **Key card access systems:** With optional biometric readers and/or personal identification number (PIN) pads, these systems control who is permitted to access which buildings and rooms. These systems are used in both preventive (by restricting access to sensitive areas) and detective (by recording subjects' movement) contexts.

>> **Escorts:** Visitors and subjects with lower security clearances may be escorted by other personnel.

>> **Guards and guard dogs:** Security personnel with optional canine assistants ensure that only authorized and/or properly escorted personnel can enter a building.

>> **Visitor logs:** Although they serve as an administrative control, visitor logs provide a business record of guests and visitors who enter and leave a facility. This control is improved somewhat through the verification of visitor identity via government-issued photo identification.

>> **Fences, walls, and gates:** These features help establish a secure physical perimeter and controlled entry/exit points around a building or facility.

>> **Mantraps and sally ports:** These combinations of passageways and entry-ways restrict access to an area, for example, with a set of interlocking doors that require one set of doors to be closed before the next set can open.

>> **Bollards and crash gates:** These features control vehicle flow approaching and near facilities.

Many other aspects of physical security are discussed in Chapter 5.

Applications

Access to applications (and their associated data) is typically controlled through an identity management system (discussed in the following section). Attackers can target vulnerabilities in applications, APIs, or — increasingly commonly — user credentials to gain access to application data. Enterprise applications such as customer relationship management (CRM), e-commerce, enterprise resource planning (ERP), and financial systems can contain sensitive information that is valuable to an attacker, including customer data, financial information, and intellectual property. These applications may be hosted in an on-premises data center and may be accessible only internally or via a virtual private network (VPN) connection, or they may be hosted as web applications that can be accessed over the Internet. These applications may also run as Infrastructure as a Service (IaaS), Platform as a Service (PaaS), or Software as a Service (SaaS) workloads hosted in a public cloud.

Manage Identification and Authentication of People, Devices, and Services

The core activity within IAM is the management of identities, including people, devices, and services. In this section, we describe the processes and technologies in use today.

CROSS REFERENCE

This section covers Objective 5.2 of the Identity and Access Management (IAM) domain in the CISSP Exam Outline (May 1, 2021).

Identity management implementation

Implementing identity management (IdM) begins with a plan. IdM is a complex, distributed system that touches systems, networks, and applications, and also controls access to assets within an organization. IdM also includes the business processes that work together with IAM technologies and personnel to get the job done.

Single-/multifactor authentication

Authentication is a two-step process that consists of identification and authentication (I&A). *Identification* is the means by which a user or system (subject) presents a specific identity (such as a username) to a system (object). *Authentication* is the process of verifying that identity. A username/password combination is one common technique (albeit a weak one) that demonstrates the concepts of identification (username) and authentication (password).

Authentication is based on any of these factors:

>> **Something you know (such as a password or PIN):** This concept is based on the assumption that only the owner of the account knows the secret password or PIN needed to access the account. Username and password combinations are the simplest, least expensive, and therefore most common

authentication mechanism implemented today. Passwords, of course, are often shared, stolen, guessed, or otherwise compromised; thus, they're among the weakest authentication mechanisms.

» **Something you have (such as a smart card, security token, or smartphone):** This concept is based on the assumption that only the owner of the account has the necessary key to unlock the account. Smart cards, USB tokens, smartphones, and key fobs are becoming more common, particularly in relatively secure organizations, such as government and financial institutions. Many online applications, such as LinkedIn and Twitter, have implemented multifactor authentication (MFA) as well. Although smart cards and tokens are somewhat more expensive and complex than other, less-secure authentication mechanisms, they're (usually) not prohibitively expensive or overly complicated to implement, administer, and use. Smartphones that can receive text messages or run soft token apps such as Google Authenticator or Microsoft Authenticator are increasingly popular because of their lower cost and convenience. Regardless of the method chosen, all forms of multifactor authentication provide a significant boost to authentication security. Tokens, smartcards, and smartphones are sometimes lost, stolen, or damaged, however.

WARNING

Because of the risks associated with text messages (such as mobile phone porting scams), the U.S. National Institute of Standards and Technology has deprecated the use of text messages for multifactor authentication.

» **Something you are (such as fingerprint, face, voice, retina, or iris characteristics):** This concept is based on the assumption that the face, finger, or eyeball attached to your body is actually yours and uniquely identifies you. (Fingers and eyes can be lost, of course.) Actually, the major drawback with this authentication mechanism is acceptance; people are sometimes uneasy about using these systems. There is also the issue of spoofing: Some biometric systems, such as fingerprint and facial recognition, are not immune to spoofing attacks. Software-based biometric systems such as facial recognition are generally inexpensive, but hardware-based biometric systems are more costly to deploy.

REMEMBER

Authentication is based on something you *know*, something you *have*, or something you *are*.

The various I&A techniques that we discuss in the following sections include passwords/passphrases and PINs (knowledge-based); biometrics and behavior (characteristic-based); and one-time passwords, tokens, and single sign-on.

The identification component is normally a relatively simple mechanism based on a username or, in the case of a system or process, on a computer or process name, Media Access Control (MAC) address, IP address, or Process ID (PID).

Identification requirements include only unique identification of the user (or system/process) without identifying that user's role or relative importance in the organization. That is, the identification shouldn't include labels such as *accounting* or *CEO*. Common, shared, and group accounts such as *root, admin,* or *system* should not be permitted. Such accounts provide no accountability and are prime targets for malicious beings.

REMEMBER

Identification is the act of claiming a specific identity. *Authentication* is the act of verifying that identity.

Single-factor authentication

Single-factor authentication requires only one of the three preceding factors discussed in the preceding section (something you *know*, something you *have*, or something you *are*) for authentication. Common single-factor authentication mechanisms include passwords and passphrases, one-time passwords, and PINs.

PASSWORDS AND PASSPHRASES

"A password should be like a toothbrush. Use it every day; change it regularly; and *don't* share it with friends." — USENET

Passwords are easily the most common — and weakest — authentication credentials in use today. Although more advanced and secure authentication technologies are available, including tokens and biometrics, organizations typically use those technologies as supplements to or in combination with — rather than as replacements for — traditional usernames and passwords.

A *passphrase* is a variation on a password that uses a sequence of characters or words rather than a single password. Generally, attackers have more difficulty breaking passphrases than breaking regular passwords because longer passphrases are generally more difficult to break than shorter, complex passwords. Passphrases also have the following advantages:

>> Users frequently use the same passwords to access numerous accounts, such as their corporate networks, their home computers, their email accounts, their eBay accounts, and their Amazon.com accounts. So, an attacker who targets a specific user may be able to gain access to their work account by going after a less secure system, such as their home computer, or by compromising an Internet account (because the user has passwords conveniently stored in that bastion of security, Internet Explorer!). Internet sites and home computers typically don't use passphrases, so you improve the chances that your users have to use different passwords/passphrases to access their work accounts.

>> Users can remember and type passphrases more easily than they can remember and type much shorter, cryptic passwords that require contorted finger acrobatics to type on a keyboard.

Passphrases also have down sides:

>> Users can find passphrases inconvenient ("You mean I need to have a 20-character password now?"), so you may find them difficult to implement.

>> Not all systems support passphrases. Such systems ignore anything longer than the system limit (such as eight characters).

>> Many command-line interfaces and tools don't support the space character that separates words in a passphrase.

>> Ultimately, a passphrase is still a password (albeit a much longer and better one) and thus has some of the problems associated with passwords.

You, as a CISSP candidate, should understand the general problems associated with passwords, as well as common password controls and management features.

Password/passphrase have the problems of being

>> **Insecure:** Passwords are generally insecure for several reasons, including

- *Human nature:* In the case of user-generated passwords, users often choose passwords that they can easily remember and, consequently, attackers can easily guess (such as a spouse's or pet's name, birthday, anniversary, or hobby). Users may also be inclined to write down passwords (particularly complex, system-generated passwords) or share their passwords with others.

- *Transmission and storage:* Many applications and protocols — such as Telnet, File Transfer Protocol (FTP), and Password Authentication Protocol (PAP) — transmit passwords in clear text. These applications and protocols may also store passwords in plaintext files or in a security database that uses a weak hashing algorithm.

>> **Easily broken:** Passwords are susceptible to brute-force and dictionary attacks by readily available programs such as John the Ripper and L0phtCrack (pronounced *loft-crack*).

>> **Easily stolen:** From phishing scams to watering-hole attacks and key loggers, users can be tricked into giving up passwords, and malware can steal those passwords as users type them. Some organizations store their users'

passwords unencrypted, hashed without salting, or encrypted with an easily discovered key; any of these methods makes it relatively easy for an intruder to obtain passwords from a poorly protected system.

>> **Inconvenient:** Easily agitated users can find entering passwords tiresome. In an attempt to bypass these controls, users may select an easily typed, weak password; they may automate logins (such as using a keyboard macro or selecting the Remember My Password check box in a browser); and they can neglect to lock their workstations or log out when they leave their desks.

>> **Refutable:** Transactions authenticated with only a password don't necessarily provide absolute proof of a user's identity. Authentication mechanisms must guarantee nonrepudiation, which is a critical component of accountability. (For more on nonrepudiation, see "Accountability" later in this chapter.)

Passwords have the following login controls and management features that you should configure in accordance with an organization's security policy and security best practices:

>> **Length:** Generally, the longer the better. A password is in effect an encryption key. Just as larger encryption keys (such as 1024-bit or 2048-bit) are more difficult to crack, so are longer passwords. You should configure systems to require a minimum password length of 10 to 15 characters. Users can easily forget long passwords, of course, or may simply find them too inconvenient, leading to some of the human-nature problems discussed earlier in this section.

>> **Complexity:** Strong passwords contain a mix of upper- and lowercase letters, numbers, and special characters such as # and $. Be aware that some systems may not accept certain special characters. Also be aware that those characters may perform special functions (such as in terminal emulation software).

>> **Expiration (or maximum password aging):** You should set maximum password aging to require password changes at regular intervals; 30-, 60-, and 90-day periods are common.

>> **Minimum password aging:** This approach prevents a user from changing their password too frequently. The recommended setting is one to ten days to prevent a user from easily circumventing password history controls (such as by changing their password five times within a few minutes and then setting it back to their original password).

>> **Reuse:** Password reuse settings (five to ten are common) allow a system to remember previously used passwords (or, more appropriately, their hashes) for a specific account. This security setting prevents users from circumventing

maximum password expiration by alternating among two or three familiar passwords when they're required to change passwords.

>> **Limited attempts:** This control limits the number of unsuccessful login attempts and consists of two components: counter threshold (such as three or five) and counter reset (such as 5 or 30 minutes). The *counter threshold* is the maximum number of consecutive unsuccessful attempts permitted before some action occurs (such as automatic disabling of the account). The *counter reset* is the amount of time between unsuccessful attempts. Three unsuccessful login attempts within a 30-minute period, for example, may result in an account lockout for a set period, such as 24 hours, but two unsuccessful attempts in 25 minutes and a third unsuccessful attempt 10 minutes later wouldn't result in an account lockout. A successful login attempt also resets the counter.

>> **Lockout duration (or intruder lockout):** When a user exceeds the counter threshold that we describe in the preceding item, the account is locked out. Organizations commonly set the lockout duration to 30 minutes, but you can set it for any duration. If you set the duration to forever, an administrator must unlock the account. Some systems don't notify the user when it locks out an account, instead quietly alerting the system administrator to a possible break-in attempt. An attacker can use the lockout duration as a simple means to perform a denial of service attack (intentionally making repeated bad login attempts to keep the user's account locked).

>> **Limited time periods:** This control restricts the time of day when a user can log in. You can effectively reduce the period when attackers can compromise your systems by limiting users' access to business hours. This type of control is becoming less common in the modern age of the workaholic and the global economy, both of which require users to perform work legitimately at all hours of the day.

>> **System messages:** System messages include the following:

- *Login banner:* Welcome messages invite criminals to access your systems. Disable any welcome message and replace it with a legal warning that requires the user to click OK to acknowledge the warning and accept the legal terms of use.

- *Last username:* Many popular operating systems display the username of the last successful account login. Users (who need only type their password) find this feature convenient — and so do attackers (who need only crack the password without worrying about matching it to a valid user account). Disable this feature.

- *Last successful login time:* After a user successfully logs in to the system, this message tells the user the last time that they logged in. If the system shows that the last successful login for a user was Saturday morning at

2 a.m., and the user knows that they couldn't possibly have logged in at that time because they have a life, they know that someone has compromised their account, and they can report the incident.

- *Last successful login location:* After a user successfully logs in to the system, this message tells the user the last geographical location used when they logged in. If the system reports that the user last logged in from some obscure, far-away country, this message can be a clue that the user's account has been compromised.

We're sure that you know many of the following widely available, well-known guidelines for creating more secure passwords, but just in case, here's a recap:

>> Use a mix of upper- and lowercase letters, numbers, and special characters (such as !@#$%).

>> Do not include your name or other personal information (such as spouse, street address, school, birthdays, and anniversaries).

>> Replace some letters with numbers (such as *e* with *3*). This technique of modifying spelling is known as *leet* or *leetspeak.*

>> Use nonsense phrases, misspellings, substitutions, or before-and-after words and phrases (combining two unrelated words or phrases, such as "Wheel of Fortune Cookies").

>> Combine multiple words by using special characters (such as sALT&pEPPER or W3'r3-n0t-in-K4ns4s-4nym0r3).

>> Create a longer password that is actually a passphrase (such as "I Love Green Bananas").

>> Use a combination of all the other tips in this list (so that "Snow White and the Seven Habits of Highly Effective People", for example, becomes SW&t7HoH3P!).

>> Do not use repeating patterns between changes (such as password1, password2, password3).

>> Do not use the same passwords for work and personal accounts.

>> Do not use passwords that are too difficult to remember.

>> Do not use any passwords you see in a book, including this one. (But you knew that.)

The problem with these guidelines is that they're *widely available and well known!* In fact, attackers use some of these same guidelines to create their aliases or handles: super-geek becomes 5up3rg33k. Also, a password such as Qwerty12!

technically satisfies these guidelines, but it's not really a good password because it's a relatively simple and obvious pattern (the first row on your keyboard). Many dictionary attacks include not only word lists, but also patterns such as this one.

TIP

You can use a software tool that helps users evaluate the quality of their passwords when they create them. These tools are commonly known as *password/passphrase generators* or *password appraisers*. Also, a password manager can be used to securely store passwords so that users are less tempted to write down or re-use their passwords.

ONE-TIME PASSWORDS

A *one-time password* is valid for one login session only. After a single login session, the password is no longer valid. Thus, if an attacker obtains a one-time password that someone has already used, that password has no value. A one-time password is a *dynamic password*, meaning that it changes at some regular interval or event. Conversely, a *static password* is a password that remains the same for each login. Similar to the concept of a one-time pad in cryptography (which we discuss in Chapter 5), a one-time password provides maximum security for access control.

Security professionals should be sure to distinguish one-time passwords from passwords that are valid for a short period. Often, what is considered to be a one-time password is actually a password that is valid for several minutes. Limited-time passwords are a big improvement in security, but they're subject to replay attacks if the attacker acts quickly.

PERSONAL IDENTIFICATION NUMBERS

A PIN in itself is a relatively weak authentication mechanism because you have only 10,000 possible combinations for a four-digit numeric PIN. Therefore, organizations usually use some other safeguard in combination with a PIN. A PIN used with a one-time token password and an account lockout policy is also very effective, allowing a user to attempt only one PIN/password combination per minute and then locking the account after three or five failed attempts, as determined by the security policy.

REMEMBER

Two examples of one-time password implementations are tokens (which we discuss in the following section) and the S/Key protocol. The *S/Key protocol*, developed by Bell Communications Research and defined in Internet Engineering Task Force (IETF) Request For Comment (RFC) 1760, is client/server-based and uses MD4 and MD5 to generate one-time passwords. *MD4* and *MD5* are algorithms used to verify data integrity by creating a 128-bit message digest from data input.

Multifactor authentication

Multifactor authentication (MFA) involves two or more of *what you know*, *what you have*, and *what you are*. MFA is more challenging for an adversary to attack, because a successful attack of MFA requires the attacker to possess the user's token or the ability to trick a biometric reader. Types of MFA discussed in this section include tokens, certificates, and biometrics.

TOKENS

Tokens are access control devices such as key fobs, dongles, USB keys, smart cards, magnetic cards, software (known as *soft tokens* and installed on a tablet, mobile device, smartphone, laptop, or PC), and keypad or calculator-type cards that store static passwords (or digital certificates) or that generate dynamic passwords. The three general types of tokens are

>> **Static password tokens:** Store a static password or digital certificate.

>> **Synchronous dynamic password tokens:** Continuously generate a new password or passcode at a fixed time interval (such as 60 seconds) or in response to an event (such as every time you press a button). Typically, the passcode is valid only during a fixed time window (say, 1 minute) and only for a single login, so if you want to log in to more than one system, you must wait for the next passcode.

>> **Asynchronous (or *challenge-response*) dynamic password tokens:** Generate a new password or passcode asynchronously by calculating the correct response to a system-generated random challenge string (known as a *nonce*) that the owner enters manually.

Tokens provide two-factor authentication (something you have and something you know) by either requiring the owner to authenticate to the token first or by requiring the owner to enter a secret PIN along with the generated password. Both RADIUS and Terminal Access Controller Access Control System (TACACS+), which we discuss later in this chapter support various token products.

WARNING

A soft token that's installed on a laptop or PC doesn't provide strong (two-factor) authentication because the "something you have" is the computer you're trying to log in to! A soft token such as Google Authenticator and Microsoft Authenticator on a smartphone, however, would provide adequate two-factor authentication, provided that the user is not trying to log in to an application from a smartphone.

You can use tokens to generate one-time passwords and provide two-factor authentication.

SMARTPHONE / SMS PASSWORDS

When a user attempts to log in to a system, a one-time or short-duration password can be sent to a smartphone or mobile device via a text message or other messaging mechanism. Upon receiving this password, the user would enter it into the system's password field and complete the login procedure.

WARNING

Because of the proliferation of subscriber identity module swap and mobile device takeover scams, Short Message System (SMS) text messages generally should *not be used* for MFA.

DIGITAL CERTIFICATES

A digital certificate can be installed on the user's device. When the user attempts to authenticate to a system, the system will query the user's device for the digital certificate to confirm the user's identity. If the digital certificate can be obtained and is confirmed to be genuine, the user is permitted to log in.

Digital certificate authentication also helps ensure that users log in by using only company-provided devices. This presupposes the fact that the user is unable to copy the digital certificate to another, perhaps personally owned, device, or that an intruder is unable to copy the certificate to his own device.

WARNING

When implementing digital certificates on devices such as laptop computers, administrators need to be sure that they implement a per-device or per-user certificate on each laptop, not a general company certificate.

BIOMETRICS

The only absolute method for positively identifying a person is basing authentication on some unique physical or behavioral characteristic of that person. Biometric identification uses physical characteristics, such as fingerprints, hand geometry, and facial features such as retina and iris patterns. Behavioral biometrics are based on measurements and data derived from an action, and they indirectly measure characteristics of the human body. Behavioral characteristics include voice, signature, and keystroke patterns.

Biometrics are based on the third factor of authentication: something you are. Biometric access control systems apply the concept of I&A slightly differently, depending on their use:

>> **Physical access controls:** The person presents the required biometric characteristic, and the system attempts to identify them by matching the input characteristic with its database of authorized personnel. This type of control is also known as a *one-to-many* search.

>> **Logical access controls:** The user enters a username or PIN (or inserts a smart card), and then presents the required biometric characteristic for verification. The system attempts to *authenticate* the user by matching the claimed identity and the stored biometric image file for that account. This type of control is also known as a *one-to-one* search.

WARNING

Biometric authentication in and of itself doesn't provide *strong* authentication because it's based on only one of the three authentication requirements: something you *are*. To be considered a truly strong authentication mechanism, biometric authentication must include something you *know* or something you *have*. (Although you might argue that your hand or eye is something you have *and* something you are, for the purposes of the CISSP exam, you'd be wrong!)

The necessary factors for an effective biometrics access control system include

>> **Accuracy:** The most important characteristic of any biometric system. The uniqueness of the body organ or characteristic that the system measures to guarantee positive identification is an important element of accuracy. In common biometric systems today, the only organs that satisfy this requirement are the fingers/hands and eyes.

Another important element of accuracy is the system's ability to detect and reject forged or counterfeit input data. The accuracy of a biometric system is normally stated as a percentage, in the following terms:

- *False Reject Rate (FRR) or Type I error:* Authorized users to whom the system incorrectly denies access, stated as a percentage. Reducing a system's sensitivity reduces the FRR but increases the False Accept Rate (FAR).

REMEMBER

The FRR Rate (or Type I error) is the percentage of authorized users to whom the system incorrectly denies access.

- *False Accept Rate (FAR) or Type II error:* Unauthorized users to whom the system incorrectly grants access, stated as a percentage. Increasing a system's sensitivity reduces the FAR but increases the FRR.

REMEMBER

The FAR (or Type II error) is the percentage of unauthorized users to whom the system incorrectly grants access.

- *Crossover Error Rate (CER):* The point at which the FRR equals the FAR, stated as a percentage. (See Figure 7-1.) Because you can adjust the FAR and FRR by changing a system's sensitivity, the CER is considered to be the most important measure of biometric system accuracy.

REMEMBER

- The CER is the point at which the FRR equals the FAR, stated as a percentage.

>> **Speed and throughput:** The length of time required to complete the entire authentication procedure. This time, measurement includes stepping up to the system, inputting a card or PIN (if required), entering biometric data (such as inserting a finger or hand into a reader, pressing a sensor, aligning an eye with a camera or scanner, speaking a phrase, or signing a name), processing the input data, and opening and closing an access door (in the case of a physical access control system). Another important measure is the initial enrollment time required to create a biometric file for a user account. Generally accepted standards are a speed of less than five seconds, a throughput rate of six to ten per minute, and enrollment time of less than two minutes.

>> **Data storage requirements:** The size of a biometric system's input files can be as small as 9 bytes or as large as 10,000 bytes, the normal range being 256 to 1,000 bytes.

>> **Reliability:** An important factor in any system. The system must operate continuously and accurately without frequent maintenance outages.

>> **Acceptability:** The biggest hurdle to widespread implementation. Certain privacy and ethics issues arise with the prospect of organizations using these systems to collect medical or other physical data about employees. Other factors that might alarm users include intrusiveness of the data collection procedure and undesirable physical contact with common system components, such as pressing an eye against a plastic cup or placing lips close to a microphone for voice recognition.

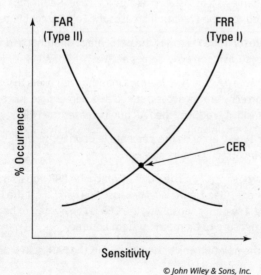

FIGURE 7-1:
Use CER to compare FAR and FRR.

© John Wiley & Sons, Inc.

Gaining user acceptance is the most common difficulty with biometric systems.

Table 7-1 summarizes the generally accepted standards for the factors described in the preceding list.

TABLE 7-1

Generally Accepted Standards for Biometric Systems

Characteristic	Standard
Accuracy	CER < 10%
Speed	5 seconds
Throughput	6–10 per minute
Enrollment time	< 2 minutes

Common types of physical biometric access control systems include

>> **Fingerprint recognition and finger scan systems:** The most common biometric systems in use today, these systems analyze the ridges, whorls, and minutiae (bifurcations and ridge endings, dots, islands, ponds and lakes, spurs, bridges, and crossovers) of a fingerprint to create a digitized image that uniquely identifies the owner of the fingerprint. A *fingerprint recognition* system stores the entire fingerprint as a digitized image. A disadvantage of this type of system is that it can require a lot of storage space and resources. More commonly, organizations use a *finger scan system,* which stores only sample points or unique features of a fingerprint and therefore requires less storage and processing resources. Also, users may more readily accept the technology because no one can re-create an entire fingerprint from the data in a finger scan system. See Table 7-2 for general characteristics of finger scan systems.

Finger scan systems, unlike fingerprint recognition systems, don't store an image of the entire fingerprint — only a digitized file describing its unique characteristics. This fact should allay the privacy concerns of most users.

>> **Facial recognition systems:** Fast becoming a popular authentication method used by Apple, Microsoft, and others, facial recognition works through recognition of the unique geometry of the user's facial features. Facial recognition software examines the face as the user looks into the device's camera and decides whether the person looking into the camera is the same person who is authorized to use the device.

TABLE 7-2   **General Characteristics of Finger Scan and Hand Geometry Systems**

Characteristic	Finger Scan	Hand Geometry
Accuracy	< 1%–5% (CER)	< 1%–2% (CER)
Speed	1–7 seconds	3–5 seconds
File size	~250–1500 bytes	~10 bytes
Advantages	Nonintrusive, inexpensive	Small file size
Disadvantages	Sensor wear and tear; accuracy may be affected by swelling, injury, or jewelry	Sensor wear and tear; accuracy may be affected by swelling, injury, or jewelry

» **Hand geometry systems:** Like finger scan systems, *hand geometry systems* are nonintrusive and therefore generally better accepted than other biometric systems. These systems generally can more accurately uniquely identify a person than finger scan systems do, and they have some of the smallest file sizes compared with other biometric system types. A digital camera simultaneously captures a vertical and a horizontal image of the subject's hand, acquiring the 3D hand geometry data. The digitized image records the length, width, height, and other unique characteristics of the hand and fingers. See Table 7-2 for general characteristics of hand geometry systems.

» **Retina pattern:** These systems record unique elements in the vascular pattern of the retina. Major concerns with this type of system are fears of eye damage from a laser (which is only a camera with a focused low-intensity light) directed at the eye and, more feasibly, privacy concerns. Certain health conditions, such as diabetes and heart disease, can cause changes in the retinal pattern, which these types of systems may detect. See Table 7-3 for general characteristics of retina pattern systems.

» **Iris pattern:** This system is by far the most accurate biometric system. The *iris* is the colored portion of the eye surrounding the pupil. The complex patterns of the iris include unique features such as coronas, filaments, freckles, pits, radial furrows, rifts, and striations. The characteristics of the iris, formed shortly before birth, remain stable throughout life. The iris is so unique that even the two eyes of a single person have different patterns. A camera directed at an aperture mirror scans the iris pattern. The subject must glance at the mirror from a distance of approximately 3 to 10 inches. It's technically feasible — but perhaps prohibitively expensive — to perform an iris scan from a distance of several feet. See Table 7-3 for general characteristics of iris pattern systems.

TABLE 7-3

General Characteristics of Retina and Iris Pattern Systems

Characteristic	Retina Pattern	Iris Pattern
Accuracy	1.5% (CER)	< 0.5% (CER)
Speed	4–7 seconds	2.5–4 seconds
File size	~96 bytes	~256–512 bytes
Advantages	Overall accuracy	Best overall accuracy
Disadvantages	Perceived intrusiveness; sanitation and privacy concerns	Subject must remain absolutely still; subject can't wear colored contact lenses or glasses (clear contacts are generally okay)

Common types of behavioral biometric systems include

» **Voice recognition:** These systems capture unique characteristics of a subject's voice and may also analyze phonetic or linguistic patterns. Most voice recognition systems are text-dependent, requiring the subject to repeat a specific phrase. This functional requirement of voice recognition systems also helps improve their security by providing two-factor authentication: something you know (a phrase) and something you are (your voice). More advanced voice recognition systems may present a random phrase or group of words, which prevents an attacker from recording a voice authentication session and later replaying the recording to gain unauthorized access. See Table 7-4 for general characteristics of voice recognition systems.

» **Signature dynamics:** These systems typically require the subject to sign their name on a signature tablet. The enrollment process for a signature dynamics system captures numerous characteristics, including the signature pattern itself, the pressure applied to the signature pad, and the speed of the signature. Signatures commonly exhibit some slight changes because of different factors, of course, and they can be forged (although the signature dynamics are difficult to forge). See Table 7-4 for general characteristics of signature dynamics systems.

» **Keystroke or typing dynamics:** These systems typically require the subject to type a password or phrase. The keystroke dynamic identification is based on unique characteristics such as how long a user holds down a key on the keyboard (dwell time) and how long it takes a user to get to and press a key (seek or flight time). These characteristics are measured by the system to form a series of mathematical data representing a user's unique typing pattern or signature, which is used to authenticate the user.

TABLE 7-4 **General Characteristics of Voice Recognition and Signature Dynamics Systems**

Characteristic	Voice Recognition	Signature Dynamics
Accuracy	< 10% (CER)	1% (CER)
Speed	10–14 seconds	5–10 seconds
File size	~1,000–10,000 bytes	~1,000–1,500 bytes
Advantages	Inexpensive; nonintrusive	Nonintrusive
Disadvantages	Accuracy, speed, file size; affected by background noise, voice changes; can be fooled by voice imitation	Signature tablet wear and tear; speed; can be fooled by a forged signature

WARNING

Digital signatures and *electronic signatures* — which are electronic copies of people's signatures — are not the same as the signatures used in biometric systems. These terms are not related and are not interchangeable.

TIP

In general, the CISSP candidate doesn't need to know the specific characteristics and specifications of the different biometric systems, but you should know how they compare. You should know, for example, that iris pattern systems are more accurate than retina pattern systems, and you should be familiar with the concepts of FRR, FAR, and CER.

Accountability

The concept of *accountability* refers to the capability of a system to associate users and processes with their actions (what they did). Audit trails and system logs are components of accountability.

Systems use audit logs and audit trails primarily as a means of troubleshooting problems and verifying events. Users should not view audit logs and audit trails as a threat or as Big Brother watching over them because they cannot be trusted. As a matter of fact, astute users consider these mechanisms to be protective, because the system not only prove what they did, but also help prove what they did *not* do. Still, it's wise for users to be mindful of the fact that the systems they use are recording their actions.

An important security concept that's closely related to accountability is nonrepudiation. *Nonrepudiation* means that a user (username Madame X) can't deny an action because her identity is positively associated with her actions. Nonrepudiation is an important legal concept. If a system permits users to log in by using a generic user account, a user account that has a widely known password, or no user account, you can't absolutely associate any user with a given (malicious) action or (unauthorized) access on that system, which makes it extremely difficult to prosecute or otherwise discipline that user.

Accounting in authentication, authorization and accounting (AAA) services records what a subject did.

REMEMBER

Nonrepudiation means that a user can't deny an action because you can irrefutably associate them with that action.

REMEMBER

Session management

A *session* is a formal term referring to an individual user's dialogue, or series of interactions, with an information system. Information systems need to track individual users' sessions to properly distinguish one user's actions from another's.

To protect the confidentiality and integrity of data accessible through a session, information systems generally use session or activity timeouts to prevent an unauthorized user from continuing a session that has been idle or otherwise inactive for a specified period.

Two primary means of implementing session timeouts are

>> **Screen savers:** Implemented by the operating system, a screen saver locks the workstation or mobile device itself and requires the user to log back into the system after a period of inactivity. The workstation's or mobile device's screen saver protects all application sessions. Make sure that this feature actually locks the screen or device, as some systems can be configured to not require a PIN or password to unlock them.

>> **Inactivity timeouts:** Individual software applications may use an auto-locking or auto-logout feature if a user has been inactive for a specific period.

If an authorized user leaves a computer terminal unlocked or a browser window on a workstation unattended, for example, an unauthorized user can simply sit down at the workstation and continue the session.

Workstation inactivity timeouts were originally called screen savers, to prevent a static image on a cathode ray tube display from being burned into the display. Although today's monitors do not require this protection, the term *screen saver* is still in common use.

TIP

Registration, proofing, and establishment of identity

Formal user registration processes are important for secure account provisioning, particularly in large organizations where it is not practical or possible to know all the workers. This type of process is particularly critical in single-sign on (SSO), federated, and PKI environments (see Chapter 5), where users will have access to multiple systems and applications.

Proof of identity often begins at the time of hire, when new workers are usually required to show government-issued identification and legal right-to-work status. These procedures should form the basis for initial user registration in information systems.

Organizations need to take several precautions when registering and provisioning users:

>> **User identity:** The organization must ensure that new user accounts are provisioned for, and given to, the correct user.

>> **Protection of privacy:** The organization should not use Social Security number, date of birth, or other sensitive private information to authenticate the user. Instead, other values should be used, such as employee number (or other values that cannot be obtained by other employees).

>> **Temporary credentials:** The organization must ensure that temporary login credentials are assigned to the correct person. Others should not be able to easily guess temporary credentials. Finally, temporary credentials should be set to expire in a short period.

>> **Birthright access:** The organization should periodically review the birthright access granted to new workers, following the principles of need to know and least privilege.

Additional considerations about user identity occur when a user is attempting to log on to a system:

>> **Geographic location:** Location can be derived from the IP address of the user. The IP address is not absolutely reliable but can be helpful for determining the user's location. Many devices, particularly smartphones and tablet computers, use GPS technology for location information, which is generally more reliable than IP address.

- **Workstation in use:** The organization may have policies about whether a user is permitted to log in on a personally owned device or from a public kiosk.

- **Elapsed time since last login:** This metric is how long it has been since the user last logged in to the system or application.

- **Login attempt after failed attempts:** This metric records whether there have been recent unsuccessful login attempts.

Depending on the preceding conditions, the system may be configured to present additional challenges to the user. These challenges ensure that the person attempting to log in actually is the authorized user, not another person or machine. This type of challenge is known as *risk-based authentication*.

Federated identity management

Federated identity management (FIM) enables multiple organizations to use one another's user identification and authentication systems to access their networks and systems. Federation of identity (FIdM) comprises the standards, technologies, and tools used to facilitate the portability of identity across separately managed organizations.

FIdM permits organizations that want to facilitate easier user access to their systems without having to create custom solutions. Instead, they need only configure existing tools and occasionally add connectors to facilitate inter-organization identity management.

Technologies in common use in federated environments (all discussed later in this chapter) include

- Single sign-on (SSO)
- Security Assertion Markup Language (SAML)
- Open Authorization (OAuth) and OpenID Connect (OIDC)

Credential management systems

Credential management systems enable an organization to centrally organize and control user IDs and passwords. This type of system should not be confused with systems used to store and manage users' professional credentials (such as the CISSP certification).

Credential management systems are available as commercial software products that can be implemented either on-premises or in the cloud.

Credential management systems create user accounts for subjects and provision those credentials as required in both individual systems and centralized identity management systems (such as LDAP or Microsoft Active Directory). Credential management systems can be either separate applications (as explained previously) or an integral part of an IAM system.

Single sign-on

The concept of single sign-on (SSO) addresses a common problem for both users and security administrators. Every account that exists in a system, network, or application is a potential point of unauthorized access. Multiple accounts that belong to a single user represent an even greater risk:

>> Users who need access to multiple systems or applications often must maintain numerous sets of credentials. Inevitably, this requirement leads to shortcuts in creating and recalling passwords. Left to their own devices, users create weak passwords that have only slight variations; worse, they use the same passwords everywhere they can. When they have multiple sets of credentials to manage, users are more likely to write them down. The problem doesn't stop at the organization's boundary: Users often use the same passwords at work that they do for their personal accounts.

>> Multiple accounts also affect user productivity (and sanity!) because the user must stop to log in to different systems. Someone must also create and maintain accounts, which involves unlocking accounts and supporting, removing, resetting, and disabling multiple sets of user IDs and passwords.

At first glance (alas), SSO seems to be the perfect solution that users and security administrators seek. SSO allows a user to present a single set of login credentials, typically to an authentication server, which then transparently logs the user into all other enterprise systems and applications in which the user is authorized. SSO does have some disadvantages, of course, which include the following:

>> **Woo-hoo!:** After you're authenticated, you have the keys to the kingdom. Read that as *access to all authorized resources!* This situation is the security professional's nightmare. If login credentials for a user's accounts are compromised, an intruder can access everything the user was authorized to access.

>> **Complexity:** Implementing SSO can be difficult and time-consuming. You have to address interoperability issues among systems and applications. But hey — that's why you get paid (or should get paid) the big bucks!

Just-in-Time

Just-in-time (JIT) access provides temporary, granular access to an application or resource (such as a virtual machine) only when it is needed to perform a specific task or function. This approach helps organizations implement and enforce the principle of least privilege (discussed in chapters 5 and 9). An organization might provide JIT access for a system administrator or a process that temporarily requires elevated privileges on a sensitive system. JIT access can be implemented in several ways:

>> **Key vault:** A centralized vault securely stores shared account credentials that are checked out for a specified, limited period by an authorized user when elevated privileges are needed. The user typically needs to provide justification for using the credentials, and access is automatically enforced via defined policies. This approach is commonly used in privileged access management (PAM) solutions.

>> **Ephemeral accounts:** These temporary, one-time accounts are created dynamically with the required privilege when needed; after use, they are immediately and automatically deprovisioned and/or deleted.

>> **Temporary elevation:** An account may be assigned elevated privileges for a specified and limited period. The elevated privileges are removed automatically when the time period expires.

Federated Identity with a Third-Party Service

Most organizations have a variety of business applications, some of which run on-premises and others of which are in the cloud. To avoid the issue of users having to manage multiple sets of user credentials, many organizations have implemented some form of third-party federated identity service. The benefits to organizations are twofold:

>> **Increased convenience:** Users have fewer sets of login credentials (as few as one) for access to business systems.

>> **Reduced risk:** Users are more likely to use stronger passwords and less likely to handle credentials unsafely (such as using sticky notes on monitors). Many organizations employ multifactor authentication, further reducing risk.

CROSS REFERENCE This section covers Objective 5.3 of the Identity and Access Management (IAM) domain in the CISSP Exam Outline (May 1, 2021).

The manner in which organizations implement a centralized IAM system depends on several factors, including

>> **Integration effort:** Newer applications have one or more interfaces available to facilitate automated account provisioning and SSO. Older applications usually lack these interfaces.

>> **Available resources:** Even for easily integrated applications, some effort is still required to perform and maintain integrations over time.

>> **Efficiency tolerance:** If the organization is intolerant of inefficiencies, such as users having to log in to business applications many times each day, they may be more likely to pursue an IAM solution.

>> **Risk tolerance:** If an organization is averse to the risks associated with users possessing multiple sets of login credentials for critical business systems, it will be more likely to implement an IAM solution.

Although each IAM platform has its own unique capabilities and architecture, an IAM system generally resembles the architecture depicted in Figure 7-2.

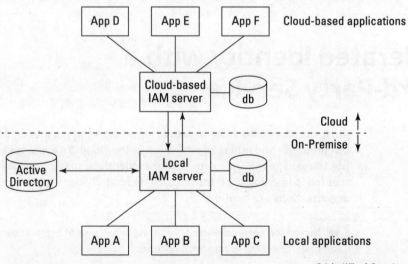

FIGURE 7-2: Typical identity and access management system architecture.

© John Wiley & Sons, Inc.

Organizations with on-premises systems often purchase and integrate identity management tools into their environments to reduce the burden of identity management, as well as improve end user experience. Where Microsoft servers are used, organizations can integrate their systems and applications with Active Directory, which is included with Microsoft server operating systems. In organizations without Microsoft servers, open source tools that use Lightweight Directory Access Protocol (LDAP) are a preferred choice. Also, several commercial on-premises identity service products can be installed and integrated with systems, devices, and software applications.

On-premises identity management tools generally have the same features as their cloud-based counterparts. Some of these tools can be implemented on-premises or based in the cloud, and a few offer solutions in which cloud-based and on-premises tools work together as a single identity access solution.

On-premises

An organization may choose to implement federated identity to manage access among multiple on-premises environments, such as when a trusted connection is needed for partners on different networks or domains.

Cloud

Because business systems and applications are increasingly cloud-based, many organizations are opting to implement cloud-based identity management and/or SSO systems to provide federated identity management.

Hybrid

Federated identity management is particularly helpful in hybrid environments comprised of on-premises data centers and multiple public or private cloud environments. Many organizations today have a multicloud strategy that uses services from different public cloud providers (such as Microsoft Azure and Amazon Web Services [AWS]) and SaaS providers (such as Concur, Salesforce, and Workday).

Implement and Manage Authorization Mechanisms

Authorization mechanisms are the portions of operating systems and applications that determine the data and functions a user is permitted to access, based on the user's identity.

CROSS REFERENCE

This section covers Objective 5.4 of the Identity and Access Management (IAM) domain in the CISSP Exam Outline (May 1, 2021).

Authorization (also referred to as *establishment*) defines the rights and permissions granted to a user account or process (what the user can do and/or what data the user can access). After a system or application authenticates a user, authorization determines what that user (subject) can do with a system or resource (object).

Data access controls protect systems and information by restricting access to system files and user data based on user identity. Data access controls also provide authorization and accountability, relying on system access controls to provide identification and authentication.

Role-based access control

Role-based access control (RBAC) is a method of managing user access controls. RBAC assigns group membership according to organizational or functional roles. People may belong to one or many groups (either acquiring cumulative permissions or limited to the most restrictive set of permissions for all assigned groups); a group may contain only a single person (corresponding to a specific organizational role assigned to that person). Access rights and permissions for objects are assigned to groups rather than (or in addition to) people. RBAC greatly simplifies the management of access rights and permissions, particularly in organizations that have large functional groups or departments or that routinely rotate personnel through various positions or otherwise experience high turnover.

The advantages of role-based access control include

>> User access tends to be more uniform.

>> Changing many users' access often involves changing the access rights for one or more roles.

Many systems that employ RBAC still permit access rights to be granted to individual end users. Still, many organizations tend to stick with the use of roles, even if in some instances only one person is a member of a role.

Figure 7-3 depicts the concept of RBAC.

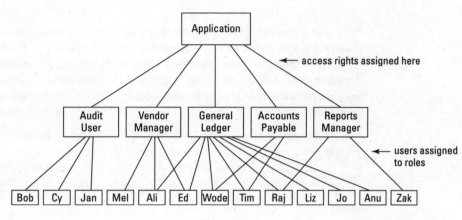

FIGURE 7-3: Role-based access control.

Role-based access control
© John Wiley & Sons, Inc.

Rule-based access control

Rule–based access control (not to be confused with RBAC) is one method of applying mandatory access control. All MAC–based systems (discussed next) implement a simple form of rule–based access control by matching an object's sensitivity label and a subject's sensitivity label to determine whether the system should grant or deny access. You can apply additional rules by using rule–based access control to further define specific conditions for access to a requested object. Other types of rules that govern access include

>> Time of day

>> Workstation or terminal in use

>> User geographical location

>> Contents of data being accessed

Mandatory access control

A *mandatory access control* (MAC) is an access policy determined by the system rather than by the owner. Organizations use MAC in multilevel systems that process highly sensitive data, such as classified government and military information. A *multilevel system* is a single computer system that handles multiple classification levels between subjects and objects. Two important concepts in MAC are

>> **Sensitivity labels:** In a MAC-based system, all subjects and objects must have assigned labels. A subject's sensitivity label specifies its level of trust. An

object's sensitivity label specifies the level of trust required for access. To access a given object, the subject must have a sensitivity level equal to or higher than the requested object. A user (subject) with a Top Secret clearance (sensitivity label) is permitted to access a file (object) that has a Secret classification level (sensitivity label) because their clearance level exceeds the minimum required for access. We discuss classification systems in Chapter 4.

>> **Data import and export:** Controlling the import of information from other systems and the export to other systems (including printers) is a critical function of MAC-based systems, which must ensure that the system properly maintains and implements sensitivity labels so that sensitive information is appropriately protected at all times.

Lattice-based access controls are another method of implementing mandatory access controls. A *lattice model* is a mathematical structure that defines greatest lower-bound and least upper-bound values for a pair of elements, such as a subject and an object. Organizations can use this model for complex access control decisions involving multiple objects and/or subjects. Given a set of files that have multiple classification levels, for example, the lattice model determines the minimum clearance level that a user requires to access all the files.

Major disadvantages of mandatory access control techniques include

>> Lack of flexibility

>> Difficulty in implementation and programming

>> User frustration

REMEMBER

In MAC, the system determines the access policy.

Discretionary access control

A *discretionary access control* (DAC) is an access policy determined by the owner of a file (or other resource). The owner decides who's allowed access to the file and what privileges they have.

REMEMBER

In DAC, the owner determines the access policy.

Two important concepts in DAC are

>> **File and data ownership:** Because the owner of the resource (which may consist of files, directories, data, system resources, and devices) determines the access policy, every object in a system must have an owner. Theoretically,

an object without an owner is left unprotected or without a user who can determine who or what can access it. Normally, the *owner* of a resource is the person who created the resource (such as a file or directory), but in certain cases, you may need to identify the owner explicitly.

>> **Access rights and permissions:** These rights and permissions are the controls that an owner can assign to individual users or groups for specific resources. Various systems (Windows-based or Unix-based) define different sets of permissions that are essentially variations or extensions of three basic types of access:

- *Read (R):* The subject can read the contents of a file or list the contents of a directory.

- *Write (W):* The subject can change the contents of a file or directory (including add, rename, create, and delete).

- *Execute (X):* If the file is a program, the subject can execute the program.

Access control lists (ACLs) provide a flexible method for applying discretionary access controls. An ACL lists the specific rights and permissions that are assigned to a subject for a given object.

Major disadvantages of discretionary access control techniques such as ACLs and RBAC include the following:

>> They lack centralized administration.

>> They depend on security-conscious resource owners.

>> Many popular operating systems default to full access for everyone if the owner doesn't set permissions explicitly.

>> Auditing is difficult, if not impossible, because of the large volume of individual permissions, as well as the log entries that can be generated.

REMEMBER

Various operating systems implement ACLs differently. Although the CISSP exam doesn't directly test your knowledge of specific operating systems or products, you should be aware of this fact. Also, understand that ACLs in this context are different from ACLs used on routers (see Chapter 5), which have nothing to do with DAC.

Attribute-based access control

Attribute-based access control (ABAC) is an access policy determined by the attributes of a subject and object, as well as environmental factors. In an ABAC-based system, the ability of a subject to access an object is based on one or more attributes of the subject (such as position title, department, or project assignment), as well as attributes of the object itself (such as name, project name, owner, or location). Further, environmental factors are used to determine whether access will be granted. Example environmental factors include the location of the subject and the time of day.

In an ABAC-based system, the access decision is made by the Policy Decision Point (PDP) and enforced by the Policy Enforcement Point (PEP).

TIP

ABAC is defined in NIST SP-162, Guide to Attribute Based Access Control (ABAC) Definition and Considerations, which is available at https://www.nist.gov.

Risk-based access control

Risk-based access controls are policy-based access controls that dynamically assess risk to determine whether access should be granted, under what conditions, and at what level. A user may only need a username and password if they are logging in from a known or trusted IP address during normal working hours, for example, but they may be prompted to use MFA if they are logging in from a different location (such as while traveling) or an unknown device (such as a desktop PC in a hotel kiosk).

Manage the Identity and Access Provisioning Life Cycle

Organizations must adopt formal policies and procedures to address account provisioning, review, and revocation. IAM is a key requirement of a zero trust strategy (discussed in Chapter 5) that requires organizations to always verify who (users) and what (devices) is connecting to their network and what they are allowed to do on the network.

CROSS REFERENCE

This section covers Objective 5.5 of the Identity and Access Management (IAM) domain in the CISSP Exam Outline (May 1, 2021).

The phases in the IAM provisioning life cycle are

> » **Role design, creation, and review:** During development, customization, or configuration of a system, each of the user roles is designed. These designs are subjected to a review process to ensure that they are appropriate.

>> **Access provisioning:** When new or temporary employees, contractors, partners, auditors, and third parties require access to an organization's systems and networks, the organization must have a formal methodology for requesting access. The steps in access provisioning are

(a) *Access request:* A requester, typically but not always the end user, makes a formal request to access specific data or a specific role in a system.

(b) *Review request:* The request is reviewed by a designated person or group for appropriateness. Sometimes, the reviewer asks the requester questions to better understand the reason for the request.

(c) *Access approval:* The request is approved by a designated person or group.

(d) *Access provisioning:* The requested and approved access is provisioned in the system.

REMEMBER

New accounts must be provisioned correctly and in a timely manner to ensure that access is ready and available when the user needs it, but not too soon (to ensure that new accounts not yet in active use are not compromised by an attacker).

>> **Privilege escalation:** User and system accounts may temporarily require a privileged access level to perform a specific task or function. Privilege escalation helps enforce the principle of least privilege (discussed in chapters 5 and 9) and may be implemented via a PAM solution or JIT access (discussed earlier in this chapter).

>> **Account access review:** User and system accounts, along with their assigned privileges, should be reviewed on a regular basis to ensure that they are still appropriate. An employee may no longer require the same privilege levels due to rotation of duties (see Chapter 9) or a transfer or promotion, for example.

>> **Role review:** Periodically, designated personnel examine the roles in a system to determine whether the access rights defined in each role are appropriate.

>> **Inactivity review:** Periodically, designated personnel examine a system to see whether users are accessing it. Users who have not accessed a system (or had a role in the system) for a given period (such as 90 days) may have their access revoked.

>> **Access termination:** Finally, when access is no longer required, accounts must be promptly disabled and deprovisioned. Deprovisioning should be automated when possible and should include moving the disabled account to a protected organizational unit (in Active Directory) or other restricted group, as well as removing the disabled account from all group memberships.

TIP

User accounts are typically locked within 24 hours of termination. In the case of dismissal, user accounts are typically locked immediately before the employee is notified.

Implement Authentication Systems

Commonly used authentication systems include OpenID Connect (OIDC)/Open Authorization (Oauth), Security Assertion Markup Language (SAML), Kerberos, and Remote Authentication Dial-In User Service (RADIUS)/Terminal Access Controller Access Control System Plus (TACACS+).

CROSS REFERENCE

This section covers Objective 5.6 of the Identity and Access Management (IAM) domain in the CISSP Exam Outline (May 1, 2021).

OpenID Connect/Open Authorization

OIDC is a standards-based authentication protocol built on the OAuth (specifically, OAuth 2.0) framework. OAuth 2.0 is used to grant an application access to another application's data. Typically, four entities are involved in an OIDC/OAuth exchange:

>> Authorization server (identity platform)

>> OAuth client (native or web app)

>> Resource server (Representational State Transfer [REST] API)

>> Resource owner (end user)

Security Assertion Markup Language

The de facto protocol for authentication, SAML is used to facilitate user authentication across systems and among organizations, through the exchange of authentication and authorization information among organizations. SAML is the glue that is used to make most SSO systems work.

As its full name suggests, SAML is an extensible markup language (SML) based standard. XML is becoming a standard method for exchanging information between dissimilar systems.

OIDC allows users to sign-in to an identity provider (IdP) to access third-party websites and apps. OAuth 2.0 allows the IdP to issue tokens to third-party applications and services on the user's behalf. SAML allows different organizations to exchange authentication and authorization data between IdPs and service providers.

TECHNICAL STUFF

Kerberos

Kerberos, commonly used in the Sun Network File System and Microsoft Windows, is perhaps the most popular ticket-based symmetric key authentication protocol in use today.

 Kerberos is named for the fierce three-headed dog that guards the gates of Hades in Greek mythology (not to be confused with *Ker-beer-os*, the fuzzy, six-headed dog sitting at the bar that keeps looking better and better!). Researchers at the Massachusetts Institute of Technology (MIT, also known as Millionaires in Training) developed this open-systems protocol in the mid-1980s.

TECHNICAL STUFF

The CISSP exam requires a general understanding of Kerberos operation. Unfortunately, Kerberos is a complex protocol that has many implementations and no simple explanation. The following step-by-step discussion is a basic description of Kerberos operation:

1. The client prompts the subject (such as a user) for an identifier and a credential (such as username and password). Using the authentication information (password), the client temporarily generates and stores a secret key for the subject by using a one-way hash function and then sends only the subject's identification (username) to the Key Distribution Center's (KDC) Authentication Server (AS). The password/secret key isn't sent to the KDC. See Figure 7-4.

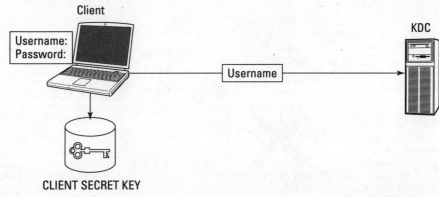

FIGURE 7-4:
Kerberos: Login
initiation (step 1).

© John Wiley & Sons, Inc.

2. The AS on the KDC verifies that the subject (known as a *principal*) exists in the KDC database. The KDC Ticket Granting Service (TGS) generates a Client/TGS Session Key encrypted with the subject's secret key, which only the TGS and the client know. The TGS also generates a Ticket Granting Ticket (TGT) consisting of the subject's identification, the client network address, the valid period of the ticket, and the Client/TGS Session Key. The TGS encrypts the TGT by using its secret key, which only the TGS knows, and then sends the Client/TGS Session Key and TGT back to the client. See Figure 7-5.

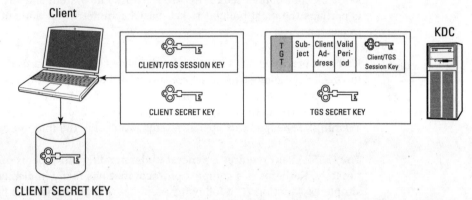

FIGURE 7-5: Kerberos: Client/ TGS session key and TGT generation (step 2).

© John Wiley & Sons, Inc.

3. The client decrypts the Client/TGS session key, using the stored secret key that it generated by using the subject's password; authenticates the subject (user); and then erases the stored secret key to avoid possible compromise. The client can't decrypt the TGT, which the TGS encrypted by using the TGS secret key. See Figure 7-6.

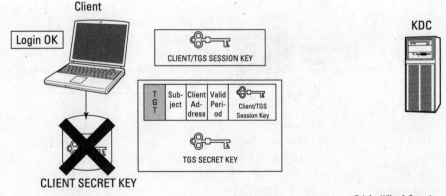

FIGURE 7-6: Kerberos: Login completion (step 3).

© John Wiley & Sons, Inc.

4. When the subject requests access to a specific object (such as a server, also known as a *principal*), it sends the TGT, the object identifier (such as a server name), and an authenticator to the TGS on the KDC. (The *authenticator* is a separate message that contains the client ID and a time stamp, and uses the Client/TGS session key to encrypt itself.) See Figure 7-7.

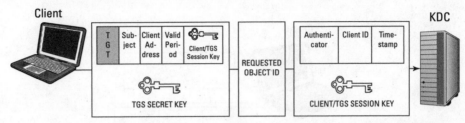

FIGURE 7-7: Kerberos: Requesting services (step 4).

© John Wiley & Sons, Inc.

5. The TGS on the KDC generates both a Client/Server session key (which it encrypts by using the Client/TGS session key) and a service ticket (which consists of the subject's identification, the client network address, the valid period of the ticket, and the Client/Server session key). The TGS encrypts the service ticket by using the secret key of the requested object (server), which only the TGS and the object know. Then the TGS sends the Client/Server session key and service ticket back to the client. See Figure 7-8.

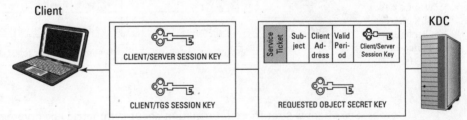

FIGURE 7-8: Kerberos: Client/ Server session key and service ticket generation (step 5).

© John Wiley & Sons, Inc.

6. The client decrypts the Client/Server session key by using the Client/TGS session key. The client can't decrypt the service ticket, which the TGS encrypted by using the secret key of the requested object. See Figure 7-9.

7. The client can communicate directly with the requested object (server). The client sends the service ticket and an authenticator to the requested object (server). The client encrypts the authenticator (comprising the subject's identification and a time stamp) by using the Client/Server session key that the TGS generated. The object (server) decrypts the service ticket by using its secret key. The service ticket contains the Client/Server session key, which allows the

object (server) to decrypt the authenticator. If the subject identification and time stamp are valid (according to the subject identification, client network address, and valid period specified in the service ticket), communication between the client and server is established. Then the Client/Server session key is used for secure communications between the subject and object. See Figure 7-10.

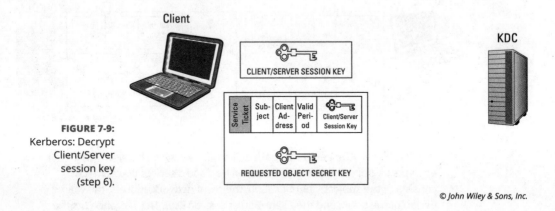

FIGURE 7-9:
Kerberos: Decrypt Client/Server session key (step 6).

© John Wiley & Sons, Inc.

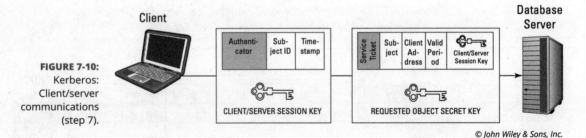

FIGURE 7-10:
Kerberos: Client/server communications (step 7).

© John Wiley & Sons, Inc.

See Chapter 5 for more information about symmetric key cryptography.

REMEMBER

In Kerberos, a *session key* is a dynamic key that is generated when needed, shared between two principals, and destroyed when it is no longer needed. A *secret* key is a static key that is used to encrypt a session key.

RADIUS and TACACS+

The *Remote Authentication Dial-In User Service* (RADIUS) protocol is an open-source, client-server networking protocol — defined in more than 25 current IETF RFCs — that provides AAA services. RADIUS is an Application Layer protocol that uses User Datagram Protocol (UDP) packets for transport. UDP is a connectionless protocol, which means that it's fast but not as reliable as other transport protocols.

RADIUS is commonly implemented in network service provider networks, as well as corporate VPNs. RADIUS is also becoming increasingly popular in corporate wireless networks. A user provides username/password information to a RADIUS client by using PAP or CHAP. The RADIUS client encrypts the password and sends the username and encrypted password to the RADIUS server for authentication.

Note: Passwords exchanged between the RADIUS client and RADIUS server are encrypted, but passwords exchanged between the workstation client and the RADIUS client are not necessarily encrypted — when PAP authentication is used, for example. If the workstation client happens also to be a RADIUS client, all password exchanges are encrypted regardless of the authentication protocol used.

TIP

Diameter is the successor protocol to RADIUS. Diameter is an Application Layer protocol that uses Transmission Control Protocol (TCP) or Stream Control Transmission Protocol (SCTP) instead of UDP for transport and uses either IP Security (IPSec) or Transport Layer Security (TLS). It supports both roaming and local AAA services and stateful or stateless modes.

The *Terminal Access Controller Access Control System* (TACACS) is a remote authentication control protocol, originally developed for the MILNET (U.S. Military Network), which provides AAA services. The original TACACS protocol has been significantly enhanced, as XTACACS (no longer used) and TACACS+ (which is the most common implementation of TACACS). But TACACS+ is a new protocol and therefore isn't backward-compatible with either TACACS or XTACACS. TACACS+ is TCP based on (port 49) and supports practically any authentication mechanism (PAP, CHAP, MS-CHAP, EAP, token cards, Kerberos, and so on). The major advantages of TACACS+ are its wide support of various authentication mechanisms and granular control of authorization parameters. TACACS+ can also use dynamic passwords; TACACS uses static passwords only.

The original TACACS protocol is often used in organizations to simplify administrative access to network devices such as firewalls and routers. TACACS facilitates the use of centralized authentication credentials that are managed centrally so that organizations don't need to manage user accounts on every device.

Chapter **8**

Security Assessment and Testing

I n this chapter, you learn about the various tools and techniques that security professionals use to continually assess and validate an organization's IT environment. This domain represents 12 percent of the CISSP certification exam.

Design and Validate Assessment, Test, and Audit Strategies

Modern security threats are rapidly and constantly evolving. Likewise, an organization's systems, applications, networks, services, and users are frequently changing. Thus, it is critical that organizations develop an effective strategy to regularly test, evaluate, and adapt their business and technology environment to reduce the probability and impact of successful attacks, as well as to achieve compliance with applicable laws, regulations, and contractual obligations.

CROSS REFERENCE

This section covers Objective 6.1 of the Security Assessment and Testing domain in the CISSP Exam Outline (May 1, 2021).

Organizations need to implement a proactive assessment, test, and audit strategy for both existing and new information systems and assets. The strategy should be an integral part of the risk management process to help the organization identify new and changing risks that are important enough to warrant analysis, decisions, and action.

Security personnel must identify all applicable laws, regulations, and other legal obligations (such as contracts) to understand what assessments, testing, and auditing are required. Further, security personnel should examine their organization's risk management framework and control framework to see what assessments, control testing, and audits are suggested or required. Then the combination of these solutions would become part of the organization's overall strategy for ensuring that all its security-related tools, systems, and processes are operating properly.

Three main perspectives come into play in planning for an organization's assessments, testing, and auditing:

>> **Internal:** Assessments, testing, and auditing performed by personnel who are part of the organization. The advantages of using internal resources for assessments, tests, and audits include lower costs and greater familiarity with the organization's practices and systems. Internal personnel may not be as objective as external parties, however.

>> **External:** Assessments, testing, and audits performed by people from an external organization or agency. Some laws and regulations, as well as contractual obligations, may require external assessments, tests, and audits of certain systems and processes. The greatest advantage of using external personnel is that they're objective. They're often more expensive, however, particularly for activities that require higher skill levels or specialized tools.

>> **Third-party:** Audits of critical business activities that have been outsourced to external service providers (third parties). The systems and personnel being examined belong to an external service provider. Depending on the requirements in applicable laws, regulations, and contracts, these assessments of third parties may be performed by internal personnel; in some cases, external personnel may be required.

TIP

To avoid multiple audits, many third-party service providers commission external audits whose audit reports can be distributed to their customers. Examples of such audits include SSAE 18, SOC-1, SOC-2, and PCI DSS. Service providers also commission security consulting firms to conduct penetration tests on systems and applications, which helps them reduce the number of customers who would want to do this testing themselves.

Conduct Security Control Testing

Security control testing employs various tools and techniques, including vulnerability assessments, penetration (or *pen*) testing, synthetic transactions, interfaces testing, and more. You learn about these and other tools and techniques in the following sections.

CROSS REFERENCE This section covers Objective 6.2 of the Security Assessment and Testing domain in the CISSP Exam Outline (May 1, 2021).

Vulnerability assessment

A vulnerability assessment is performed to identify, evaluate, quantify, and prioritize security weaknesses in an application or system. Additionally, a vulnerability assessment provides remediation steps to mitigate specific vulnerabilities that are identified in the environment.

There are three general types of vulnerability assessments:

» Port scan (not intensive)

» Vulnerability scan (more intensive)

» Penetration test (most intensive)

Generally, automated network-based scanning tools are used to identify vulnerabilities in applications, systems, and network devices in a network. Sometimes, system-based scanning tools are used to examine configuration settings to identify exploitable vulnerabilities. Often, network- and system-based tools are used together to build a complete picture of vulnerabilities in an environment.

The various types of vulnerability assessments fit into a fundamental activity in information security known as *vulnerability management,* which is a formal process of assessment, vulnerability identification, and remediation within specific timeframes. The purpose of vulnerability management often includes *attack surface reduction,* the quest to reduce the number of systems, devices, and components that are potentially exploitable.

Port scanning

A *port scan* uses a tool that communicates over the network with one or more target systems on various Transmission Control Protocol/Internet Protocol (TCP/IP) ports. A port scan can discover ports that probably should be disabled (because they serve no useful or necessary purpose on a particular system).

Vulnerability scans

Network-based vulnerability scanning tools send network messages to systems in a network to identify any utilities, programs, or tools that may be configured to communicate over the network. These tools attempt to identify the version of any utilities, programs, and tools; often, it is enough to know the versions of the programs that are running, because scanning tools often contain a database of known vulnerabilities associated with program versions. Scanning tools may also send specially crafted messages to running programs to see whether those programs contain any exploitable vulnerabilities.

Tools are also used to identify vulnerabilities in software applications. Generally, these tools are divided into two types: dynamic application security testing (DAST) and static application security testing (SAST). DAST executes an application and then uses techniques such as fuzzing in an attempt to identify exploitable vulnerabilities that could permit an attacker to compromise a software application and then alter or steal data or take control of the system. SAST examines an application's source code for exploitable vulnerabilities. Neither DAST nor SAST can find all vulnerabilities, but when these tools are used together by skilled personnel, many exploitable vulnerabilities can be found.

Examples of network-based vulnerability scanning tools include Nessus, Rapid7, and Qualys. Flexera (formerly Secunia) PSI is an example of a system-based vulnerability scanning tool. Example application scanning tools include HCL AppScan, Fortify WebInspect, Acunetix, and Burp Suite.

Unauthenticated and authenticated scans

Vulnerability scanning tools (those used to examine systems and network devices, as well as those that examine applications) generally perform two types of scans: unauthenticated and authenticated. In an *authenticated scan,* the scanning tool is configured with login credentials and attempts to log in to the device, system, or application to identify vulnerabilities that are not discoverable otherwise. In an *unauthenticated scan,* the scanning tool does not attempt to log in; hence, it can discover only vulnerabilities that would be exploitable by someone who does not possess valid login credentials.

Vulnerability scan reports

Generally, all the types of scanning tools discussed in this section create a report that contains summary and detailed information about the scan that was performed and vulnerabilities that were identified. Many of these tools produce a good amount of detail, including steps used to identify each vulnerability, the severity of each vulnerability, and steps that can be taken to remediate each vulnerability.

Some vulnerability scanning tools employ a proprietary methodology for vulnerability identification, but most scanning tools include a common vulnerability scoring system score for each identified vulnerability. Application security is discussed in more detail in Chapter 10.

Vulnerability assessments are a key part of risk management (discussed in Chapter 3).

Penetration testing

Penetration testing (*pen testing* for short) is the most rigorous form of vulnerability assessment. The level of effort required to perform a penetration test is far higher than that required for a port scan or vulnerability scan. Typically, an organization will employ a penetration test on a target system or environment when it wants to simulate an attack by an adversary.

Network penetration testing

A *network penetration test* of systems and network devices generally begins with a port scan and/or a vulnerability scan. This scan gives the pen tester an inventory of the attack surface of the network and the systems and devices connected to the network. The pen test continues with extensive use of manual techniques to identify and/or exploit vulnerabilities. In other words, the pen tester uses both automated and manual techniques to identify and confirm vulnerabilities.

Occasionally, a pen tester exploits vulnerabilities during a penetration test. Pen testers generally tread carefully because they must be acutely aware of the target environment. If a pen tester is testing a live production environment, for example, exploiting vulnerabilities could result in malfunctions or outages in the target environment. In some cases, data corruption or data loss could also result.

When performing a penetration test, the pen tester may take screen shots showing the exploited system or device because some system/device owners don't believe that their environments contain exploitable vulnerabilities. By including screen shots in the final report, the pen tester is "proving" that vulnerabilities exist and are exploitable.

THE COMMON VULNERABILITY SCORING SYSTEM

The common vulnerability scoring system (CVSS) is an industry-standard method used to numerically score vulnerabilities according to severity. Numeric scores for vulnerabilities help security personnel prioritize remediation, generally by fixing vulnerabilities with higher scores before tackling those with lower scores.

The formula for arriving at a CVSS score for a given vulnerability is fairly complicated, but all you need to understand is its basic structure. A CVSS score examines several aspects of a vulnerability, including the following:

- **Access vector:** How a vulnerability is exploited

- **Access complexity:** How easy or difficult it is to exploit a vulnerability

- **Authentication:** Whether an attacker must authenticate to a target system to exploit it

- **Confidentiality:** The potential impact on the confidentiality of data on the target system if it is exploited

- **Integrity:** The potential impact on the integrity of data on the target system if it is exploited

- **Availability:** The potential impact on the availability of data or applications on the target system if it is exploited

Generally speaking, the higher a CVSS score, the easier it is to exploit a given vulnerability and the greater the impact on the target systems.

Pen testers often include details for reproducing exploits in their reports. These details are helpful to system or network engineers who want to reproduce the exploit so that they can see for themselves that the vulnerability does in fact exist. They're also helpful when engineers or developers make changes to mitigate the vulnerabilities; they can use the same techniques to see whether their fixes closed the vulnerabilities.

In addition to scanning networks, techniques that are generally included in the topic of network penetration testing include the following:

>> **Wardialing:** Hackers use *wardialing* to sequentially dial all phone numbers in a range to discover any active modems; then they attempt to compromise any connected systems or networks via the modem connection. This attack is old-school, but it's still used occasionally.

>> **Wardriving:** *Wardriving* is the 21st-century version of wardialing. Someone with a laptop computer literally drives around a densely populated area, looking for unprotected (or poorly protected) wireless access points.

>> **Radiation monitoring:** *Radio frequency* (RF) emanations are the electromagnetic radiation emitted by computers and network devices. *Radiation monitoring* is similar to packet sniffing and wardriving, in that someone uses sophisticated equipment to try to determine what data is being displayed on monitors, transmitted on local area networks (LANs), or processed in computers.

>> **Eavesdropping:** Eavesdropping is as low-tech as dumpster diving but a little less (physically) dirty. Basically, an *eavesdropper* takes advantage of one or more people who are talking or using a computer and paying little attention to whether someone else is listening to their conversations or watching them work with discreet over-the-shoulder glances. (The technical term for the latter approach is *shoulder surfing*.)

>> **Packet sniffing:** A *packet sniffer* is a tool that captures all TCP/IP packets on a network, not just those being sent to the system or device doing the sniffing. An Ethernet network is a shared-media network (see Chapter 6), which means that any or all devices on the LAN can (theoretically) view all packets. Switched-media LANs are more prevalent today, however, and sniffers on switched-media LANs generally pick up only packets intended for the device running the sniffer.

TIP

A network adapter that operates in *promiscuous mode* accepts all packets, not just the packets destined for the system, and sends them to the operating system.

PACKET SNIFFING ISN'T ALL BAD

Packet sniffing isn't just a tool that hackers use to pick up user IDs and passwords from a LAN; it has legitimate uses as well. Primarily, you can use it as a diagnostic tool to troubleshoot network devices, such as a firewall (to see whether the desired packets get through), routers, switches, and virtual LANs (VLANs).

The obvious danger of the packet sniffer's falling into the wrong hands is that it provides the capability to capture sensitive data, including user IDs and passwords. Equally perilous is the fact that packet sniffers can be difficult to detect on a network.

Application penetration testing

An *application penetration test* is used to identify vulnerabilities in a software application. Although the principles of an application penetration test are the same as those of a network penetration test, the tools and skills are somewhat different. Someone performing an application penetration test generally has an extensive background in software development. Indeed, the best application pen testers are often former software developers or software engineers.

Physical penetration testing

Penetration tests are also performed on the controls protecting physical premises to see whether it is possible for an intruder to bypass security controls such as locked doors and keycard-controlled entrances. Sometimes, pen testers employ various social engineering techniques to gain unauthorized access to work centers and sensitive areas within work centers, such as computer rooms and file storage rooms. Often, they plant evidence, such as a business card or other object, to prove that they were successful.

TIP

Hacking For Dummies, 6th Edition (John Wiley & Sons, Inc.), by Kevin Beaver, explores penetration testing and other techniques in more detail.

In addition to breaking into facilities, physical pen testers practice dumpster diving. *Dumpster diving* is low-tech penetration testing at its best (or worst) and is exactly what it sounds like. This test can be an extraordinarily fruitful way to obtain information about an organization. Organizations in highly competitive environments also need to be concerned about where their trash and recycled paper go.

GET OUT OF JAIL FREE

Penetration testers who are hired to target physical premises often ask for a signed letter printed on company letterhead that authorizes them to use various techniques to break into physical premises. Pen testers carry these letters (often called "get out of jail free" letters) in case on-site personnel call security or law enforcement.

This safeguard usually helps keep a pen tester out of trouble, but not always. Knowing this technique, cybercriminals may use these letters to try to fool personnel into leaving them alone. A key feature of these letters is contact information for one or more senior officials in the organization whom security or law enforcement can call to verify that the pen tester is legitimate. But this feature is not foolproof either; cybercriminals can cite a real name but provide the phone number of one of their accomplices.

Social engineering

Social engineering is any testing technique that employs some means of tricking people into performing some action or providing some information that enables the pen tester to break into an application, system, or network. Social engineering involves such low-tech tactics as pretending to be a support technician, calling an employee, and asking for their password. You'd think that most people would be smart enough not to fall for this trick, but people are people (and Soylent Green is people)! Some of the ruses used in social engineering tests include the following:

>> **Phishing messages:** Email messages purporting to be something they're not, sent in an attempt to lure someone into opening a file or clicking a link. Test phishing messages are harmless, of course, but they're used to see how many personnel fall for the ruse.

>> **Telephone calls:** Calls made to various workers inside an organization to trick them into performing tasks. A call to the service desk, for example, might attempt to reset a user's account (possibly enabling the pen tester to log in to that user's account).

>> **Tailgating:** Attempts to enter a restricted work area by following legitimate personnel as they pass through a controlled doorway. Sometimes, the tester carries boxes in the hopes that an employee will hold the door open for them or poses as a delivery or equipment repair person.

PHISHING AND ITS VARIANTS

Phishing messages pretend to be something they're not. There are several specific forms of phishing, including

- **Pharming:** This attack results in users visiting an imposter website instead of the site they intend to visit. Pharming can be accomplished through an attack on a system's hosts file, an organization's Domain Name System (DNS), or a domain homograph attack.

- **Spearphishing:** These phishing messages target a single organization (or part of an organization) with highly customized messaging.

- **Whaling:** These phishing messages are sent to executives in a target organization.

- **Smishing:** These phishing messages are delivered through Short Message Service (SMS), also known as texting.

Log reviews

Reviewing your various security logs on a regular basis (ideally, daily) is a critical step in security control testing. Unfortunately, this important task often ranks only slightly higher than updating documentation on many administrators' to-do lists. Log reviews often happen only after an incident has occurred, but that's not the time to discover that your logging is incomplete or insufficient.

Logging requirements (including any regulatory or legal mandates) need to be clearly defined in an organization's security policy, including the following:

>> What gets logged, such as

- Events in network devices, such as firewalls, intrusion prevention systems (IPSes), web filters, and data loss prevention (DLP) systems

- Events in server and workstation operating systems

- Events in subsystems, such as web servers, database management systems, and application gateways

- Events in applications

>> What's in the logs, such as

- Date/time of the event

- Source (and destination, if applicable), protocol, and IP addresses

- Device, system, and/or User ID

- Event ID and category

- Event details

>> When and how often the logs are reviewed

>> The level of logging (how verbose the logs are)

>> How and where the logs are transmitted, stored, and protected:

- Are the logs stored on a centralized log server or on the local system hard drives?

- Which secure transmission protocol is used to ensure the integrity of the logging data in transit?

- How are date and timestamps synchronized (for example, using a network time protocol (NTP) server)?

- Is encryption of the logs required?

- Who is authorized to access the logs?

- Which safeguards are in place to protect the integrity of the logs?
- How is access to the logs logged?

» How long the logs are retained

» Which events in logs are triggered to generate alerts and to whom alerts are sent

TIP

Various log management tools, such as security information and event management (SIEM) systems (discussed in Chapter 9), may be used to help with real-time monitoring, parsing, anomaly detection, and generation of alerts to key personnel.

Synthetic transactions

Synthetic transactions are real-time actions or events that execute on monitored objects automatically. A tool might be used to regularly perform a series of scripted steps on an e-commerce website, for example, to measure performance, identify impending performance issues, simulate the user experience, and confirm calculations. Thus, synthetic transactions can help an organization proactively test, monitor, and ensure integrity and availability (refer to the C-I-A triad in Chapter 3) for critical systems and monitor service-level agreement (SLA) guarantees.

NOBODY REVIEWS LOGS ANYMORE

Systems create event logs that are sometimes the only indicator that something is amiss. Originally, logs were designed for either of two purposes: for periodic reviews, as a way of looking for unwanted events, or for forensic purposes in case of an incident or breach, so that investigators can piece together the clues.

Back in the day, sysadmins would check logs first thing in the morning to see what was amiss. But as sysadmins got busier, guess what was the first daily task to fall by the wayside? You got it: reviewing logs. Soon after, the mere existence of logs was practically forgotten. Logs had become only forensic resources. But for logs to be useful, you must know that an unwanted event has occurred.

Enter the security information and event management (SIEM) system. A SIEM system does what no sysadmin could ever do: monitors log entries from all systems and network devices in real time, correlates events from various systems and devices, and automatically creates actionable alerts on the spot when unwanted events occur.

Not every organization has a SIEM system, and many organizations that don't have one don't review logs either. We strongly discourage this form of negligence. It's essential for an organization to be aware of what's happening in its environment.

Application performance monitoring tools traditionally produce such metrics as system uptime, correct processing, and transaction latency. Although uptime certainly is an important aspect of availability, it is only one component. Increasingly, reachability (which is a more user- or application-centric metric) is becoming the preferred metric for organizations that focus on customer experience. After all, it doesn't do your customers much good if your web servers are up 99.999 percent of the time but Internet connections in their region of the world are slow, DNS doesn't resolve quickly, or web pages take 5 or 6 seconds to load in an online world that measures responsiveness in milliseconds. Hence, other key metrics for applications are correct processing (perhaps expressed as a percentage, which should be close to 100 percent) and *transaction latency* (the length of time it takes for specific types of transactions to complete). These metrics help operations personnel spot application problems.

Code review and testing

Code review and testing (sometimes known as *peer review*) involves systematically examining application source code to identify bugs, mistakes, inefficiencies, and security vulnerabilities in software programs. Online software repositories, such as Mercurial and Git, enable software developers to manage source code in a collaborative development environment. A code review can be accomplished manually, by carefully examining code changes visually, or by using automated code reviewing software (such as HCL AppScan Source, HP Fortify, and CA Veracode). Different types of code review and testing techniques include

>> **Pair programming:** *Pair* (or *peer) programming* is a technique commonly used in agile software development and extreme programming (both discussed in Chapter 10), in which two developers work together and alternate between writing and reviewing code line by line.

>> **Lightweight code review:** Often performed as part of the development process, this technique consists of conducting informal walk-throughs and email pass-around, tool-assisted, and/or over-the-shoulder (not recommended for the rare introverted or paranoid developer) reviews.

>> **Formal inspections:** Structured processes such as the Fagan inspection are used to identify defects in design documents, requirements specifications, test plans, and source code throughout the development process.

TIP

Code review and testing can be invaluable for identifying software vulnerabilities such as buffer overflows, script injection vulnerabilities, memory leaks, and race conditions (see Chapter 10).

Misuse case testing

The opposite of use case testing (in which normal or expected behavior in a system or application is defined and tested), *abuse/misuse case testing* is the process of performing unintended and malicious actions in a system or application to produce abnormal or unexpected behavior and thereby identify potential vulnerabilities.

After misuse case testing identifies a potential vulnerability, a use case can be developed to define new requirements for eliminating or mitigating similar vulnerabilities in other programs and applications.

A common technique used in misuse case testing is *fuzzing*, which involves the use of automated tools that can produce dozens (or hundreds, or even more) of combinations of input strings to be fed to a program's data input fields to elicit unexpected behavior. Fuzzing is used, for example, in an attempt to attack a program by using *script injection*, a technique that tricks a program into executing commands in various languages, mainly JavaScript and SQL. Tools such as HP WebInspect, IBM AppScan, Acunetix, and Burp Suite have built-in fuzzing and script injection tools that are pretty good at identifying script injection vulnerabilities in software applications.

WHY WOULD SOMEONE TYPE *THAT?*

Since time immemorial, while writing programs that interfaced with people, programmers thought of all the valid use cases for input fields. An input field that asked for an amount of currency, for example, would be programmed to accept proper numeric input. The program would expect and accept a numeric value in that field, and process it accordingly. Programmers focused on valid input, and that was that.

Fast-forward to the web with HTML, in which programs with their input fields were exposed to the world. Hackers soon found that interesting things could be typed in input fields to provide interesting results. We know these results today as SQL injection, JavaScript injection, cross-site scripting, cross-site request forgery, and buffer overflow.

These attacks caught nearly the entire programming community off guard. Simply put, a programmer's perspective was to ask "Why would someone type binary code in a numeric field?" Abuse simply had not occurred to many programmers. Fortunately, today a multitude of code libraries sanitize input to make sure that only proper characters are input, thereby blunting the effects of SQL injection and other attacks. And organizations like the Open Web Application Security Project (OWASP) produce learning content so that programmers can more easily write better, more secure programs that are less susceptible to input field attacks.

Test coverage analysis

Test coverage analysis (also called *code coverage analysis)* measures the percentage of source code that is tested by a given test or validation suite. Basic coverage criteria typically include

- ≫ **Branch coverage** (every branch at a decision point is executed as TRUE or FALSE, for example)
- ≫ **Condition (predicate) coverage** (each Boolean expression is evaluated as both TRUE and FALSE, for example)
- ≫ **Function coverage** (every function or subroutine is called, for example)
- ≫ **Statement coverage** (every statement is executed at least once for example)

A security engineer might use a dynamic application security testing tool (DAST) such as AppScan or WebInspect to test a travel booking program to determine whether the program has any exploitable security defects. Tools such as these are powerful, using a variety of methods to "fuzz" input fields in attempts to discover flaws. But the other thing these tools need to do is fill out forms in every conceivable combination so that all the program's code will be executed. In this example of a travel booking tool, these combinations would involve every way in which flights, hotels, or cars could be searched, queried, examined, and finally booked. In a complex program, this test can be really daunting. Highly systematic analysis would be needed to make sure that every possible combination of conditions is tested so that all of a program's code is tested.

Interface testing

Interface testing focuses on the interface between different systems and components. It ensures that functions (such as data transfer and control between systems or components) perform correctly and as expected. Interface testing also verifies that any execution errors are handled properly and do not expose any potential security vulnerabilities. Examples of interfaces tested include

- ≫ Application programming interfaces (APIs)
- ≫ Web services
- ≫ Transaction processing gateways
- ≫ Physical interfaces, such as keypads, keyboard/mouse/display, and device switches and indicators

TIP

APIs, web services, and transaction gateways can often be tested with automated tools such as HP WebInspect, IBM AppScan, and Acunetix, which are also used to test the human-input portion of web applications.

Breach attack simulations

Organizations that regularly employ penetration testing in their environments can do still more to understand the effectiveness of their protective safeguards. A *breach attack simulation* is an attack on an organization that includes

» Penetration testing

» An intrusion objective such as the theft of specific data

» A test of security event monitoring to recognize the attack

» Security incident response

The value of breach attack simulation comes from an exercise of not only defensive safeguards, but also detective safeguards and the steps taken after personnel recognize that an attack has occurred.

Compliance checks

In many industries, it's not enough to be secure; it's also necessary to be compliant with various laws, standards, and other types of obligations. For IT, security, and privacy-related matters, information security personnel often perform various types of compliance checks to ensure that organizations are doing what is specifically required of them. The Payment Card Industry Data Security Standard (PCI DSS), for example, requires specific safeguards to be implemented regardless of whether they are justified through risk management. HIPAA, NYDFS, and others are similar in that they prescribe specific safeguards that must be tested periodically to ensure that organizations are not only secure, but also compliant with applicable laws and regulations.

Collect Security Process Data

Assessments of security management processes and systems help organizations determine the efficacy of their key processes and controls. Periodic testing of key activities is an important part of management and regulatory oversight, confirming the proper functioning of key processes and identifying improvement areas.

CROSS REFERENCE

This section covers Objective 6.3 of the Security Assessment and Testing domain in the CISSP Exam Outline (May 1, 2021).

Several factors must be considered in determining who will perform this testing, including

>> **Regulations:** Various regulations specify which parties must perform testing, such as qualified internal staff or outside consultants.

>> **Staff resources and qualifications:** Regulations and other conditions permitting, an organization may have adequately skilled and qualified staff members who can perform some or all of its testing.

>> **Independence:** Although an organization may have the resources and expertise to test its management processes, organizations often elect to have a qualified outside organization perform testing. Independent outside testing helps prevent bias.

These factors also determine required testing methods, including the tools used, testing criteria, sampling, and reporting. In a U.S. public company, an organization is required to self-evaluate its information security controls in specific ways and with specific auditing standards under the auspices of the Sarbanes–Oxley (SOX) Act of 2002, also known as the Public Company Accounting Reform and Investor Protection Act.

The types of testing that can be performed include

>> **Document review:** An auditor will examine process or control procedure documentation to get an understanding of the activity and how it is performed.

>> **Walk-through:** An auditor will interview a process or control owner to hear in their own words how a process or control is performed and the nature of any business records that are created. The auditor will also note any variations between what they are told and what is written in process or control procedures.

>> **Records review:** An auditor will examine the business records that are created by the process or control to see whether they are consistent with process or control documentation and what they heard in a walk-through.

>> **Corroboration:** After a walk-through with a process or control owner, an auditor will ask others about the same process or control to see whether there are variations or inconsistencies in the descriptions of the process or control.

>> **Reperformance:** Here, an auditor will follow the steps in process or procedure documentation to see whether they obtain the same results as workers.

Account management

Management must regularly review user and system accounts, as well as related business processes and records, to ensure that user privileges are provisioned and deprovisioned appropriately and with proper approvals. The types of reviews include the following:

>> All user account provisioning was properly requested, reviewed, approved, and executed.

>> All internal personnel transfers resulted in timely termination of access that was no longer needed.

>> All personnel terminations resulted in timely termination of all access.

>> All users who hold privileged account access still require it, and their administrative actions are logged.

>> All user accounts can be traced back to a proper request, review, and approval.

>> All unused user accounts are evaluated to see whether they can be deactivated.

>> All users' access privileges are certified regularly as necessary.

Account management processes are discussed in more detail in Chapter 9.

Management review and approval

Management provides resources and strategic direction for all aspects of an organization, including its information security program. As a part of its overall governance, management needs to review key aspects of the security program. There is no single way that this review is done; instead, in the style and with the same rigor that it reviews other key activities in an organization, management reviews the security program. In larger organizations, this review will likely be quite formal, with executive-level reports created periodically for senior management, including key activities, events, and metrics. (Think *eye candy* here.) In smaller organizations, this review will probably be a lot less formal. In the smallest organizations, as well as organizations with lower security maturity levels, there may be no management review.

Management review often includes these activities:

>> Review of recent security incidents

>> Review of recent and anticipated security-related spending

>> Review (and ratification) of recent policy changes

>> Review (and ratification) of risk treatment decisions

>> Review (and ratification) of major changes to security-related processes and the security-related components of other business processes

>> Review of operational- and management-level metrics and risk indicators

The internationally recognized standard ISO/IEC 27001, "Information technology — Security techniques — Information security management systems — Requirements," requires an organization's management to determine what activities and elements in the information security program need to be monitored, the methods to be used, and the people or teams that will review them. ISO/IEC 27001 formally defines the structure and activities of an *information security management system* (ISMS), the set of all high-level activities that constitute a complete information security program.

Key performance and risk indicators

The leaders of an organization's information security program are generally required to report to upper management some key indicators that depict the health and effectiveness of the security program. These indicators include

>> **Key performance indicator (KPI):** A measurable value that depicts the level of effectiveness or success of a process or procedure

>> **Key risk indicator (KRI):** A measurement of a process or procedure that depicts the level of risk

KPIs and KRIs are meaningful measurements of key activities in an information security program that can be used to help management at every level better understand how well the security program and its components are performing. This process is easier said than done, however, and here are a few reasons why:

>> No single set of universal metrics applies to every organization.

>> There are different ways to measure performance and risk.

>> Executives will want key activities to be measured in specific ways.

>> Maturity levels vary from organization to organization.

Organizations typically develop metrics and KRIs for their key security-related activities to ensure that security processes are operating as expected. Metrics help identify improvement areas by alerting management to unexpected trends.

Focus areas for security metrics include the following:

>> **Vulnerability management:** Operational metrics include the numbers of scans performed, numbers of vulnerabilities identified (ranked by severity), and numbers of patches applied. KRIs focus on the coverage of scans and elapsed time between the public release of a vulnerability and the completion of patching.

>> **Incident response:** Operational metrics focus on the numbers and categories of incidents and on whether trends suggest new weaknesses in defenses. KRIs focus on the time required to realize that an incident is in progress (known as *dwell time)* and the time required to contain and resolve the incident.

>> **Security awareness training:** Operational metrics and KRIs generally focus on the completion rate over time.

>> **Logging and monitoring:** Operational metrics generally focus on the numbers and types of events that occur. KRIs focus on the proportion of assets whose logs are being monitored and the elapsed time between the start of an incident and the time when personnel begin to take action.

KRIs are so called because they are harbingers of information risk in an organization. Although the development of operational metrics is not all that difficult, security managers often struggle with the problem of developing KRIs that make sense to executive management. The vulnerability management process, for example, involves using one or more vulnerability scanning tools and subsequent remediation efforts. In this example, some good operational metrics include the numbers of scans performed, the numbers of vulnerabilities identified, and the time required to remediate identified vulnerabilities. These metrics will make no sense to management because they lack business context, but at least one good KRI can be derived from data in the vulnerability management process. "Percentage of servers supporting manufacturing whose critical security defects are not remediated within 10 days," for example, is a great KRI. This metric directly helps management understand how well the vulnerability management process is performing in a specific business context. It is also a good leading indicator of the risk of a breach that exploits an unpatched, vulnerable server that could affect business operations (manufacturing, in this case).

Backup verification data

Organizations need to routinely review and test system and data backups, as well as recovery procedures, to ensure that they are accurate, complete, and readable. They also need to regularly test the ability to recover data from backup media to ensure that they can do so in the event of a ransomware attack, hardware malfunction, or disaster event that damages information systems or facilities.

On the surface, this process seems easy enough. But as they say, the devil's in the details. Several gotchas and considerations exist, including the following:

>> **Data recovery versus disaster recovery:** There are two main reasons for backing up data:

- *Data recovery:* When various circumstances require the recovery of data from a past state

- *Disaster recovery:* When an event has resulted in damage to primary processing systems, necessitating recovery of data to alternative processing systems

For data recovery, you want your backup media (in whatever form) to be logically and physically near your production systems so that the logistics of data recovery are simple. Disaster recovery, however, requires backup media to be far from the primary processing site so that it is not involved in the same natural disaster. These two processes are at odds. Organizations sometimes solve this dilemma by creating two sets of backup media, one that stays in the primary processing center, and one that is stored at a secure offsite facility.

>> **Data integrity:** To respond to requests to roll back data to an earlier date and time, it is vital to know exactly what data needs to be recovered. Database management systems enforce a rule known as *referential integrity,* which means that a database cannot be recovered to a state in which relationships between indexes, tables, and foreign keys would be broken. This issue often comes into play in large distributed systems with multiple databases on different servers, sometimes owned by different organizations.

>> **Version control:** To respond to requests to recover data to an earlier state, personnel also need to be mindful of all changes to programs and database design that are dependent on one another. Rolling data back to a point in time last week, for example, may also require that rolling back the associated computer programs if changes in those applications last week involved both code and data changes. Further, rolling back to an earlier point in time could involve other components such as run-time libraries, subsystems such as Java, and even operating system versions and patches.

>> **Staging environments:** Depending on the reason for recovering data from a point in time in the past, it may be appropriate to recover data in a separate environment. If certain transactions in an e-commerce environment were lost, it may make sense to recover data, including the lost transactions, to a test server so that those transactions can be found. If older data was recovered to the primary production environment, transactions from that time up to the present would effectively be wiped out.

Organizations have several choices for backup media, including

>> **Magnetic tape:** For decades, organizations have used various forms of magnetic tape (*magtape*), which is reliable and can last for years when properly stored.

>> **Optical disc:** Media such as CD-ROM and DVD-ROM held promise as the heirs apparent to magtape, as they are impervious to magnetic fields and are thought to last longer.

>> **Virtual tape library (VTL):** A VTL is a disk-based storage library that simulates a tape library. The advantage of VTL is its write and read speed, which is far higher than that of magtape or optical disc.

>> **Redundant storage system:** This second storage system is usually located tens or hundreds of miles from the primary storage system. Data is copied to a redundant storage system during backups or during real-time replication.

>> **Electronic vaulting:** Electronic vaulting (*e-vaulting*) consists of using a cloud-based data storage repository that functions as a data recovery or disaster recovery archive.

Training and awareness

Organizations need to measure participation in and effectiveness of security training and awareness programs to ensure that people at all levels of the organization understand how to respond to new and evolving threats and vulnerabilities.

Key characteristics for examining training and awareness programs include

>> **Relevance:** Is the training content relevant to the workforce and the organization?

>> **Methods of delivery:** In what ways is security awareness knowledge imparted to the workforce? Using a variety of methods is more effective than using one method.

>> **Specialized content:** Do technical workers such as system administrators and DBAs receive additional training that is relevant to their responsibilities?

>> **Competency testing:** Does security awareness training include any competency testing to see whether workers are learning from the lessons?

Security awareness training is discussed in more detail in Chapter 3.

Disaster recovery and business continuity

Disaster recovery (DR) and business continuity (BC) planning enable an organization to be more resilient, even when disrupting events occur. Business continuity planning ensures that critical business processes will continue operating, whereas disaster recovery planning ensures the restoration of critical assets. Organizations need to periodically review and test their disaster recovery and business continuity plans to determine whether recovery plans are up to date and will result in the successful continuation of critical business processes in the event of a disaster.

Techniques used for testing DR and BC include

>> **Document review:** DR and BC documents can be examined for completeness, relevance, and date of last review and update.

>> **Test-report review:** DR and BC test reports can be examined to see how well recent tests went, whether improvement opportunities were identified, and whether those improvements have been made.

>> **Plan testing:** Various types of BC and DR tests can be performed, including tabletop testing, parallel testing, and cutover testing. Tabletop testing is a group walk-through of recovery procedures; parallel and cutover testing involve the use of primary and recovery systems and procedures.

DR and BC plan development and testing are discussed in detail in Chapters 3 and 9.

TIP

Information security continuous monitoring (ISCM) is defined in NIST SP 800-137 as "maintaining ongoing awareness of information security, vulnerabilities, and threats to support organizational risk management decisions." An ISCM strategy helps the organization systematically maintain an effective security management program in a dynamic environment.

Analyze Test Output and Generate Reports

Various systems and tools are capable of producing volumes of log and testing data. Without proper analysis and interpretation, these reports are useless or may be used out of context. Security professionals must be able to analyze log and test data, and to report this information in meaningful ways, so that senior management can understand organizational risks and make informed security decisions.

CROSS REFERENCE

This section covers Objective 6.4 of the Security Assessment and Testing domain in the CISSP Exam Outline (May 1, 2021).

Often, this process requires developing test output and reports for different audiences with information in a form that is useful to them. The output of a vulnerability scan report, with its lists of IP addresses, DNS names, and vulnerabilities with their respective common vulnerabilities and exposures codes, would be useful to system engineers and network engineers, who would use such reports as lists of individual defects to be fixed. But give that report to a senior executive, and they'll have little idea what it's about or what it means in business terms. For senior executives, vulnerability scan data would be rolled up into meaningful business metrics and KRIs to inform senior management of any appreciable changes in risk levels.

The key for information security professionals is knowing the meaning of *data* and transforming it for various purposes and different audiences. Security professionals who perform this task well are better able to obtain funding for additional tools and staff because they're able to state the need for resources in business terms.

Remediation

It is said in our industry that the real work begins when the assessment is completed. In other words, assessments of systems and processes are likely to find problems and improvement areas, and we're duty-bound to fix those problems. Management knows the game: They're going to ask about remediation, and we'd better have an answer. Take the following steps upon receipt of a security assessment:

1. Validate.

The important first step to take when you receive a final assessment report is to validate the findings. This step is especially important when an outside firm performed the assessment; that firm doesn't know the organization as you do, and its understanding of the matter is likely to be incomplete. Validation is a sanity check to confirm the validity of the findings.

2. Prioritize.

Initial prioritization is needed to give the organization an idea of which assessment findings will receive early attention and which ones can wait. This initial prioritization is done at face value before the details are known, and sometimes, priorities will change. Priorities may be based on risk level, visibility, the perception of quick wins, or a combination of all three.

3. **Identify a remediation owner.**

   When the context of the issue is identified, management assigns the task of remediation to a remediation owner. In some cases, the remediation owner performs the remediation; in other cases, the owner manages or supervises those who will do the work. Still, one person is identified as being responsible for seeing remediation through.

4. **Develop a work plan.**

   Subject-matter experts get down to business, developing a detailed work plan to change the process or system so as to resolve the issue found in the assessment. The work plan tells management how much effort (and what kinds of effort), cost, and time are required to remediate the finding. Depending on the nature of the finding, the work plan could be one line in a project plan or a thousand lines.

5. **Reprioritize.**

   When the work plan has been created, you'll know whether the cost, effort, and time are within initial estimates. Sometimes, remediation is easier than you initially thought, and sometimes, it's harder. Reprioritizing the effort can be the right thing to do.

6. **Remediate.**

   This work itself that resolves the issue. Whether remediation means making changes in a business process, an information system, or both, the changes are intended to resolve the issue that was identified in the assessment. In some cases, an assessment highlights the absence of something. Remediation consists of building the process or the system (or both) and putting it in place.

7. **Close the issue.**

   When the remediation owner has completed the remediation effort, the work is confirmed, and the issue is marked as closed and completed.

Management is likely to request a periodic report on remediation progress. Whoever produces this report will need to stay in contact with remediation owners to maintain an up-to-date status on all issues, providing management an accurate depiction of progress.

Exception handling

Virtually all organizations have cases in which it's infeasible (or impossible) to comply with every policy, control, and standard in every business process and information system. An account reconciliation procedure may require a manager to sign off on the process, for example, but the small size of the department may mean that there is no one to do the sign-off. Or an information system may be unable to enforce password complexity standards.

When these situations arise, the methodical approach is as follows:

1. Analyze the situation.

Study the situation to validate the assertion and explore reasonable options.

2. Analyze risk.

Study the risk levels of various options.

3. Get approval.

The exception can be approved or denied. In case of a denial, a different approach is generally prescribed.

4. Enter the records.

Enter the request, its analysis, and final disposition into an exception register.

Exception approvals should be time-bound, not perpetual. We suggest that exceptions be granted for no more than one year, after which time the matter will be reopened and reconsidered. Much may have changed in that year to put the risks in a new light.

Ethical disclosure

Now and again, personnel who conduct security assessments encounter wrongdoing. The nature of the wrongdoing may range from incompetence to intentional malice, and it may be difficult for the assessor to know the difference. Discerning intent can be especially difficult when the assessor is an employee of the firm that was hired to perform the assessment. Further, it can be difficult to know who else may be involved in the wrongdoing — who ordered it, who knows about it, and who is covering it up.

In situations like these, assessors have three options:

>> **Include the finding in the report.** Although including the finding is ethically sound, management (if involved) may deliberately bury the finding, resulting in no change in the situation. The assessor may not be invited back to perform subsequent assessments now that the misbehavior has been found.

>> **Notify law enforcement.** If an actual crime is being committed, auditors may be compelled to notify law enforcement. This approach may backfire, however, as the law enforcement organization's caseload may result in unwillingness to pursue the case.

>> **Notify the board of directors.** Bypassing management (who may be complicit in the misbehavior) and going directly to the board may be a viable option. Board members have a fiduciary responsibility and are legally bound to act on a notification of wrongdoing.

Assessors in these situations need to document their findings carefully, as those findings are likely to be challenged, considered, or even submitted as evidence in legal proceedings.

TIP

Wrongdoing discovered within an information system, particularly when a crime has been committed, may require forensic analysis and a chain of custody.

Conduct or Facilitate Security Audits

Auditing is the process of examining systems and/or business processes to ensure that they've been designed properly, are being used properly, and are considered to be effective. Audits are frequently performed by an independent third party or an independent group within an organization, which helps ensure that the audit results are accurate and not biased due to organizational politics or other circumstances.

CROSS REFERENCE

This section covers Objective 6.5 of the Security Assessment and Testing domain in the CISSP Exam Outline (May 1, 2021).

Audits are frequently performed to ensure that an organization is in compliance with business or security policies and with other requirements to which the business may be subject. These policies and requirements can include laws and regulations, legal contracts, industry or trade group standards, and best practices.

The major factors in play for internal and external audits include

>> **Purpose and scope:** The reason for an internal or external audit, and the scope of the audit, need to be fully understood by both management in the audited organization and those who will be performing the audit. Scope may include one or more of the following factors:

- Organization business units and departments

- Geographic locations

- Business processes, systems, and networks
- Time periods

» **Applicable standards or regulations:** Often, an audit is performed under the auspices of a law, regulation, or standard, which determines such matters as who may perform the audit, auditor qualifications, the type and scope of the audit, and the obligations of the audited organization at the conclusion.

» **Qualifications of auditors:** The personnel who perform audits may be required to have specific work experience, possess specific training and/or certifications, or work in certain types of firms.

» **Types of auditing:** Several activities comprise an audit, including

- *Observation:* Auditors passively observe activities performed by personnel and/or information systems.

- *Inquiry:* Auditors ask questions of control or process owners to understand how key activities are performed.

- *Inspection:* Auditors inspect documents, records, and systems to verify that key controls or processes are operating properly.

- *Reperformance:* Auditors perform tasks or transactions on their own to see whether the results are correct.

» **Sampling:** The process of selecting items in a large population is known as *sampling*. Regulations and standards often specify the types and rates of sampling that are required for an audit.

» **Management response:** In some types of audits, management in the auditee organization is permitted to write a statement in response to an auditor's findings. Management response may range from "We already fixed it" to "We will fix it" to "We don't think this is an issue."

There are three main contexts for audits of information systems and related processes:

» **Internal audit:** Personnel in the organization conduct an audit on selected information systems and/or business processes.

» **External audit:** Auditors from an outside firm conduct an audit on one or more information systems and/or business processes.

>> **Third-party audit:** Auditors, internal or external, perform an audit of a third-party service provider that is performing services on behalf of the organization. An organization may outsource a part of its software development to another company, for example. From time to time, the organization audits the software development company to ensure that its business processes and information systems are in compliance with applicable regulations and business requirements. Alternatively, the third party may hire its own audit firm and make audit reports available to its customers. This approach is common among service providers. SOC 1, SOC 2, and SOC 3 audits are often used for this purpose.

Business–critical systems need to be subject to regular audits as dictated by regulatory, contractual, or trade group requirements.

WARNING

For organizations that are subject to regulatory requirements, such as Sarbanes-Oxley (discussed in Chapter 3), it's all too easy and far too common to make the mistake of focusing on audits and compliance rather than on implementing a truly effective and comprehensive security strategy. Compliance *does not* equal security. Compliance isn't optional, but neither is security. Don't assume that achieving compliance will automatically achieve effective security (or vice versa). Fortunately, security and compliance aren't mutually exclusive, but you need to ensure that your efforts truly achieve both objectives.

Chapter **9**

Security Operations

T he Security Operations domain covers lots of essential security concepts and builds on many of the other security domains, including Security and Risk Management (Chapter 3), Asset Security (Chapter 4), Security Architecture and Engineering (Chapter 5), and Communication and Network Security (Chapter 6). Security operations represents routine operations that occur across many of the CISSP domains. This domain represents 13 percent of the CISSP certification exam.

Understand and Comply with Investigations

Conducting investigations for various purposes is an important function for security professionals. You must understand evidence collection and handling procedures, reporting and documentation requirements, various investigative processes, and digital forensics tools and techniques. Successful conclusions in investigations depend heavily on proficiency in these skills.

CROSS REFERENCE

This section covers Objective 7.1 of the Security Operations domain in the CISSP Exam Outline (May 1, 2021).

Evidence collection and handling

Evidence is information presented in a court of law to confirm or dispel a fact that's under contention, such as the commission of a crime, the violation of policy, or an ethics matter. A case can't be brought to trial or other legal proceeding without sufficient evidence to support the case. Thus, gathering and protecting evidence properly is one of the most important and most difficult tasks that an investigator must master.

Important evidence collection and handling topics covered on the CISSP exam include the types of evidence, rules of evidence, admissibility of evidence, chain of custody, and the evidence life cycle.

Types of evidence

Sources of legal evidence that you can present in a court of law generally fall into one of four major categories:

>> **Direct evidence:** Oral testimony or a written statement based on information gathered through a witness's five senses (in other words, an eyewitness account) that proves or disproves a specific fact or issue.

>> **Real (or physical) evidence:** Tangible objects from the actual crime, such as the tools or weapons used and any stolen or damaged property; may also include visual or audio surveillance tapes generated during or after the event. Physical evidence from a computer crime is not always available.

>> **Documentary evidence:** Includes originals and copies of business records, computer-generated and computer-stored records, manuals, policies, standards, procedures, and log files. Most evidence presented in a computer crime case is documentary evidence. The *hearsay rule* (which we discuss in "Hearsay rule" later in this chapter) is an extremely important test of documentary evidence that must be understood and applied to this type of evidence.

>> **Demonstrative evidence:** Used to aid the court's understanding of a case. Opinions are considered to be demonstrative evidence and may be *expert* (based on personal expertise and facts) or *nonexpert* (based on facts only). Other examples of demonstrative evidence include models, simulations, charts, and illustrations.

Other types of evidence that may fall into one or more of the major categories include

>> **Best evidence:** Original, unaltered evidence, which courts prefer over secondary evidence. Read more about this evidence in "Best evidence rule" later in this chapter.

>> **Secondary evidence:** A duplicate or copy of evidence, such as a tape backup, screen capture, or photograph.

>> **Corroborative evidence:** Evidence that supports or substantiates other evidence presented in a case.

>> **Conclusive evidence:** Incontrovertible and irrefutable evidence — you know, the smoking gun.

>> **Circumstantial evidence:** Relevant facts that you can't directly or conclusively connect to other events, but about which a reasonable person can make a reasonable inference.

Rules of evidence

Important rules of evidence for computer crime cases include the *best evidence rule* and the *hearsay evidence rule*. The CISSP candidate must understand both of these rules and their applicability to evidence in computer crime cases.

BEST EVIDENCE RULE

The best evidence rule, defined in the U.S. Federal Rules of Evidence, states that "to prove the content of a writing, recording, or photograph, the original writing, recording, or photograph is [ordinarily] required."

The Federal Rules of Evidence, however, define an exception to this rule as "[i]f data are stored in a computer or similar device, any printout or other output readable by sight, shown to reflect the data accurately, is an 'original.'"

Thus, data extracted from a computer — if that data is a fair and accurate representation of the original data — satisfies the best evidence rule and may be introduced into court proceedings as such.

HEARSAY RULE

Hearsay evidence is evidence that's not based on personal, firsthand knowledge of a witness but comes from other sources. Under the Federal Rules of Evidence, hearsay evidence normally is not admissible in court. This rule exists to prevent unreliable testimony from improperly influencing the outcome of a trial.

Business records, including computer records, have traditionally, and perhaps mistakenly, been considered hearsay evidence by most courts because these records cannot be proved to be accurate and reliable. One of the most significant obstacles for a prosecutor to overcome in a computer crime case is getting computer records admitted as evidence.

TIP

A prosecutor may be able to introduce computer records as best evidence, rather than hearsay evidence.

Several courts have acknowledged that the hearsay rules are applicable to *computer-stored* records containing human statements but are not applicable to *computer-generated* records untouched by human hands.

Perhaps the most successful and commonly applied test of admissibility for computer records, in general, has been the *business records exception*, established in the Federal Rules of Evidence for records of regularly conducted activity that meet the following criteria:

>> Made at (contemporaneously) or near the time when the act occurred

>> Made by a person who has knowledge of the business process or from information transmitted by a person who has knowledge of the business process

>> Made and relied on during the regular conduct of business or in the further-ance of the business, as verified by the custodian or other witness who is familiar with the records' use

>> Kept for motives that tend to ensure their accuracy

>> In the custody of the witness on a regular basis (as required by the chain of evidence)

The chain of evidence establishes accountability for the handling of evidence throughout the evidence life cycle. See "Chain of custody and the evidence life cycle" later in this chapter.

Admissibility of evidence

Because computer-generated evidence can be easily manipulated, altered, or tampered with, and because it's not easily and commonly understood, this type of evidence is usually considered to be suspect in a court of law. To be admissible, evidence must be

>> **Relevant:** It must tend to prove or disprove facts that are relevant and material to the case.

>> **Reliable:** It must be reasonably proved that what is presented as evidence is what was originally collected and that the evidence itself is reliable. This proof is established in part through proper evidence handling and the chain of custody. (We discuss this topic in the upcoming section "Chain of custody and the evidence life cycle.")

>> **Legally permissible:** It must be obtained through legal means. Evidence that's not legally permissible may include evidence obtained through the following means:

- *Illegal search and seizure:* Law enforcement personnel must obtain a court order. But non–law enforcement personnel, such as a supervisor or system administrator, may be able to conduct an authorized search under some circumstances.

- *Illegal wiretaps or phone taps:* Anyone conducting wiretaps or phone taps must obtain a court order.

- *Entrapment or enticement: Entrapment* encourages a person to commit a crime that they may have had no intention of committing. Conversely, *enticement* lures a person toward certain evidence (a honeypot, if you will) after they have already committed a crime. Enticement isn't necessarily illegal, but it does raise certain ethical arguments and may not be admissible in court.

- *Coercion:* Coerced testimony or confessions are not legally permissible. Coercion involves compelling a person to provide evidence involuntarily through the use of threats, violence (torture), bribery, trickery, or intimidation.

- *Unauthorized or improper monitoring:* Active monitoring must be properly authorized and conducted in a standard manner; users must be notified that they may be subject to monitoring.

Chain of custody and the evidence life cycle

The *chain of custody* (or *chain of evidence*) provides accountability and protection for evidence throughout its entire life cycle and includes the following information, which is normally kept in an evidence log:

>> **People involved (who):** Identify any and all people who discovered, collected, seized, analyzed, stored, preserved, transported, or otherwise controlled the evidence; also identify any witnesses or other people who were present during any of these activities

>> **Description of evidence (what):** Ensure that all evidence is described completely and uniquely

>> **Location of evidence (where):** Provide specific information about the evidence's location when it is discovered, analyzed, stored, or transported

>> **Date/time (when):** Record the date and time when evidence is discovered, collected, seized, analyzed, stored, or transported; also record date and time information for any evidence log entries associated with the evidence

>> **Methods used (how):** Provide specific information about how evidence was discovered, collected, stored, preserved, or transported

Any time evidence changes possession or is transferred to a different media type, it must be recorded properly in the evidence log to maintain the chain of custody.

Law enforcement officials must strictly adhere to chain-of-custody requirements, and this adherence is highly recommended for anyone else who is involved in collecting or seizing evidence. Security professionals and incident response teams must fully understand and follow chain-of-custody principles and procedures, no matter how minor or insignificant a security incident may initially appear to be. In both cases, chain of custody serves to prove that digital evidence has not been modified at any point in the forensic examination and analysis.

Even properly trained law enforcement officials sometimes make crucial mistakes in evidence handling and safekeeping. Most attorneys won't understand the technical aspects of the evidence that you may present in a case, but they will definitely know evidence-handling rules and will most certainly scrutinize your actions in this area. Improperly handled evidence, no matter how conclusive or damaging, will likely be inadmissible in a court of law.

The *evidence life cycle* describes the various phases of evidence, from its initial discovery to its final disposition. The evidence life cycle has the following five stages:

>> Collection and identification

>> Analysis

>> Storage, preservation, and transportation

>> Presentation in court

>> Final disposition, such as return to owner or destroy (for copies)

The following sections explain more about these stages.

COLLECTION AND IDENTIFICATION

Collecting evidence involves taking that evidence into custody. Unfortunately, evidence can't always be collected and must instead be seized. Many legal issues are involved in seizing computers and other electronic evidence. The publication *Searching and Seizing Computers and Obtaining Evidence in Criminal Investigations*, 3rd Edition (2009), published by the U.S. Department of Justice Computer Crime and Intellectual Property Section, provides comprehensive guidance on this subject. This publication is available for download at https://www.justice.gov/sites/default/files/criminal-ccips/legacy/2015/01/14/ssmanual2009.pdf.

In general, law enforcement officials can search and/or seize computers and other electronic evidence under any of four circumstances:

>> **Voluntary or consensual:** The owner of the computer or electronic evidence can freely surrender the evidence.

>> **Subpoena:** A court issues a subpoena to a person, ordering that person to deliver the evidence to the court.

>> **Search warrant or Anton Piller order:** A *search warrant* is issued to a law enforcement official by the court, allowing that official to search and seize specific evidence. An *Anton Piller order* allows the premises to be searched and evidence seized without warning, usually to prevent possible destruction of evidence.

>> **Exigent circumstances:** If probable cause exists and the destruction of evidence is imminent, that evidence may be searched or seized without a warrant.

When evidence is collected, it must be marked and identified properly to ensure that it can be presented in court properly as actual evidence gathered from the scene or incident. The collected evidence must be recorded in an evidence log with the following information:

>> A description of the piece of evidence, including specific information such as make, model, serial number, physical appearance, material condition, and preexisting damage

>> The name(s) of the person or people who discovered and collected the evidence

>> The exact date and time, specific location, and circumstances of the discovery/collection

Additionally, the evidence must be marked according to the following guidelines:

>> **Mark the evidence.** If possible, without damaging the evidence, mark the piece of evidence with the collecting person's initials, the date, and the case number (if known). Seal the evidence in an appropriate container, and again mark the container with the same information.

>> **Use an evidence tag.** If the actual evidence cannot be marked, attach an evidence tag with the information in the preceding item, seal the evidence and tag it in an appropriate container, and mark the container with the same information.

>> **Seal the evidence.** Seal the container with evidence tape, and mark the tape in a manner that will clearly indicate any tampering or altering of the evidence.

>> **Protect the evidence.** Use extreme caution when collecting and marking evidence to ensure that it's not damaged. If you're using plastic bags for evidence containers, make sure that they're static-free to protect magnetic media.

Always collect and mark evidence in a consistent manner so that you can easily identify evidence and describe your collection and identification techniques to an opposing attorney in court, if necessary.

ANALYSIS

Analysis involves examining the evidence for information pertinent to the case. Analysis should be conducted with extreme caution — and only by experienced, properly trained personnel — to ensure the evidence is not altered, damaged, or destroyed.

STORAGE, PRESERVATION, AND TRANSPORTATION

All evidence must be stored properly in a secure facility and preserved to prevent damage or contamination from various hazards, including intense heat or cold, extreme humidity, water, magnetic fields, and vibration. Evidence that's not properly protected may be inadmissible in court, and the party responsible for collection and storage may be liable. Care must also be exercised during transportation to ensure that evidence is not lost, temporarily misplaced, damaged, or destroyed.

PRESENTATION IN COURT

Evidence to be presented in court must continue to follow the chain of custody and be handled with the same care as at all other times in the evidence life cycle. This process continues throughout the trial until all testimony related to the evidence is completed and the trial has concluded, or the case is settled or dismissed.

FINAL DISPOSITION

After the conclusion of the trial or other disposition, evidence is normally returned to its proper owner. Under some circumstances, however, certain evidence may be ordered destroyed, such as contraband, drugs, or drug paraphernalia. Any evidence obtained through a search warrant is legally under the control of the court, possibly requiring the original owner to petition the court for its return.

Reporting and documentation

As described in the preceding section, complete and accurate recordkeeping is critical to each investigation. An investigation's report is intended to be a complete record of an investigation and usually includes the following:

>> Incident investigators, including their qualifications and contact information

>> Names of the parties interviewed, including their roles, involvement, and contact information

>> List of all evidence collected, including chain(s) of custody

>> Tools used to examine or process evidence, including versions

>> Samples and sampling methodologies used, if applicable

>> Computers used to examine, process, or store evidence, including a description of configuration

>> Root-cause analysis of incident, if applicable

>> Conclusions and opinions of investigators

>> Hearings or proceedings

>> Parties to whom the report is delivered

Investigative techniques

An investigation should begin immediately upon report of an alleged computer crime, policy violation, or incident. Any incident should be handled, at least initially, as a computer crime investigation or policy violation until a preliminary investigation determines otherwise. Different investigative techniques may be required, depending on the goal of the investigation or applicable laws and regulations. Incident handling, for example, requires expediency to contain any potential damage as quickly as possible. A root-cause analysis requires in-depth examination to determine what happened, how it happened, and how to prevent the same thing from happening again. In all cases, proper evidence collection and handling are essential. Even if a preliminary investigation determines that a security incident was not the result of criminal activity, you should always handle any potential evidence properly, in case further legal proceedings are anticipated or a crime is uncovered during the course of a full investigation. The CISSP candidate should be familiar with the general steps of the investigative process:

1.   Detect and contain an incident.

Early detection is critical to a successful investigation. Unfortunately, computer-related incidents usually involve passive or reactive detection techniques (such as the review of audit trails and accidental discovery), which often leave a cold evidence trail. Containment minimizes further loss or damage. The computer incident response team (CIRT), which we discuss later in this chapter, normally is responsible for conducting an investigation. The team should be notified or activated as quickly as possible after a computer crime is detected or suspected.

2.   Notify management.

Management must be notified of any investigations as soon as possible. Knowledge of the investigations should be limited to as few people as possible and on a need-to-know basis. Out-of-band communication methods (reporting in person) should be used to ensure that an intruder does not intercept sensitive communications about the investigation.

3.   Conduct a preliminary investigation.

This preliminary investigation determines whether an incident or crime actually occurred. Most incidents turn out to be honest mistakes rather than malicious conduct. This step includes reviewing the complaint or report, inspecting damage, interviewing witnesses, examining logs, and identifying further investigation requirements.

4. Determine whether the organization should disclose that the crime occurred.

First, and most important, determine whether laws or regulations require the organization to disclose a crime or incident. Next, by coordinating with a public relations or public affairs official of the organization, determine whether the organization wants to disclose this information.

5. Conduct the investigation.

Conducting the investigation involves three activities:

(a) Identify potential suspects.

Potential suspects include organization insiders and outsiders. One standard discriminator that helps identify and eliminate potential suspects is the MOM test: Did the suspect have the motive, opportunity, and means? The *motive* might relate to financial gain, revenge, or notoriety. A suspect had *opportunity* if they had access, whether as an authorized user for an unauthorized purpose or as an unauthorized user (due to a security weakness or vulnerability) for an unauthorized purpose. *Means* relates to whether the suspect had the necessary tools and skills to commit the crime.

(b) Identify potential witnesses.

Determine whom you want to interview and who should conduct the interviews. Be careful not to alert any potential suspects to the investigation; focus on obtaining facts, not opinions, in witness statements.

(c) Prepare for search and seizure.

Identify the types of systems and evidence that you plan to search or seize, designate and train the search and seizure team members (normally, members of the CIRT), obtain and serve proper search warrants (if required), and determine the potential risk to the system during a search-and-seizure effort.

6. Report your findings.

The results of the investigation, including evidence, should be reported to management and turned over to proper law enforcement officials or prosecutors as appropriate.

MOM stands for *motive, opportunity,* and *means.* Motive refers to the reason or incentive for a suspect to commit a crime, such as financial gain or revenge. Opportunity refers to the suspect having a chance or opening to commit the crime — for example, if they had access (either authorized or unauthorized) to a system that was breached. Means refers to a suspect having the ability to commit a crime — for example, they had the skills and tools to commit the crime.

Digital forensics tools, tactics, and procedures

Digital forensics is the science of conducting a computer incident investigation to determine what has happened and who is responsible, and to collect legally admissible evidence for use in subsequent legal proceedings, such as a criminal investigation, internal investigation, or lawsuit.

Proper forensic analysis and investigation requires in-depth knowledge of hardware (such as endpoint devices and networking equipment), operating systems (including desktop, server, mobile device, and other device operating systems, such as routers, switches, and load balancers), applications, databases, and software programming languages, as well as knowledge and experience using sophisticated forensics tools and tool kits.

The types of forensic data-gathering techniques include

>> **Hard drive forensics:** Specialized tools are used to create one or more forensically identical copies of a computer's hard drive. A device called a *write blocker* is typically used to prevent any possible alterations to the original drive. Cryptographic checksums can be used to verify that a forensic copy is an exact duplicate of the original. Then tools are used to examine the contents of the hard drive to determine the following:

   • Last known state of the computer

   • History of files accessed

   • History of files created

   • History of files deleted

   • History of programs executed

   • History of websites visited by a browser

   • History of attempts by the user to remove evidence

>> **Live forensics:** Specialized tools are used to examine a running system, including

- Running processes
- Currently open files
- Contents of main storage (RAM)
- Keystrokes
- Communications traffic in and out of the computer

Live forensics are difficult to perform because the tools used to collect information can affect the system being examined.

Artifacts

Key artifacts that may be collected during an investigation may include computers, mobile devices, servers (physical or virtual), network equipment (such as routers and switches), and security equipment (such as a firewall). These artifacts may contain indicators of compromise (IoC) that can be preserved as evidence to support an investigation.

Conduct Logging and Monitoring Activities

Event logging is an essential part of an organization's IT operations. Increasingly, organizations are implementing centralized log collection systems that often serve as security information and event management (SIEM) platforms.

CROSS REFERENCE

This section covers Objective 7.2 of the Security Operations domain in the CISSP Exam Outline (May 1, 2021).

Intrusion detection and prevention

Intrusion detection is a passive technique used to detect unauthorized activity on a network. An intrusion detection system is frequently called an *IDS*. Three types of IDSes used today are

>> **Network-based:** Consists of a separate device attached to a network that listens to all network traffic by using various methods (which we describe later in this section) to detect anomalous activity

>> **Host-based:** A subset of network-based IDS in which only the network traffic destined for a particular host is monitored

>> **Wireless:** Another type of network intrusion detection that focuses on wireless intrusion by scanning for rogue access points

Both network- and host-based IDSes use a couple of methods:

>> **Signature-based:** A *signature-based* IDS compares network traffic that is observed with a list of patterns in a signature file. A signature-based IDS detects any of a known set of attacks, but if an intruder is able to change the patterns that they use in the attack, the attack may be able to slip by the IDS without being detected. The other downside of signature-based IDS is that the signature file must be updated frequently.

>> **Reputation-based:** Closely akin to signature-based alerting, reputation-based alerting detects when communications and other activities involve known-malicious domains and IP networks. Some IDSes update themselves several times daily, including adding to a list of known-malicious domains and IP addresses. Then, when any activities are associated with a known-malicious domain or IP address, the IDS can create an alert that lets personnel know about the activity.

>> **Anomaly-based:** An *anomaly-based* IDS monitors all the traffic over the network and builds traffic profiles. Over time, the IDS will report deviations from the profiles that it has built. The upside of anomaly-based IDSes is that there are no signature files to update periodically. The downside is that you may have a high volume of false positives. Behavior-based and heuristics-based IDSes are similar to anomaly-based IDSes and have many of the same advantages. Rather than detect anomalies in normal traffic patterns, behavior-based and heuristics-based systems attempt to recognize and remember potential attack patterns.

Intrusion detection doesn't stop intruders, but intrusion *prevention* does . . . or at least, it slows them down. *Intrusion prevention systems* (IPSes) are newer and more common than IDSes, and IPSes are designed to detect and block intrusions. An IPS is simply an IDS that can take action, such as dropping a connection or blocking a port, when an intrusion is detected.

REMEMBER

Intrusion detection looks for known attacks and/or anomalous behavior on a network or host.

See Chapter 6 for more on IDSes and IPSes.

Security information and event management

Security information and event management (SIEM) solutions provide real-time collection, analysis, correlation, and presentation of security logs and alerts generated by various network sources (such as firewalls, IDSes/IPSes, routers, switches, servers, and workstations).

A SIEM solution can be software- or appliance-based, and may be hosted and managed internally or by a managed security service provider.

A SIEM requires a lot of up-front configuration and tuning, so only the most important, actionable events are brought to the attention of staff members in the organization. The effort is worthwhile, however: A SIEM combs through millions or billions of events daily and presents only the most important few actionable events so that security teams can take appropriate action.

Many SIEM platforms also have the ability to accept threat intelligence feeds from various vendors, including the SIEM manufacturers. This approach permits the SIEM to adjust its detection and blocking capabilities automatically for the most up-to-date threats.

Security orchestration, automation, and response

A *security orchestration, automation, and response* (SOAR) solution takes a SIEM one step further through the automation of repeatable tasks as a result of an event that has been detected. A SIEM might produce an alert regarding suspicious activity originating from the Internet, and an analyst might investigate the alert by performing various tasks to obtain additional information about the originating system or the contents of the alert. Such investigating may take several minutes, during which time an intruder may be performing reconnaissance or stealing information. With SOAR, these steps may take only seconds, providing the analyst key actionable information. A SOAR platform can also direct response steps, such as black-holing a domain or IP address, locking a user or administrator account, or enacting a firewall rule to contain an intrusion. In other cases, a SOAR can prepare for action and perform that action when an analyst approves it.

Continuous monitoring

Continuous monitoring technology collects and reports security data in near real time. Continuous monitoring components may include

- >> **Discovery:** Ongoing inventory of network and information assets, including hardware, software, and sensitive data
- >> **Assessment:** Automatic scanning and baselining of information assets to identify and prioritize vulnerabilities
- >> **Threat intelligence:** Feeds from one or more outside organizations that produce high-quality, actionable data
- >> **Audit:** Nearly real-time evaluation of device configurations and compliance with established policies and regulatory requirements
- >> **Scanning:** Automatic scanning of systems and networks to discover new vulnerabilities
- >> **Patching:** Automatic security patch installation and software updating
- >> **Reporting:** Aggregating, analyzing, and correlating log information and alerts

Egress monitoring

Egress monitoring (or *extrusion detection*) is the process of monitoring outbound traffic to discover potential data leakage (or loss). Modern cyberattacks employ various stealth techniques to avoid detection as long as possible for the purpose of data theft. These techniques may include the use of encryption, such as secure sockets layer/transport layer security (SSL/TLS) and steganography (discussed in Chapter 4).

Data loss prevention (DLP) systems are often used to detect the exfiltration of sensitive data, such as personally identifiable information (PII) or protected health information (PHI)in email messages, data uploads, PNG or JPEG images, and other forms of communication. These technologies often perform deep packet inspection to decrypt and inspect outbound traffic that is TLS encrypted.

DLP systems can also be used to disable the use of removable media drive interfaces on servers and workstations, as well as to encrypt data written to removable media so that only systems with the same organization's DLP agent can read the contents of the removable media drive.

Static DLP tools are used to discover sensitive and proprietary data in databases, file servers, and other data storage systems.

Log management

Log data is (or should be) collected from practically every application, server, network, and security device, and user device in an organization's environment. Without complete log data, it can be nearly impossible to determine exactly what happened in the event of an attack.

To the greatest extent possible, log information should be synchronized to a network time server to ensure that log data from disparate sources can be correlated accurately. Logs should be stored centrally and securely to ensure that the data collected is immutable and can be readily ingested into various security analytics platforms, SIEM solutions, and other security tools for log aggregation, analysis, and correlation. Appropriate retention periods for log information should also be defined and implemented based on legal or regulatory compliance requirements.

Threat intelligence

Threat intelligence involves collecting and analyzing data about attempted or successful intrusions. Global threat intelligence is usually provided automatically to subscriber organizations via a threat intelligence feed. You might subscribe to numerous threat feeds from your firewall, endpoint protection (antimalware), and other security vendors that collect and analyze threat information from numerous sources, typically including their globally deployed customer bases. Security teams use the information from these threat intelligence feeds to proactively hunt and identify active threats and bad actors within their organization's environment.

A key element in threat intel feeds is the *indicator of compromise* (IoC), which is digital information that an organization can use to determine whether specific types of intrusion or malware are present in an organization's environment. An IoC typically takes the form of a virus signature, IP address, MD5 hash of a malware file, URL, or domain name. Security analysts may use tools to proactively search for IoCs in an activity known as *threat hunting*.

Machine-readable threat intel feeds use any of several formats, including CSV (comma-separated values), STIX (Structured Threat Information Expression), XML (Extensible Markup Language), JSON (JavaScript Object Notation), OpenIOC (Open Indicators of Compromise), and TAXII (Trusted Automated Exchange of Indicator Information).

TIP

Threat intel tools enable an organization to detect the tactics, techniques, and procedures that threat actors use to attack networks and systems.

User and entity behavior analysis

User and entity behavior analysis (UEBA) is used in various security tools to create a baseline of "normal" user, device, or other activity on the network. Anomalous behavior that deviates from this baseline, beyond a defined threshold, may be an indicator of threat activity and can be alerted on or trigger an automated response. A user who normally logs in between 9 a.m. and 5 p.m. on the East Coast of the United States, for example, may be required to use an authorized (managed or domain-joined) device and multifactor authentication (MFA) if they attempt to log into the network after 10 p.m. from Europe. The user may, in fact, be on a business trip and therefore legitimately needs to log in outside their "normal" hours, in which case they can still log in with some additional precautions to verify their identity. Or they may be sound asleep at home in New York, and the login attempt is coming from a threat actor somewhere in Europe, in which case the additional precautions will help prevent a successful intrusion using the unsuspecting user's credentials.

Perform Configuration Management

An organization's information architecture changes all the time. As a result, its security posture changes all the time. Provisioning and decommissioning various information resources can have significant effects, both direct and indirect, on the organization's security posture. An application, for example, might directly introduce new vulnerabilities into an environment or integrate with a database in a way that compromises the integrity of the database. Or a system administrator might create a new virtual machine (VM) in a public cloud environment and forget a step in the VM's security settings.

CROSS REFERENCE

This section covers Objective 7.3 of the Security Operations domain in the CISSP Exam Outline (May 1, 2021).

Security planning and analysis must be an integral part of every organization's resource provisioning processes, as well as throughout the life cycle of all resources. Important security considerations include

>> **Provisioning:** Security should be consulted any time the organization is considering introducing new equipment, such as a Wi-Fi access point or network router from a manufacturer whose products have not previously been deployed in the environment. This approach ensures that security can assess any known risks associated with the new equipment and its impact on the organization's overall security posture.

>> **Asset management (or inventory):** Maintaining a complete, accurate inventory is critical to ensure that all potential vulnerabilities and risks in an environment can be identified, assessed, and addressed. Indeed, so many other critical security processes depend on sound asset inventory that asset inventory is one of the most important (if most mundane) activities in IT organizations.

TIP

Asset management is covered in the first two controls of the Center for Internet Security (CIS) 20 Controls and the first category of the first (Identify) core function of the National Institute of Standards and Technology's (NIST) Cybersecurity Framework (CSF). Asset management appears first in these and other)important frameworks because it is fundamental to most other security controls.

>> **Baselining:** Establishing a baseline helps security teams tune security events and alerts that are received and can also be used to feed user and entity behavior capabilities (discussed earlier in this chapter) in security tools deployed throughout the environment.

>> **Change management:** Change management processes are used to strictly control changes to systems in production environments so that only duly requested and approved changes are made.

>> **Configuration management:** Configuration management processes need to be implemented and strictly enforced to ensure that information resources are operated in a safe and secure manner. Organizations typically implement an automated configuration management database (CMDB) that is part of a system configuration management system used to manage asset inventory data. Often, this database is also used to manage the configuration history of systems.

>> **Physical assets:** Physical assets must be protected against loss, damage, or theft. Valuable or sensitive data stored on a physical asset may far exceed the value of the asset itself.

>> **Virtual assets:** VM sprawl has increasingly become an issue for organizations with the popularity of virtualization technology and software defined networks (SDN). VMs can be (and often are) provisioned in a matter of minutes but aren't always properly decommissioned when they are no longer needed. Dormant VMs aren't always backed up and can go unpatched for many months, exposing the organization to increased risk from unpatched security vulnerabilities.

Of particular concern to security professionals is the implementation of VMs without proper review and approvals. This problem didn't exist before virtualization, as organizations had other checks and balances in place to prevent the implementation of unauthorized systems (namely, the purchasing process). But VMs can be implemented unilaterally, often without the knowledge or involvement of other personnel within the organization.

- » **Cloud assets:** As more organizations adopt cloud strategies, including Software as a Service (SaaS), Platform as a Service (PaaS), and Infrastructure as a Service (IaaS) solutions, it's important to keep track of these assets. Ultimately, an organization is responsible for the security and privacy of its applications and data — not the cloud service provider. Issues of data residency and transborder data flow need to be considered.

  A new class of security tools known as *cloud access security brokers* (CASB) can detect access to and use of cloud-based services. These tools give the organization more visibility into its sanctioned and unsanctioned use of cloud services. Many of these systems, in cooperation with cloud services, can be used to control the use of cloud services.

- » **Applications:** This category includes commercial and custom applications, private clouds, web services, SaaS products, and the interfaces and integrations among application components. Securing the provisioning of these assets requires strict access controls; only designated administrators should be able to deploy and configure them.

- » **Automation:** IT organizations, including DevOps teams, increasingly use automation and orchestration to deploy information resources at scale. This massive level of activity requires coordination with security teams to ensure that information resources can be accounted for properly and security risks identified. It is increasingly common, for example, for an application deployed in a microservices architecture to provision and deprovision thousands of containerized ephemeral microservices, deployed across a multi-cloud environment, with a life span of a few minutes or even seconds.

Apply Foundational Security Operations Concepts

Fundamental security operations concepts that CISSP candidates need to understand and manage well include the principles of need to know and least privilege, separation of duties and responsibilities, monitoring of special privileges, job rotation, information life cycle management, and service-level agreements.

CROSS REFERENCE

This section covers Objective 7.4 of the Security Operations domain in the CISSP Exam Outline (May 1, 2021).

Need-to-know and least privilege

The *need-to-know* concept states that only people with a valid business justification should have access to specific information or functions. In addition to having a need to know, a person must have an appropriate security clearance level to be granted access. Conversely, a person who has the appropriate security clearance level but no need to know should not be granted access.

One of the most difficult challenges in managing need to know is the use of controls that enforce it. Also, information owners need to be able to distinguish *I need to know* from *I want to know, I want to feel important,* and *I'm just curious.*

Need-to-know is closely related to the concept of least privilege and can help organizations implement least privilege in a practical manner.

The principle of *least privilege* states that people should have the capability to perform only the tasks (or have access to only the data) required for their primary jobs — no more.

Giving a person more privileges and access than required invites trouble. Offering the capability to perform more than the job requires may become a temptation that results, sooner or later, in an abuse of privilege.

Giving a user full permissions on a network share, for example, rather than just read and modify rights to a specific directory, opens the door not only to abuse of those privileges (such as reading or copying other sensitive information on the network share), but also to costly mistakes (such as accidentally deleting a file or the entire directory). As a starting point, organizations should approach permissions with a "deny all" mentality and add needed permissions as required.

TIP

Least privilege is closely related to separation of duties and responsibilities, described in the following section. Distributing the duties and responsibilities for a given job function among several people means that those people require fewer privileges on a system or resource.

REMEMBER

The principle of least privilege states that people should have the fewest privileges necessary to allow them to perform their tasks.

Several important concepts associated with need to know and least privilege include

>> **Entitlement:** When a new user account is provisioned in an organization, the permissions granted to that account must be appropriate for the level of access required by the user. In too many organizations, human resources simply instructs the IT department to give a new user "whatever so-and-so (another user in the same department) has access to." Instead, entitlement needs to be based on the principle of least privilege.

>> **Aggregation:** When people transfer between jobs and/or departments within an organization (see "Job rotation" later in this chapter), they often need different access and privileges to do their new jobs. Far too often, organizational security processes do not adequately ensure that access rights that are no longer required are actually revoked. Instead, people accumulate privileges, and over a period of many years, an employee can have far more access and privileges than they actually need. This situation is known as *aggregation,* and it's the antithesis of least privilege. *Privilege creep* and *accumulation of privileges* are common terms.

>> **Transitive trust:** Trust relationships (in the context of security domains) are often established within and between organizations to facilitate ease of access and collaboration. A trust relationship enables subjects (such as users or processes) in one security domain to access objects (such as servers or applications) in another security domain. (See chapters 5 and 7 for more about objects and subjects.) A transitive trust extends access privileges to the subdomains of a security domain, analogous to inheriting permissions to subdirectories within a parent directory structure. Instead, a nontransitive trust should be implemented by requiring access to each subdomain to be granted explicitly based on the principle of least privilege rather than inherited.

Separation of duties and responsibilities

The concept of *separation* (or *segregation*) *of duties* (SoD) *and responsibilities* ensures that no single person has complete authority and control of a critical system or process. This practice promotes security in the following ways:

>> **Reduces opportunities for fraud or abuse:** For fraud or abuse to occur, two or more people must collude or be complicit in the performance of their duties.

>> **Reduces high-impact mistakes:** Because two or more people perform the process, mistakes are less likely to occur, or are more quickly detected and corrected.

>> **Reduces dependence on a single person:** Critical processes are accomplished by groups of people. Multiple people should be trained on different parts of the process (such as through job rotation, discussed in the following section) to help ensure that the absence of one person doesn't unnecessarily delay or impede the successful completion of a step in the process.

Here are some common examples of separation of duties and responsibilities within organizations:

>> A bank assigns the first three numbers of a six-number safe combination to one employee and the second three numbers to another employee. A single employee isn't permitted to have all six numbers, so a lone employee is unable to gain access to the safe and its contents.

>> An accounting department might separate record entry and internal auditing functions or accounts payable and check disbursing functions.

>> A system administrator is responsible for setting up new accounts and assigning permissions, which a security administrator then verifies.

>> A programmer develops software code, but a separate person is responsible for testing and validation, and yet another person is responsible for loading the code on production systems.

>> Destruction of classified materials may require two people to complete or witness the destruction.

>> Disposal of assets may require an approval signature by the office manager and verification by building security.

In smaller organizations, separation of duties and responsibilities can be difficult to implement because of limited personnel and resources.

Privileged account management

Privileged entity controls are the mechanisms, generally built into computer operating systems and network devices, that provide and monitor privileged access to hardware, software, and data. In Unix and Windows, the controls that permit privileged functions reside in the operating system. Operating systems for servers, desktop computers, and many other devices use the concept of *modes* of execution to define privilege levels for various user accounts, applications, and processes that run on a system. For instance, the Unix root account and Windows

Server Enterprise, Domain, and Local Administrator account roles have elevated rights that allow those accounts to install software, view the entire file system and in some cases access the OS kernel and memory directly.

Specialized tools are used to monitor and record activities performed by privileged and administrative users. This approach helps ensure accountability on the part of each administrator and aids in troubleshooting through the ability to view actions performed by administrators.

System or network administrators typically use privileged accounts to perform operating system and utility management functions. Supervisor or Administrator mode should be used only for system administration purposes. Unfortunately, many organizations allow system and network administrators to use these privileged accounts or roles as their normal user accounts even when they aren't doing work that requires this level of access. Yet another horrible security practice allows administrators to share a single administrator or root account.

Privileged access management (PAM) solutions help security teams organize administrative accounts and service accounts in an environment. Some PAM solutions can permit an administrator to check out temporary privileged access credentials to access a network device or operating system and can even record the administrative session. Although administrators may feel threatened by this level of monitoring, they should remember that a PAM solution that records their session can help exonerate them by implicitly proving that they did not perform some malicious (or accidental) task.

WARNING

System or network administrators occasionally grant root or administrator privileges to normal applications as a matter of convenience, rather than spend the time to figure out exactly what privileges the application requires and then create an account role for the application with only those privileges. Allowing a normal application these privileges is a serious mistake, because applications that run in privileged mode bypass some or all security controls, which could lead to unexpected application behavior. Any user of a payroll application, for example, could view or change anyone's data if the application running in privileged mode was never told *no* by the operating system. Further, if an application running in privileged mode is compromised by an attacker, the attacker may inherit privileged access to the entire system.

TIP

Hackers specifically target Supervisor and other privileged modes because those modes have a great deal of power over systems. The use of Supervisor mode should be limited wherever possible, especially on user workstations.

MONITORING (EVERYBODY'S SPECIAL!)

Monitoring the activities of an organization's users, particularly those who have special (such as administrator) privileges, is an important security operations practice.

User monitoring can include casual or direct observation, analysis of security logs, inspection of workstation hard drives, random drug testing (in certain job functions and in accordance with applicable laws), audits of attendance and building access records, review of call logs and transcripts, and other activities.

User monitoring and its purposes should be fully addressed in an organization's written policy manuals. Information systems should display a login warning that clearly informs the user that their activities may be monitored and for what purposes. The login warning should also clearly indicate who owns the information and information assets processed on the system or network and that the user has no expectation of privacy with regard to information stored or processed on the system. The login process should require users to affirmatively acknowledge the login warning by clicking OK or I Agree to gain access to the system.

An organization should conduct user monitoring in accordance with its written policies and applicable laws. Also, only personnel who are authorized to do so (such as security, legal, or human resources) should perform this monitoring, and only for authorized purposes. User and entity behavior analytics is a process that is helpful for detecting potential breaches, intrusions, or other malicious activity by using monitoring data to establish baselines of normal behavior or activity and analyzing anomalies.

Job rotation

Job rotation (or *rotation of duties*) is another effective security control that offers many benefits to an organization. Similar to the concept of separation of duties and responsibilities, job rotation involves regularly or randomly transferring key personnel to different positions or departments within an organization, with or without notice. Job rotations accomplish several important organizational objectives:

>> **Reduce opportunities for fraud or abuse:** Regular job rotations can accomplish this objective in the following two ways:

- People hesitate to set up the means for periodically or routinely stealing corporate information because they know that they could be moved to another shift or task at almost any time.

- People don't work with one another long enough to form collusive relationships that could damage the company.

>> **Eliminate single points of failure:** By ensuring that numerous people within an organization or department know how to perform several job functions, an organization can reduce dependence on certain people and thereby eliminate single points of failure when a person is absent, incapacitated, no longer employed with the organization, or otherwise unavailable to perform a critical job function.

>> **Promote professional growth:** Through cross-training opportunities, job rotations can help an employee's professional growth and career development, and reduce monotony and/or fatigue.

Job rotations can also include changing workers' workstations and work locations, which can also keep would-be saboteurs off balance and less likely to commit wrongful acts.

MANDATORY AND PERMANENT VACATIONS: JOB ROTATIONS OF A DIFFERENT SORT

Mandatory vacations and termination of employment are two important security operations topics that warrant a few paragraphs. You might think of a mandatory vacation as being a short (one- or two-week) job rotation and a termination as being a permanent vacation!

Requiring employees to take one or more weeks of their vacation in a single block of time gives an organization an opportunity to uncover potential fraud or abuse. Employees who engage in illegal or prohibited activities are sometimes reluctant to be away from the office, concerned that these activities will be discovered in their absence as a result of an actual audit or investigation or when someone else who performs that person's normal day-to-day functions in their absence uncovers an irregularity. Less ominously, mandatory vacations may help in other ways:

- Reducing stress and therefore reducing opportunities for mistakes or coercion by others

- Discovering inefficient processes when a substitute performs a job function faster or discovers a better way to get something done

- Revealing single points of failure, shadow processes, and opportunities for job rotation (and separation of duties and responsibilities) when a process or job function idles because the only person who knows how to perform that function is lying on a beach somewhere

As with the practice of separation of duties, small organizations can have difficulty implementing job rotations.

Finally, it is vital to lock down or revoke local and remote access for a terminated employee as soon as possible, especially when the employee is being fired or laid off. The potential consequences associated with continued access by an angry employee are serious enough to warrant emergency procedures for immediate termination of access.

Service-level agreements

Users of business- or mission-critical information systems need to know whether their systems or services will function when they need them, and users need to know more than "Is it up?" or "Is it down *again?*" Their customers, and others, hold users accountable for getting their work done in a timely and accurate manner, so consequently, those users need to know whether they can depend on their systems and services to help them deliver as promised.

The service-level agreement (SLA) is a quasilegal document (a real legal document when it is included in or referenced by a contract) pledging that the system or service performs to a set of minimum standards, such as

>> **Hours of availability:** The wall-clock hours that the system or service will be available for users, which could be 24 x 7 (24 hours per day, 7 days per week) or something more limited, such as daily from 4 a.m. to 12 p.m. Availability specifications may also cite *maintenance windows* (such as Sundays from 2–4 a.m.) when users can expect the system or service to be down for testing, upgrades, and maintenance.

>> **Average and peak number of concurrent users:** The maximum number of users who can use the system or service at the same time.

>> **Transaction throughput:** The number of transactions that the system or service can perform or support in a given period. Usually, *throughput* is expressed as transactions per second, per minute, or per hour.

>> **Transaction accuracy:** The accuracy of transactions that the system or service performs. Generally, this figure is related to complex calculations (such as sales tax) and accuracy of location data.

>> **Data storage capacity:** The amount of data that users can store in the system or service (such as cloud storage). Capacity may be expressed in raw terms (megabytes or gigabytes) or in numbers of transactions.

>> **Response times:** The maximum periods of time (in seconds) that key transactions take. Response times for long processes such as nightly runs and batch jobs also should be covered in the SLA.

>> **Service desk response and resolution times:** The amount of time (usually in hours) that a service or help desk will take to respond to requests for support and resolve any issues.

>> **Mean time between failures:** The amount of time, typically measured in (thousands of) hours, that a component (such as a server hard drive) or system is expected to operate continuously before experiencing a failure.

>> **Mean time to restore service:** The amount of time, typically measured in minutes or hours, that it is expected to take to restore a system or service to normal operation after a failure has occurred.

>> **Security incident response times:** The amount of time (usually in hours or days) between the realization of a security incident and any required notifications to data owners and other affected parties, commonly known as *dwell time*.

>> **Escalation process during times of failure:** When things go wrong, how quickly the service provider will contact the customer, as well as what steps the provider will take to restore service.

REMEMBER

Availability is one of the three tenets of information security (confidentiality, integrity, and availability, discussed in Chapter 3). Therefore, SLAs are important security documents.

Because an SLA is a quantified statement, the service provider and the user alike can take measurements to see how well the service provider is meeting the SLA's standards. This measurement, which is sometimes accompanied by analysis, is frequently called a *scorecard*.

TIP

Operational-level agreements (OLAs) and underpinning contracts (UCs) are important SLA supporting documents. An OLA is essentially an SLA between the interdependent groups that are responsible for the terms of the SLA, such as a service desk and the desktop support team. UCs are used to manage third-party relationships with entities that help support the SLA, such as an external service provider or vendor.

Finally, for an SLA to be meaningful, it needs to have teeth! How will the SLA be enforced, and what will happen when violations occur? What are the escalation procedures? Will any penalties or service credits be paid in the event of a violation? If so, how will penalties or credits be calculated?

TIP

Internal SLAs and OLAs, such as those between an IT department and its users, typically don't provide penalties or service credits for service violations. Internal SLAs are structured more as a commitment between IT and the user community and are useful for managing service expectations. Clearly defined escalation procedures (who gets notified of a problem; when, how, and when it goes up the chain of command) are critical in an internal SLA.

TIP

SLAs rarely, if ever, provide meaningful financial penalties for service violations. An hour of Internet downtime might legitimately cost an e-commerce company $10,000 of business, for example, but most service providers typically only credit an equivalent to the amount paid for the lost hour of Internet service (a few hundred dollars). This amount may seem to be incredibly disproportionate, but consider things from the service provider's perspective. That same credit has to be given to *all* customers that experienced the outage. Thus, an outage could cost the service provider hundreds of thousands of dollars. If service providers were legally obligated to reimburse every customer for their actual losses, it's fair to guess that no one would be in the business of providing Internet service (or that an MPLS circuit would cost a few thousand dollars a month). Instead, look for such penalties as an early-termination clause that lets you get out of a long-term contract if your service provider repeatedly fails to meet its service-level obligations.

HOW MANY NINES?

Availability is often expressed in a percentage of uptime, usually in terms of "How many nines?" In other words, an application, server, or site may be available 99 percent of the time, 99.9 percent of the time, or as much as 99.999 percent of the time. Approximate amounts of downtime per year are shown in the following table:

Percentage	Number of Nines	Downtime Per Year (24/7/365)
99	Two	88 hours
99.9	Three	9 hours
99.99	Four	53 minutes
99.999	Five	5 minutes

Apply Resource Protection

Resource protection is the broad category of controls that protect information assets and information infrastructure.

CROSS REFERENCE

This section covers Objective 7.5 of the Security Operations domain in the CISSP Exam Outline (May 1, 2021).

Resources that require protection include

>> **Communications hardware and software:** Routers, switches, firewalls, load balancers, IPSes, fax machines, virtual private network (VPN) servers, and so on, as well as the software that these devices use

>> **Computers and their storage systems:** All corporate servers and client workstations, storage area networks, network-attached storage, direct-attached storage, near-line, and offline storage systems, cloud-based storage, and backup devices

>> **Business data:** All stored information, such as financial data, sales and marketing information, personnel and payroll data, customer and supplier data, proprietary product or process data, and intellectual property

>> **System data:** Operating systems, utilities, user IDs and password files, audit trails, and configuration files

>> **Backup media:** Tapes, tape cartridges, removable disks, and offsite replicated disk systems

>> **Software:** Application source code, programs, tools, libraries, vendor software, and other proprietary software

Media management

Media management refers to a broad category of controls that are used to manage information classification and physical media. *Data classification* refers to the tasks of marking information according to its sensitivity, as well as the subsequent handling, storage, transmission, and disposal procedures that accompany each classification level. Physical media is similarly marked; likewise, controls specify handling, storage, and disposal procedures. See Chapter 4 for more about data classification.

Sensitive information such as financial records, employee data, and information about customers must be clearly marked, properly handled and stored, and appropriately destroyed in accordance with established organizational policies, standards, and procedures:

>> **Marking:** Marking is the process of affixing human-readable classification labels on documents and data files, whether those files are electronic or hard copy. A marking might read PRIVILEGED AND CONFIDENTIAL. See Chapter 4 for a detailed discussion of data classification.

>> **Tagging:** Tagging is the process of affixing machine-readable classification labels on documents and data files.

>> **Handling:** The organization should have established procedures for handling sensitive information. These procedures detail how employees can transport, transmit, and use such information, as well as any applicable restrictions.

>> **Protection:** Protection involves two components:

- Physical protection of the actual media, such as locked cabinets and secured vehicles

- Logical protection of information on media, such as encryption

>> **Storage and backup:** Similar to handling, the organization must have procedures and requirements specifying how sensitive information must be stored and backed up.

>> **Retention:** Most organizations are bound by various laws and regulations to collect and store certain information, as well as keep it for minimum and/or maximum specified periods. An organization must be aware of legal requirements and ensure that it's in compliance with all applicable regulations. Records retention policies should cover any electronic records that may be located on file servers, document management systems, databases, email systems, archives, and records management systems, as well as paper copies and backup media stored at offsite facilities. Organizations that want to retain information longer than required by law should firmly establish why such information should be kept longer. Nowadays, just having information can be a liability, so keeping information longer should be the exception rather than the norm.

>> **Destruction:** Sooner or later, an organization must destroy sensitive information. The organization must have procedures detailing how to destroy sensitive information that was previously retained, whether the data is in hard-copy form or an electronic file.

WARNING

At the opposite end of the records retention spectrum, many organizations destroy records (including backup media) as soon as legally permissible to limit the scope and cost of any *future* discovery requests or litigation. Before implementing draconian retention policies that severely restrict your organization's retention periods, you should fully understand the negative implications of a policy for your disaster recovery capabilities. Also, consult your organization's legal counsel to ensure that you're in full compliance with all applicable laws and regulations.

Although extremely short retention policies and practices may be prudent for limiting future discovery requests or litigation, they're *illegal* for limiting *pending* discovery requests or litigation (or even records that you have a reasonable expectation may become the subject of future litigation). In such cases, don't destroy pertinent records; if you do, you go to jail. You go directly to jail! You don't pass Go, you don't collect $200, and (oh, yeah) you don't pass the CISSP exam, either — or even remain eligible for CISSP certification.

Media protection techniques

Media protection techniques span a broad array of technologies and approaches, depending on the media. A mobile device may encrypt data and automatically back up to the cloud. An organization might also use mobile device management (MDM) software to enforce additional protections and controls on the device. Similarly, user endpoints such as desktop and laptop PCs may have a trusted platform module (TPM) chip installed to provide hardware-based encryption on the device (discussed in Chapter 5). The operating system may provide additional protections such as disk- or file-level encryption and permissions control. Data loss prevention (DLP) tools may be used to disable an endpoint device's USB port to prevent data from being copied to a removable USB storage device. Servers and storage arrays employ media protection techniques that may include redundant array of independent disks (RAID) protection, snapshots, and replication. Removable media containing sensitive or valuable information may be placed in locking cabinets or stored at a secure off-site storage facility.

Conduct Incident Management

The formal process of detecting, responding to, and fixing a security problem is known as *incident management* (but more properly referred to as *security incident management*).

CROSS REFERENCE

This section covers Objective 7.6 of the Security Operations domain in the CISSP Exam Outline (May 1, 2021).

Do not confuse the concept of incident management, described herein, with the more general concept of incident management as defined by the Information Technology Infrastructure Library's (ITIL) Service Management best practices.

WARNING

Several incident response frameworks include minor variations in the following phases. NIST Special Publication (SP) 800-61 Revision 2, *Computer Security Incident Handling Guide*, lists the following phases of incident handling: Preparation; Detection and Analysis; Containment, Eradication, and Recovery; and Post-Incident Activity.

Incident management includes the following steps:

» **Preparation:** Incident management begins before an incident occurs. Preparation is the key to quick and successful incident management. A well-documented and regularly practiced incident management plan ensures effective preparation. The plan should include

- *Response procedures:* Include detailed procedures that address different contingencies and situations. Some organizations structure detailed procedures in a set of playbooks.

- *Response authority:* Clearly define roles, responsibilities, and levels of authority for all members of the CIRT.

- *Available resources:* Identify people, tools, and external resources (consultants and law enforcement agents) that are available to the CIRT. Training should include the use of these resources when possible.

- *Legal review:* The incident response plan should be evaluated by appropriate legal counsel to determine compliance with applicable laws and to determine whether they're enforceable and defensible.

» **Detection:** Detecting a security incident or event is the first and, often, most difficult step in incident management. Detection may occur through automated monitoring and alerting systems or as the result of a reported security incident (such as a lost or stolen mobile device). Under the best of circumstances, detection may occur in real time as soon as a security incident occurs, such as malware that is discovered by antimalware software on a computer. More often, a security incident may not be detected for quite some time (months or years), as in the case of a sophisticated "low and slow" cyberattack. Determining whether a security incident has occurred is similar to the detection and containment step in the investigative process (discussed earlier in this chapter) and includes defining what constitutes a security incident for your organization.

» **Response:** Upon determination that an incident has occurred, it's important to begin immediate, detailed documentation of every action taken throughout the incident management process. You should also identify the appropriate alert level. Ask questions such as "Is this an isolated incident or a systemwide event?", "Has personal or sensitive data been compromised?", and "What laws may have been violated?" The answers will help you determine who to notify and whether to activate the entire incident response team or only certain members. Next, notify the appropriate people about the incident — both incident response team members and management. All contact information should be documented before an incident, and all notifications and contacts during an incident should be documented in the incident log.

- » **Mitigation:** The purpose of this step is to contain the incident and minimize further loss or damage. You may need to eradicate a virus, deny access, or disable services to halt the incident in progress.

- » **Reporting:** This step requires assessing the incident and reporting the results to appropriate management personnel and authorities (if applicable). The assessment includes determining the scope and cause of damage, as well as the responsible (or liable) party.

- » **Recovery:** Recovering normal operations involves eradicating any components of the incident (such as removing malware from a system or disabling email service on a stolen mobile device). Think of recovery as returning a system to its pre-incident state, with any changes required to prevent incident recurrence.

- » **Remediation:** Remediation may include rebuilding systems, repairing vulnerabilities, improving safeguards, and restoring data and services. Do this step in accordance with a business continuity plan that properly identifies recovery priorities.

- » **Lessons learned:** The final phase of incident management requires evaluating the effectiveness of your incident management plan and identifying any lessons learned, which should include not only what went wrong, but also what went right. Organizations often perform an after-action review (AAR) to discuss and understand the steps taken during a recent incident to identify potential improvements in detection or response.

REMEMBER

Investigations and incident management follow similar steps but have different purposes. The distinguishing characteristic of an investigation is the gathering of evidence for possible prosecution, whereas incident management focuses on containing the damage and returning to normal operations.

Operate and Maintain Detective and Preventative Measures

Detective and preventative security measures include various security technologies and techniques.

CROSS
REFERENCE

This section covers Objective 7.7 of the Security Operations domain in the CISSP Exam Outline (May 1, 2021).

Important examples of detective and preventative measures include

>> **Firewalls:** Firewalls are typically deployed at the network or data center perimeter and at other network boundaries, such as between zones of trust. Increasingly, host-based firewalls are being deployed to protect endpoints and virtual servers throughout the data center. Firewalls are discussed in more detail in Chapter 6.

>> **IDSes/IPSes:** IDSes passively monitor traffic in a network segment or to and from a host and provide alerts of suspicious activity. An IPS can detect and either block an attack or drop the network packets from the attack source. IDSes and IPSes are discussed earlier in this chapter and in Chapter 6.

>> **Whitelisting and blacklisting:** *Whitelisting* involves explicitly allowing some action, such as email delivery from a known sender, traffic from a specific IP address range, or execution of a trusted application. *Blacklisting* explicitly blocks specific actions.

>> **Third-party security services:** Third-party security services cover a wide spectrum of possible security services, such as

- *Managed security services (MSS),* which typically involves a service provider that monitors an organization's IT environment for malfunctions and incidents. Service providers can also perform management of infrastructure devices, such as network devices and servers.

- *Vulnerability management services,* where a service provider periodically scans internal and external networks and then reports vulnerabilities to the customer organization for remediation.

- *SIEM,* discussed earlier in this chapter.

- *SOAR,* discussed earlier in this chapter.

- *IP reputation services,* usually in the form of a threat intelligence feed to an organization's IDSes, IPSes, and firewalls.

- *Web content filtering,* in which an on-premises appliance or a cloud-based service limits or blocks user access to banned categories of websites (think gambling or pornography), as well as websites known to contain malicious software.

- *Data loss prevention (DLP),* capabilities where the storage, movement, and use of sensitive information is monitored and (sometimes) blocked, based on organization policy. DLP is described earlier in this chapter.

- *Cloud-based malware detection,* offered as a service that provides real-time scanning of files in the cloud and uses the speed and scale of the cloud to detect and prevent zero-day threats faster than traditional on-premises antimalware solutions.

- *Cloud-based spam filtering,* offered as a service that blocks or quarantines spam and phishing emails before they reach the corporate network, thereby significantly reducing the volume of email traffic and performance overhead associated with transmitting and processing unwanted and potentially malicious email.

- *Distributed denial of service (DDoS) mitigation,* typically deployed in an upstream network to drop or reroute DDoS traffic before it affects the customer's network, systems, and applications.

» **Sandboxing:** A *sandbox* enables untrusted or unknown programs to be executed in a separate, isolated operating environment, so any security threats or vulnerabilities can be safely analyzed. Sandboxing is used in many types of systems today, including antimalware, web filtering, and IPSes.

» **Honeypots and honeynets:** A *honeypot* is a decoy system that is used to attract attackers so that their methods and techniques can be observed (somewhat like a Trojan horse for the good guys!). A *honeynet* is a network of honeypots.

» **Antimalware:** *Antimalware* (also known as *antivirus*) software intercepts operating system routines that store and open files. The antimalware software compares the contents of the file being opened or stored against a database of malware signatures. If a malware signature is matched, the antimalware software prevents the file from being opened or saved and (usually) alerts the user. Enterprise antimalware software typically sends an alert to a central management console so that the organization's security team is alerted and can take the appropriate action. Advanced antimalware tools use various advanced techniques such as machine learning to detect and block malware from executing on a system.

» **Machine learning and artificial intelligence (AI) tools:** Machine learning and AI technologies are increasingly used to perform analytics, identify anomalous and potentially malicious behavior, and automate and orchestrate security responses, such as disconnecting a compromised host from the network or terminating a session from a known-malicious IP source address or domain.

Implement and Support Patch and Vulnerability Management

Software bugs and flaws inevitably exist in operating systems, database management systems, tools, and applications, and are continually discovered by researchers. Many of these bugs and flaws are security vulnerabilities that could permit an attacker to control a target system and subsequently access sensitive data or critical functions. Patch and vulnerability management is the process of regularly assessing, testing, installing and verifying fixes and patches for software bugs and flaws as they are discovered.

CROSS REFERENCE

This section covers Objective 7.8 of the Security Operations domain in the CISSP Exam Outline (May 1, 2021).

To perform patch and vulnerability management, follow these basic steps:

1. **Subscribe to security advisories from vendors and third-party organizations.**

2. **Perform risk analysis on each advisory and patch to determine its applicability and risk to your organization.**

3. **Develop a plan to install the security patch or to perform another workaround, if any is available.**

   You should base your decision on which solution best eliminates the vulnerability or reduces risk to an acceptable level.

4. **Proactively apply security patches to systems, devices, and applications based on risk and after appropriate testing.**

   Testing ensures that stated functions still work properly and that no unexpected side effects arise as a result of installing the patch or workaround.

5. **Verify that the patch is properly installed and that systems still perform properly.**

6. **Update all relevant documentation to include any changes made or patches installed.**

7. **Perform periodic security scans of internal and external infrastructure to identify systems and applications with unsecure configuration and missing patches.**

   Security scans serve as quality assurance to make sure that proactive patching and configuration are effective.

Understand and Participate in Change Management Processes

Change management is the business process used to control architectural and configuration changes in a production environment. Instead of just making changes in systems and the way that they relate to one another, change management is a formal process of request, design, review, approval, implementation, and recordkeeping.

CROSS REFERENCE

This section covers Objective 7.9 of the Security Operations domain in the CISSP Exam Outline (May 1, 2021).

Configuration management is the closely related process of actively managing the configuration of every system, device, and application and then thoroughly documenting those configurations.

REMEMBER

>> *Change management* is the approval-based process that ensures that only approved changes are implemented.

>> *Configuration management* is the control that records all the soft configuration (settings and parameters in the operating system, database, and application) and software changes that are performed with approval from the change management process.

Implement Recovery Strategies

Developing and implementing effective backup and recovery strategies are critical for ensuring the availability of systems and data. Other techniques and strategies are commonly implemented to ensure the availability of critical systems, even in the event of an outage or disaster.

CROSS REFERENCE

This section covers Objective 7.10 of the Security Operations domain in the CISSP Exam Outline (May 1, 2021).

Backup storage strategies

Backups are performed for a variety of reasons that center on a basic principle: Sometimes, things go wrong, and we need to get our data back. To cover all reasonable scenarios, backup storage strategies often involve the following:

>> **Secure offsite storage:** Store backup media at a remote location, far enough away so that the remote location is not directly affected by the same events (weather, natural disasters, or human-made disasters), but close enough that backup media can be retrieved in a reasonable period. This approach is also known as *e-vaulting* or *remote backup*.

>> **Transport via secure courier:** This approach can discourage or prevent theft of backup media while it is in transit to a remote location.

>> **Backup media encryption:** This approach helps prevent any unauthorized third party from recovering data from backup media.

>> **Data replication:** Send data to an offsite or remote data center, or cloud-based storage provider, in near real time.

Recovery site strategies

These strategies include hot sites (fully functional data centers or other facilities that are always up and ready with near-real-time replication of production systems and data), cold sites (data centers or facilities that have some recovery equipment available but not configured and no backup data onsite), and warm sites (data centers or facilities that have some hardware and connectivity prepositioned and configured, as well as an offsite copy of backup data).

Selecting a recovery site strategy has everything to do with cost and service level. The faster you want to recover data processing operations in a remote location, the more you will have to spend to build a site that is ready to go at the speed you require. In a nutshell, *speed costs*.

Multiple processing sites

Many large organizations operate multiple data centers for critical systems with real-time replication and load balancing between the various sites. This approach is the ultimate solution for large commercial sites that have little or no tolerance for downtime. Indeed, a well-engineered multisite application can suffer even significant whole-data-center outages without customers even knowing that anything is wrong.

System resilience, high availability, quality of service, and fault tolerance

The *resilience* of a system is a measure of its ability to keep running, even under less-than-ideal conditions. Resilience is important at all levels, including network, operating system, subsystem (such as database management system or web server), and application.

Resilience can mean a lot of different things. Here are some examples:

» **Filtering malicious input:** The system can recognize and reject input that may be an attack. Examples of suspicious input include what you get typically in an injection attack, buffer-overflow attack, or DoS attack.

» **Data replication:** The system copies critical data to a separate storage system in the event of component failure.

» **Redundant components:** The system contains redundant components that permit it to continue running even when hardware failures or malfunctions occur. Examples of redundant components include multiple power supplies, multiple network interfaces, redundant storage techniques (such as RAID), and redundant server architecture techniques (such as clustering).

» **Maintenance hooks:** Hidden, undocumented features in software programs that are intended to inappropriately expose data or functions for illicit use.

» **Security countermeasures:** Knowing that systems are subject to frequent or constant attack, systems architects need to include several security counter-measures to minimize system vulnerability. Such countermeasures include the following:

  ● Revealing as little information about the system as possible. For example, don't permit the system to display the version of operating system, database, or application software that's running.

  ● Limiting access to those people who must use the system to perform needed organizational functions.

  ● Disabling unnecessary services to reduce the number of attack targets.

  ● Using strong authentication to make it as difficult as possible for outsiders to access the system.

System resilience, high availability, quality of service (QoS), and fault tolerance are similar characteristics that are engineered into a system to make it as reliable as possible:

» **System resilience:** Includes eliminating single points of failure in system designs and building fail-safes into critical systems.

» **High availability (HA):** Typically consists of clustered systems and databases configured in an active–active (both systems are running and immediately available) or active–passive (one system is active, while the other is in standby but can become active, usually within a matter of seconds). Clusters in active–passive mode have the failover mechanism used to automatically switch the active role from one server in the cluster to another.

» **Quality of service:** Refers to a mechanism in which systems that provide various services prioritize certain services to ensure that they're always available or perform at a certain level. Voice over Internet Protocol (VoIP) systems, for example, typically are prioritized to ensure that sufficient network bandwidth is always available to prevent any traffic delay or degradation of voice quality. Other services that are not as sensitive to delays (such as web browsing or file downloads) will be prioritized at a lower level in such cases.

» **Fault tolerance:** Includes engineered redundancies in critical components, such as multiple power supplies, multiple network interfaces, and RAID-configured storage systems.

HOW VIRTUALIZATION MAKES HIGH AVAILABILITY A REALITY

Server virtualization is a rapidly growing and popular trend that has come of age in recent years. Virtualization allows organizations to build more resilient, highly efficient, cost-effective technology infrastructures to support their business-critical systems and applications. Popular virtualization solutions include VMware vSphere, VirtualBox, and Microsoft Hyper-V. Although virtualization has many benefits, here's a quick look at the high-availability benefit.

Virtual systems can be replicated or moved between separate physical systems, often without interrupting server operations or network connectivity. This task can be accomplished over a local area network when two physical servers (hosting multiple virtual servers) share common storage (a storage-area network). If Physical Server #1 fails, all the virtual servers on that physical server can quickly be moved to Physical Server #2. In an alternative scenario, if a virtual server on Physical Server #1 reaches a predefined performance threshold (such as processor, memory, or bandwidth use), the virtual server can be moved" — automatically and seamlessly — to Physical Server #2.

For business continuity or disaster recovery purposes (discussed in the next section and in Chapter 3), virtual servers can also be pre-staged in separate geographic locations, ready to be activated or booted up when needed. Using a third-party application, critical applications and data can be continuously replicated to a disaster recovery site or secondary data center in near real time so that normal business operations can be restored as quickly as possible.

Implement Disaster Recovery Processes

A variety of disasters can beset an organization's business operations. They fall into two main categories: natural and human-made.

CROSS REFERENCE

This section covers Objective 7.11 of the Security Operations domain in the CISSP Exam Outline (May 1, 2021).

In many cases, formal methodologies are used to predict the likelihood of a particular disaster. The term *50-year flood plain*, for example, is one that you've probably heard to describe the maximum physical limits of a river flood that's likely to occur once in a 50-year period. The likelihood of each of the following disasters depends greatly on local and regional geography:

>> Fires and explosions

>> Earthquakes

>> Storms (snow, ice, hail, prolonged rain, wind, dust, solar)

>> Floods

>> Hurricanes, typhoons, and cyclones

>> Volcanic eruptions and lava flows

>> Tornadoes

>> Landslides

>> Avalanches

>> Tsunamis

>> Pandemics

Many of these occurrences may have secondary effects; often, these secondary effects have a bigger impact on business operations, sometimes in a wider area than the initial disaster. A landslide in a rural area, for example, can topple power transmission lines, resulting in a citywide blackout. Some of these effects are

>> **Utility outages:** Electric power, natural gas, water, and so on

>> **Communications outages:** Telephone, cable, wireless, TV, and radio

>> **Transportation outages:** Road, airport, train, and port closures

>> **Evacuations/unavailability of personnel:** From both home and work locations

As if natural disasters weren't enough, human-made disasters can also disrupt business operations as a result of deliberate and accidental acts:

» **Accidents:** Hazardous-materials spills, power outages, communications failures, and floods due to water-supply accidents

» **Crime and mischief:** Arson, vandalism, and burglary

» **War and terrorism:** Bombings, sabotage, and other destructive acts

» **Cyberattacks/cyberwarfare:** DoS attacks, malware, data destruction, and similar acts

» **Civil disturbances:** Riots, demonstrations, strikes, sickouts, and other such events

DISASTER RECOVERY PLANNING AND TERRORIST ATTACKS

The 2001 terrorist attacks in New York, Washington, D.C., and Pennsylvania — and the subsequent collapse of the World Trade Center buildings — had disaster recovery planning and business continuity planning officials all over the world scrambling to update their plans.

This kind of planning is still a highly relevant topic more than 20 years later. The attacks redefined the limits of extreme, deliberate acts of destruction. Previously, the most heinous attacks imaginable were large-scale bombings, such as the 1993 attack on the World Trade Center and the 1995 bombing of the Alfred P. Murrah Federal Building in Oklahoma City.

The collapse of the World Trade Center towers resulted in the loss of life of 40 percent of the employees of the Sandler O'Neill & Partners investment bank. Bond broker Cantor Fitzgerald lost 658 employees in the attack — nearly its entire workforce. The sudden loss of a large number of employees had rarely been figured into business continuity and disaster recovery plans before. Businesses suddenly had to figure into contingency and recovery plans the previously unheard-of scenario, "What do we do if significant numbers of employees are suddenly lost?"

Traditional plans nearly always assumed that a business still had plenty of workers around to keep the business rolling; those insiders might be delayed by weather or other events, but eventually, they'd be back to continue running the business. The attacks on September 11, 2001, changed all that forever. Organizations need to include the possibility of the loss of a significant portion of their workforces into their business continuity plans. They owe this inclusion to their constituents and to their investors.

TIP

For a complete reference on disaster recovery planning, we recommend *IT Disaster Recovery Planning For Dummies*, by Peter H. Gregory (John Wiley & Sons, Inc.).

Disasters can affect businesses in many ways, some obvious and others not so obvious:

>> **Damage to business buildings:** Disasters can damage or destroy a building or make it uninhabitable.

>> **Damage to business records:** Along with damaging a building, a disaster may damage a building's contents, including business records, whether they are in the form of paper, microfilm, or electronic files.

>> **Damage to business equipment:** A disaster may be capable of damaging business equipment, including computers, copiers, and all sorts of other machinery. Anything electrical or mechanical, from calculators to nuclear reactors, can be damaged in a disaster.

>> **Damage to communications:** Disasters can damage common-carrier facilities, including telephone networks (both landline and cellular), data networks, and even wireless and satellite-based systems. Even if a business's buildings and equipment are untouched by a disaster, communications outages can be crippling. Further, damaged communications infrastructure in other cities can be capable of knocking out many businesses' voice and data networks. The September 11, 2001, attacks had an immediate impact on communications over a wide area of the northeastern United States, where several telecommunications providers had strategic regional facilities.

>> **Damage to public utilities:** Power, water, natural gas, and steam services can be damaged by a disaster. Even if a business's premises are undamaged, a utility outage can cause significant business disruption.

>> **Damage to transportation systems:** Freeways, roads, bridges, tunnels, railroads, and airports can all be damaged in a disaster. Damaged transportation infrastructure in other regions (where customers, partners, and suppliers are located, for example) can cripple organizations that depend on the movement of materials, goods, or customers.

>> **Injuries and loss of life:** Violent disasters in populated areas often cause casualties. When employees, contractors, or customers are killed or injured, businesses are affected in negative ways. There may be fewer customers or fewer available employees to deliver goods and services, for example. Losses don't need to be the employees or customers themselves; when family members are injured or in danger, employees will usually stay home to care for them and return to work only when those situations have stabilized.

>> **Indirect damage: suppliers and customers:** If a disaster strikes a region where key suppliers or customers are located, the effect on businesses can be almost as serious as though the business itself suffered damage.

This list isn't complete, but it should help you think about all the ways that a disaster can affect your organization.

Response

Emergency response teams must be prepared for every reasonably possible scenario. Members of these teams need a variety of specialized training to deal with such things as water and smoke damage, structural damage, flooding, and hazardous materials.

Organizations must document all the types of responses so that the response teams know what to do. The emergency response documentation consists of two major parts: response to each type of incident, and the most up-to-date facts about the facilities and equipment that the organization uses.

In other words, you want your teams to know how to deal with water damage, smoke damage, structural damage, hazardous materials, and many other things. Your teams also need to know everything about every company facility — where to find utility entrances, electrical equipment, heating/ventilation/air conditioning (HVAC) equipment, fire control, elevators, communications, data closets, and so on; which vendors maintain and service them; and so on. And you need experts who know about the materials and construction of the buildings themselves. Those experts might be your own employees, outside consultants, or a little of both.

REMEMBER

It is the disaster response planning team's responsibility to identify the experts needed for all phases of emergency response.

Responding to an emergency branches into two activities: salvage and recovery. A tangential activity is preparing financially for the costs associated with salvage and recovery.

Salvage

The salvage team is concerned with restoring full functionality to the damaged facility. This restoration includes several activities:

>> **Damage assessment:** Arrange a thorough examination of the facility to identify the full extent and nature of the damage. Frequently, outside experts, such as structural engineers, perform this inspection.

>> **Salvage assets:** Remove assets, such as computer equipment, records, furniture, inventory, and so on, from the facility.

>> **Cleaning:** Thoroughly clean the facility to eliminate smoke damage, water damage, debris, and more. Outside companies that specialize in these services frequently perform this job.

>> **Restoring the facility to operational readiness:** Complete repairs, and restock and reequip the facility to return it to pre-disaster readiness. At this point, the facility is ready for business functions to resume.

REMEMBER

The salvage team is primarily concerned with the restoration of a facility and its return to operational readiness.

Recovery

Recovery comprises equipping the business continuity team with any logistics, supplies, or coordination needed to get alternative functional sites up and running. This activity should be heavily scripted, with lots of procedures and checklists to ensure that every detail is handled.

Financial readiness

The salvage and recovery operations can cost a lot of money. The organization must prepare for potentially large expenses (at least several times the normal monthly operating cost) to restore operations to the original facility.

Financial readiness can take several forms, including

>> **Insurance:** An organization may purchase an insurance policy that pays for the replacement of damaged assets and perhaps even some of the other costs associated with conducting emergency operations.

>> **Cash reserves:** An organization may set aside cash to purchase assets for emergency use, as well as to use for emergency operations costs.

>> **Line of credit:** An organization may establish a line of credit before a disaster to be used to purchase assets or pay for emergency operations should a disaster occur.

>> **Pre-purchased assets:** An organization may choose to purchase assets to be used for disaster recovery purposes in advance and store those assets at or near a location where they will be used in the event of a disaster.

>> **Letters of agreement:** An organization may want to establish legal agreements that would be enacted in a disaster. These agreements may range from the use of emergency work locations (such as nearby hotels), use of fleet vehicles, and appropriation of computers used by lower-priority systems.

>> **Standby assets:** An organization can use existing assets as items to be repurposed in the event of a disaster. A computer system that is used for software testing could be quickly reused for production operations if a disaster strikes, for example.

Personnel

People are the most important resource in any organization. As such, disaster response must place human life above all other considerations when developing disaster response plans and when emergency responders are taking action after a disaster strikes. In terms of life safety, organizations can ensure the safety of personnel in several ways:

>> **Evacuation plans:** Personnel need to know how to safely evacuate a building or work center. Signs should be clearly posted and drills held routinely so that personnel can practice exiting the building or work center calmly and safely. For organizations with large numbers of customers or visitors, additional measures need to be taken so that people who are unfamiliar with evacuation routes and procedures can exit the facilities safely.

>> **First aid:** Organizations need to have plenty of first-aid supplies on hand, including longer-term supplies in the event that a natural disaster prevents paramedics from being able to respond. Personnel need to be trained in first aid and cardiopulmonary resuscitation (CPR) in the event of a disaster, especially when communications and/or transportation facilities are cut.

>> **Emergency supplies:** For disasters that require personnel to shelter in place, organizations need to stock emergency water, food, blankets, and other necessities in the event that employees are stranded at work locations for more than a few hours.

REMEMBER

Personnel are the most important resource in any organization.

Communications

A critical component of disaster recovery planning is the communications plan. Employees need to be notified about closed facilities and any special work instructions, such as an alternative location for work. The planning team needs to realize that one or more of the usual means of communication may be adversely affected by the same event that damaged business facilities. If a building has been damaged, for example, the voicemail system that people would try to call to check messages and get workplace status might not be working.

Organizations need to anticipate the effects of an event when considering emergency communications. You need to establish two or more ways to locate each important staff member. These ways may include landlines, cellphones, spouses' cellphones, and alternative contact numbers (such as neighbors and relatives).

TIP

Text messaging is often an effective means of communication, even when mobile communications systems are congested. Text messages require fewer resources than live calls.

Many organizations' emergency operations plans include the use of audio conference bridges so that personnel can discuss operational issues hour by hour throughout the event. Instead of relying on a single provider (which you might not be able to reach because of communications problems or because it's affected by the same event), organizations should have a second and maybe even a third audioconference provider established. Emergency communications documentation needs to include dial-in information for both (or all three) conference systems.

In addition to internal communications, the disaster recovery plan must address external communications to ensure that customers, investors, government, and media are provided accurate and timely information.

Assessment

When a disaster strikes, an organization's disaster recovery plan needs to include procedures to assess damage to buildings and equipment.

First, the response team needs to examine buildings and equipment to determine which assets are a total loss, which are repairable, and which are still usable (although not necessarily in their current locations).

For such events as floods, fires, and earthquakes, a professional building inspector usually needs to examine a building to see whether it is fit for occupation. If not, the next step is determining whether a limited number of personnel will be permitted to enter the building to retrieve needed assets.

When the assessment has been completed, assets can be divided into three categories:

>> **Salvage:** Assets that are a total loss and cannot be repaired. In some cases, components can be removed to repair other assets.

>> **Repair:** Assets that can be repaired and returned to service.

>> **Reuse:** Undamaged assets that can be placed back in service, although they may need to be moved to an alternative work location if the building can't be occupied.

Restoration

The ultimate objective of the disaster recovery team is the restoration of work facilities with their required assets so that business can return to normal. Depending on the nature of the event, restoration may take the form of building repair, building replacement, or permanent relocation to a different building.

Similarly, assets used in each building may need to undergo their own restoration, whether that restoration takes the form of replacement, repair, or return to service in the chosen location.

Before full restoration, business operations may be conducted in temporary facilities, possibly by alternative personnel, who may be other employees or contractors hired to and help out. These temporary facilities may be located near the original facilities or some distance away. The circumstances of the event will dictate some of these matters, as well as the organization's plans for temporary business operations.

Training and awareness

An organization's ability to respond effectively to a disaster is highly dependent on its preparations. In addition to the development of high-quality, workable disaster recovery and business continuity plans that are kept up to date, the most important part is making sure that employees and other personnel are trained periodically in response and continuity procedures. Training and practice help reinforce understanding of proper response procedures, giving the organization its best chance at surviving a disaster.

An important part of training is participation in various types of testing, as discussed in the upcoming section, "Test Disaster Recovery Plans."

Lessons learned

Every disaster is different and requires the best-laid plans to be adapted to address unique circumstances as they arise. It is important to document any changes and adaptations that need to be made to the plan, including what worked well and what didn't work, as well as other key information collected during an actual disaster. This information will help drive a program of continuous improvement to ensure that the organization is better prepared for future events.

Test Disaster Recovery Plans

By the time an organization has created a disaster recovery plan, it's probably spent hundreds of hours and possibly tens or hundreds of thousands of dollars on consulting fees. You'd think that after making such a big investment, they'd test the plan to make sure that it really works when an actual disaster strikes!

CROSS REFERENCE

This section covers Objective 7.12 of the Security Operations domain in the CISSP Exam Outline (May 1, 2021).

The following sections outline testing methods.

Read-through or tabletop

A *read-through* (or *tabletop) test* is a detailed review of plan documents, performed by employees on their own. The purpose of a read-through test is to identify inaccuracies, errors, and omissions in documentation.

It's easy to coordinate this type of test, because each person who performs the test does it when their schedule permits (provided that they complete it before any deadlines).

By itself, a document review is an insufficient way to test a disaster recovery plan, but it's a logical starting place. You should perform one or more of the other tests described in the following sections shortly after you do a read-through test.

Walkthrough

A *walkthrough* (or *structured walkthrough*) *test* is a team approach to the read-through test in which several business and technology experts in the organization gather to walk through the plan. A moderator or facilitator leads participants to discuss each step in the plan so that they can identify issues and opportunities for making the plan more accurate and complete. Group discussions usually help identify issues that people won't find when working on their own. Often, the participants want to perform the review at a fancy mountain or oceanside retreat, where they can think much more clearly! (Yeah, right.)

During a walkthrough test, the facilitator writes down parking-lot issues (items to be considered later but written down now so they won't be forgotten) on a whiteboard or flip chart while the group identifies those issues. These action items serve to make improvements to the plan. Each action item needs to have an accountable person assigned, as well as a completion date so that the action items will be completed in a reasonable time. Depending on the extent of the changes, a follow-up walkthrough may need to be conducted.

TIP

A walkthrough test usually requires two or more hours to complete.

Simulation

In a *simulation test,* all the designated disaster recovery personnel practice going through the motions associated with a real recovery. In a simulation, the team doesn't \perform any recovery or alternative processing.

An organization that plans to perform a simulation test appoints a facilitator who develops a disaster scenario, using a type of disaster that's likely to occur in the region. An organization in San Francisco might choose an earthquake scenario, for example, and an organization in Miami could choose a hurricane.

In a simple simulation, the facilitator reads announcements as though they're news briefs. Such announcements describe an unfolding scenario and can include information about the organization's status at the time. An example announcement might read like this:

> It is 8:15 a.m. local time, and a magnitude 7.1 earthquake has just occurred 15 miles from company headquarters. Building 1 is heavily damaged, and some people are seriously injured. Building 2 (the one containing the organization's computer systems) is damaged, and personnel are unable to enter the building. Electric power is out, and the generator has not started because of an unknown problem that may be earthquake-related. Executives Jeff Johnson and Sarah Smith (CIO and CFO) are backpacking on the Appalachian Trail and cannot be reached.

The disaster-simulation team, meeting in a conference room, discusses emergency response procedures and how the response might unfold. They consider the conditions described to them and identify any issues that could affect an actual disaster response.

The simulation facilitator makes additional announcements throughout the simulation. Just as in a real disaster, the team doesn't know everything right away; instead, news trickles in. In the simulation, the facilitator reads scripted statements that . . . um, simulate the way that information flows in a real disaster.

A more realistic simulation can be held at the organization's emergency response center, where some resources that support emergency response may be available. Another idea is to hold the simulation on a day that is not announced ahead of time so that responders will be genuinely surprised and possibly be less prepared to respond.

TIP

Remember to test your backup media to make sure that you can actually restore data from backups!

Parallel

A *parallel test* involves performing all the steps of a real recovery except that you keep the real live production systems running. The actual production systems run in parallel with the disaster recovery systems. The parallel test is very time-consuming, but it does test the accuracy of the applications because analysts compare data on the test recovery systems with production data.

The technical architecture of the target application determines how a parallel test needs to be conducted. The general principle of a parallel test is that the *disaster recovery system* (meaning the system that remains on standby until a real disaster occurs, at which time the organization presses it into production service) runs process work at the same time that the primary system continues its normal work. Precisely how this is accomplished depends on technical details. For a system that operates on batches of data, those batches can be copied to the disaster recovery system for processing there, and results can be compared for accuracy and timeliness.

Highly interactive applications are more difficult to test in a strictly parallel test. Instead, it might be necessary to record user interactions on the live system and then play back those interactions using an application testing tool. Then responses, accuracy, and timing can be verified after the test to verify whether the disaster recovery system worked properly.

Although a parallel test may be difficult to set up, its results can provide a good indication of whether disaster recovery systems will perform during a disaster. Also, the risks associated with a parallel test are low, since a failure of the disaster recovery system will not affect real business transactions.

The parallel test includes loading data onto recovery systems without taking production systems down.

Full interruption (or cutover)

A *full interruption* (or *cutover*) *test* is similar to a parallel test except that in a full interruption test, a function's primary systems are shut off or disconnected. A full interruption test is the ultimate test of a disaster recovery plan because one or more of the business's critical functions depends on the availability, integrity, and accuracy of the recovery systems.

A full interruption test should be performed only after successful walk-throughs and at least one parallel test. In a full interruption test, backup systems are processing the full production workload and all primary and ancillary functions, including user access, administrative access, integrations with other applications, support, reporting, and whatever else the main production environment needs to support.

REMEMBER

A full interruption test is the ultimate test of the ability for a disaster recovery system to perform properly in a real disaster, but it's also the test with the highest risk and cost.

Participate in Business Continuity Planning and Exercises

Business continuity and disaster recovery planning are closely related but distinctly different activities. As described in Chapter 3, business continuity focuses on keeping a business running after a disaster or other event has occurred; disaster recovery deals with restoring the organization and its affected processes and capabilities back to normal operations.

CROSS REFERENCE

This section covers Objective 7.13 of the Security Operations domain in the CISSP Exam Outline (May 1, 2021).

If you don't recall the similarities and differences between business continuity and disaster recovery planning, we strongly recommend that you refer to Chapter 3!

TIP

Security professionals need to take an active role in their organization's business continuity planning activities and related exercises. As a CISSP, you'll be a recognized expert in the area of business continuity and disaster recovery, and you'll need to contribute your specialized knowledge and experience to help your organization develop and implement effective, comprehensive business continuity and disaster recovery plans.

Implement and Manage Physical Security

Physical security is yet another important aspect of the security professional's responsibilities. Important physical security concepts and technologies are covered extensively in chapters 5 and 7.

CROSS
REFERENCE

This section covers Objective 7.14 of the Security Operations domain in the CISSP Exam Outline (May 1, 2021).

As with other information security concepts, ensuring physical security requires appropriate controls at the physical perimeter (including the building exterior, parking areas, and common grounds) and internal security controls to (most important) protect personnel, as well as to protect other physical and information assets from various threats, such as fire, flooding, severe weather, civil disturbances, terrorism, criminal activity, and workplace violence. Physical security is discussed further in Chapter 5.

Address Personnel Safety and Security Concerns

Security professionals contribute to the safety and security of personnel by helping their organizations develop and implement effective personnel security policies (discussed in Chapter 3) and through physical security measures (discussed in the preceding section, as well as chapters 5 and 7).

CROSS
REFERENCE

This section covers Objective 7.15 of the Security Operations domain in the CISSP Exam Outline (May 1, 2021).

Several important aspects of personnel safety and security that need to be understood and addressed include

>> **Travel:** Personnel may be at greater risk when traveling. They may unwittingly travel into or get lost in a high-crime area of a city, for example. Even at the airport, criminals are looking for travelers, whether for business or leisure, who are easy marks. When employees are traveling, the organization should ensure that its personnel are appropriately briefed on important safety precautions, including local laws and customs (particularly for international travel), travel advisories, and high-crime areas to avoid (if possible). Organizations should develop a means for being able to contact traveling employees in the event of an emergency. Traveling in groups, checking in frequently with other members of the travel group, and using laptop security cables are a few examples of security precautions to consider.

>> **Security training and awareness:** Security training and awareness should include not only information security, but also personal safety and security. Providing your employees' safety and security information that they can use to protect themselves and their families helps promote a positive culture that

extends to every important aspect of our lives. Teaching your employees not to post personal data, travel plans, and other sensitive information on social media, for example, can potentially help them avoid identity theft, fraud, burglary, blackmail, extortion, and other crimes that may be perpetrated by violent criminals (not just faraway cybercriminals) in their homes and communities. Security training and awareness programs are discussed further in Chapter 3.

>> **Emergency management:** Saving human lives is the first priority in any emergency situation. Personnel should understand the basics such as calling 911 (or another local emergency number or system), first-aid care, CPR, use of an automated emergency defibrillator, fire evacuation routes, and other topics.

>> **Duress:** Personnel may be subjected to or threatened by coercion or violence for various purposes. Your employees should know what resources are available to them from your organization, such as legal or financial assistance, employee assistance programs, grief counseling, and suicide prevention. In public areas, organizations should implement duress alarms so that personnel can summon assistance. Personnel security and safety awareness training should cover not only what resources are available and how to use them, but also the tactics, scams, and schemes that threat actors may use to achieve their deviant purposes.

REMEMBER

Saving human lives is the first priority in any life-threatening situation.

security into the software
development life cycle

» **Identifying and applying security
controls in software development
ecosystems**

» **Assessing the effectiveness of
software security**

» **Assessing the security impact of
acquired software**

» **Defining and applying secure coding
guidelines and standards**

Chapter **10**

Software Development Security

You must understand the principles of software security controls, software
development, and software vulnerabilities. Software and data are the foundation of information processing; software can't exist apart from software
development. Understanding the software development process is essential for
creating and maintaining appropriate, reliable, and secure software. This domain
represents 10 percent of the CISSP certification exam.

Understand and Integrate Security in the Software Development Life Cycle

The *software development life cycle* (SDLC, also known as the *systems development life cycle* and the *software development methodology*) refers to all the steps required to develop software and systems from conception through implementation, support, and (ultimately) retirement. In other words, the entire life of software and systems, from birth to death, and everything in between (like adolescence, going off to college, getting married, and retirement)!

CROSS REFERENCE

This section covers Objective 8.1 of the Software Development Security domain in the CISSP Exam Outline (May 1, 2021).

The life cycle is a development process designed to achieve two objectives: software and systems that perform their intended functions correctly and securely, and a development or integration project that's completed on time and on budget.

TIP

As we point out numerous times in this chapter, the term *software development life cycle* is giving way to *systems development life cycle* because the process applies to more than just software development; it also broadly applies to systems development, which can include networks, servers, database management systems, and more.

Development methodologies

Popular development methodologies include waterfall, Agile, DevOps, and DevSec-Ops, as discussed in the following sections.

Agile

Agile development involves a more iterative, less formal approach to software and systems development than more traditional methodologies, such as waterfall (discussed in the next section). As its name implies, Agile development focuses on speed in support of rapidly, and often constantly, evolving business requirements.

The Manifesto for Agile Software Development (https://agilemanifesto.org/) describes the underlying philosophy of Agile development as follows:

>> *Individuals and interactions* over processes and tools

>> *Working software* over comprehensive documentation

>> *Customer collaboration* over contract negotiation

>> *Responding to change* over following a plan

The manifesto doesn't disregard the importance of the items on the right (such as processes and tools), but it focuses more on the italicized items on the left.

Specific implementations of agile development take many forms. One common approach is the *Scrum* methodology. Typical activities and artifacts in this methodology include

>> **Product backlog:** A prioritized list of customer requirements, commonly known as *user stories*, maintained by the *product owner*, a business or customer representative who communicates with the Scrum team on behalf of the project stakeholders.

>> **User stories:** Formal requirements written as brief, customer-centric descriptions of the desired feature or function. User stories usually take the form "As a [role], I want to [feature/function] so that I can [purpose]." An example would be "As a customer service representative, I want to be able to view full credit card information so that I can process customer refunds."

WARNING

The user story in the preceding example should be raising all sorts of red flags and sounding alarms in your head! It illustrates why security professionals need to be involved in the development process, particularly when Agile development methods are used; requirements are developed on the fly and may not be well thought out or part of a well-documented, comprehensive security strategy. The user in this example may simply be trying to perform a legitimate job function and may have limited understanding of the potential security risks that this request introduces. If the developer is not security-focused and doesn't challenge the requirement, the feature may be delivered as requested. In the developer's mind, a feature was rapidly developed as requested and delivered to the customer error-free. Still, major security risks may have been unintentionally and unwittingly made an inherent part of the software! Someone in security (maybe you!) needs to attend development meetings to make sure that risky features aren't being developed.

>> **Sprint planning:** During sprint planning, the entire team meets for two hours to selects the product backlog items that team members believe they can deliver during the upcoming *sprint* (also known as an *iteration*) — typically, a two-week time-boxed cycle. During the next two hours of the sprint planning meeting, the development team breaks the product backlog items selected during the first two hours into discrete tasks and plans the work that will be required during the sprint (including who will do what).

>> **Daily standup:** Team members hold a daily 15-minute standup meeting (called a *scrum)* throughout the two-week sprint, and each team member answers the following three questions:

- What did I accomplish yesterday?

- What will I accomplish today?

- What obstacles or issues exist that may prevent me from meeting the sprint goal?

The daily standup is run by the *scrum master,* who is responsible for tracking and reporting the sprint's progress and resolving any obstacles or issues identified during the daily standup.

>> **Sprint review and retrospective:** At the end of each two-week sprint, the team holds a sprint review meeting (typically, for two hours) with the product owner and stakeholders to present (or demonstrate) the work that was completed during the sprint and review any work that was planned but not completed during the sprint.

The sprint retrospective typically is a 90-minute meeting. The team identifies what went well during the sprint and what can be improved in the next sprint.

WARNING

The preceding process is a very high-level overview of one possible Scrum methodology. There are as many iterations of Agile software development methods as there are iterations of software development. For a more complete discussion of the Agile and Scrum methodologies, we recommend *Agile Project Management For Dummies,* by Mark Layton, and *Scrum For Dummies,* 2nd edition, by Mark Layton and David Morrow (both John Wiley & Sons, Inc.). Another thing you can do is perform an Internet search for "pigs and chickens" to learn about the folklore behind the Scrum methodology. You'll probably find it interesting. Make sure that you find the accompanying joke about the pig and the chicken who discussed opening a restaurant together.

Security concerns to be addressed within any Agile development process can include a lack of formal documentation or comprehensive planning. In more traditional development approaches, such as waterfall, extensive up-front planning is done before any actual development work begins. This planning can include creating formal test acceptance criteria, security standards, design and interface specifications, detailed frameworks and modeling, and certification and accreditation requirements. The general lack of such formal documentation and planning in the Agile methodology isn't a security issue itself. Still, it means that security needs to be front of mind for everyone involved in the Agile development process throughout the project's life cycle.

Waterfall

In the *waterfall* model of software (or system) development, each of the stages in the life cycle progress is like a series of waterfalls.

The stages are performed sequentially, one at a time. Typically, these stages consist of the following:

>> **Conceptual definition:** A high-level description of the software or system deliverable. This description generally contains no details; it's the sort of description you want to give the business and finance people (the folks who fund your projects and keep you employed). You don't want to scare them with details, and they probably wouldn't understand the details anyway.

>> **Functional requirements:** The required characteristics of the software or system deliverable (basically, a list). Rather than being a design, the functional requirements are a collection of things that the software or system must do. Although functional requirements don't give you design-level material, this description contains more details than the conceptual definition. Functional requirements usually include a *test plan,* a detailed list of software or system functions and features that must be tested. The test plan describes how each test should be performed and the expected results. Generally, you have at least one test in the test plan for each requirement in the functional requirements. Functional requirements also must contain expected security requirements for the software or system.

>> **Nonfunctional requirements:** The required characteristics of the software or system, thought of as "under the covers" properties. Examples include encryption algorithms, architecture, capacity, throughput, resilience, and portability.

>> **Functional specifications:** The software development department's version of functional requirements. Rather than being a list of have-to-have and nice-to-have items, the functional specification is more a statement of what it is (we hope) or what we think we can build statement. (To this point, the MoSCoW (M – Must have, S – Should have, C – could have, W – Won't have) prioritization method can be used to prioritize requirements.) Functional specifications aren't quite a design, but a list of characteristics that the developers and engineers think they can create in the real world. From a security perspective, the functional specifications for an operating system or application should contain all the details about authentication, authorization, access control, confidentiality, transaction auditing, integrity, and availability.

>> **Design:** The process of developing the highest-detail designs. In the application software world, design includes entity-relationship diagrams, data-flow diagrams, database schemas, and over-the-wire protocols. For networks, this stage includes the design of local area networks (LANs), wide area networks (WANs), subnets, and the devices that tie them together and provide needed security.

>> **Design review:** The last step in the design process, in which a group of experts (some on the design team and some not) examines the detailed designs. Members who are not on the design team give the design a set of fresh eyes and a chance to catch a design flaw or two.

>> **Coding:** The phase that software developers and engineers yearn for. Most software developers would prefer to skip all the preceding steps and start coding right away, even before the formal requirements are known! It's scary to think about how much of the world's software was created with coding as the first activity. (Would you fly in an airplane that the machinists built before the designers produced engineering drawings? Didn't think so.) Coding and systems development usually include *unit testing,* which is the process of verifying all the modules and other pieces built in this phase.

>> **Code review:** The phase in which developers examine one another's program code and get into philosophical arguments about levels of indenting and the correct use of curly braces. Seriously, though, engineers can discover mistakes during code review that would cost you a lot of money if you had to fix them later in the implementation process or in maintenance mode. You can use several good static and dynamic code analysis tools use to automatically identify security vulnerabilities and other errors in software code. Many organizations use these tools to ferret out programming errors that would otherwise result in vulnerabilities that attackers might exploit. You can review code review in Chapter 8.

>> **Configuration review:** A phrase in systems development, such as operating systems and networks, that involves performing system or device configuration checks and similar activities. This important step helps verify that individual components were built properly and saves time in the long run, because errors found at this stage ensure that subsequent steps will go more smoothly and that errors in subsequent steps will be somewhat easier to troubleshoot. After all, the configuration of individual components will have already taken place.

>> **Unit test:** When portions of an application or other system have been developed, testing the pieces separately is often possible. This is called *unit testing.* Unit testing allows a developer, engineer, or tester to verify the correct functioning of individual modules in an application or system. Unit testing is usually done during coding and other component development. It doesn't always show up as a separate step in process diagrams.

>> **Integration test:** As software modules and components are developed and unit tested, they can next be tested as a group in an integration test, which ensures that various modules and components communicate with each other correctly.

>> **System test:** A system test occurs when all the components of the entire system have been assembled, and the entire system is tested from *end to end*. The test plan that was developed in the functional requirements step is carried out here. The system test includes testing all the system's security functions, of course, because the program's designers included those functions in the test plan. (Right?) You can find some great tools to rigorously test for vulnerabilities in software applications, as well as operating systems, database management systems, network devices, and other things. Many organizations consider it necessary to use such tools in system tests to ensure that the system has no exploitable vulnerabilities.

>> **Certification and accreditation:** *Certification* is the formal evaluation of the application or system: Every intended feature performs as planned, and the system is declared fully functional. *Accreditation* means that the powers that be have said it's okay to put the system into production by issuing an Authority to Operate (ATO). An ATO could mean offer it for sale, build it and ship it, or whatever "put into production" means in your organization.

>> **Implementation:** The phase when all testing and required certifications and accreditations are completed and the software can be released to production. This phase usually involves a formal mechanism whereby the software developers create a release package for operations. The release package contains the new software and any instructions for operations personnel so that they know how to implement it and verify that it was implemented correctly. An implementation plan usually includes backout instructions to revert the software (and any other changes) to its pre-change state.

>> **Maintenance:** The maintenance phase is a system's "golden years." Then customers start putting in change requests because . . . well, because that's what people do! Change management and configuration management are the processes used to control (and document all changes to) the software or system over its lifetime. Change and configuration management are discussed later in this chapter.

WARNING

You need good documentation, in the form of those original specification and design documents, because the developers who wrote this software or built the system have probably moved on to some other cool projector even another organization, and new people are left to maintain it.

DevOps

DevOps is a life cycle software development methodology that can be thought of as a merger of software development and operations. DevOps was inspired by the Agile methodology and the "Plan – Do – Check – Act" Deming Cycle.

CIS SYSTEM AND DEVICE HARDENING STANDARDS

The systems development life cycle generally has to do with the design and development of information systems, which may include many components, including server operating systems, network devices, database management systems, embedded systems, and other components. Similar to OWASP (described later in this chapter for software developers), many good system and device hardening standards are available. Of particular note is the vast collection of hardening standards from the Center for Internet Security. No, we're not talking about criminal investigators, but an organization dedicated to developing high-quality documents that provide detailed, step-by-step instructions on how to build and configure a system that will be highly resistant to attack. Best of all, these standards are free. You can find out more at https://www.cisecurity.org.

DevOps is a popular trend that represents the fusion of development and operations. It extends Agile development practices to the entire IT organization. Perhaps it's not as exciting as an Asian–Italian fusion restaurant that serves a gourmet sushi calzone, but hey, this is software and systems development, not fine dining! (Sorry.)

DevOps aims to improve communication and collaboration between software/ systems developers and IT operations teams to facilitate the rapid, efficient deployment of software and infrastructure.

As with Agile development methodologies, however, inherent security challenges must be addressed in a DevOps environment. Traditional IT organizations and security best practices have maintained strict separation between development and production environments. Although these distinctions remain in a DevOps environment, they are a little less absolute. Therefore, this situation can introduce additional risks to the production environment. These risks must be adequately addressed with proper controls and accountability throughout the IT organization.

To learn more about DevOps, pick up a copy of either *The Phoenix Project* (IT Revolution Press) or *The Visible Ops Handbook* (Information Technology Process Institute), written by Kevin Behr, Gene Kim, and George Spafford. These books are considered to be must-reads in many IT organizations.

The steps in the DevOps life cycle are

» **Dev:** The steps in the development portion of DevOps are

(a) *Plan:* This step potentially contains several substeps, including the development of functional requirements, nonfunctional requirements, design, and design review.

(b) *Code:* In this step, developers write or update application code. Potentially, this step also includes developing or updating infrastructure configuration. The concept of *infrastructure as code* embodies the idea that application developers create software application source code and the configuration of the underlying database management systems, operating systems, and even network infrastructure.

(c) *Build:* This step relies on automation in the form of a build system that compiles and integrates code and infrastructure developed in the code step.

(d) *Test:* Various types of testing are performed, including security testing, regression testing, performance testing, and user acceptance testing.

(e) *Release:* In this step, the software release package is built so that the software (and potentially infrastructure) can be deployed in the next step.

» **Ops:** The steps in the operations portion of DevOps are

(a) *Deploy:* The updated software and infrastructure are moved into production. Sometimes, this step includes running utility programs that make one-time changes to database management systems when the underlying data model is being changed (such as adding new tables or new fields to a table).

(b) *Operate:* The system is in production operation, including users who use the system and all manual and automated tasks performed by users and IT operations personnel.

(c) *Monitor:* Various monitoring tools are used to observe the running application to ensure that the underlying hardware is functioning correctly, sufficient resources are available, and the application is running correctly.

As with the Agile methodology, an organization can have several concurrent sprints in play and more in the pipeline.

DevOps is often depicted in a "figure 8" life cycle, as shown in Figure 10-1.

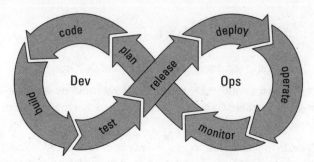

FIGURE 10-1:
The DevOps life cycle process.

© John Wiley & Sons, Inc.

DevSecOps

DevSecOps is a life cycle development process much like DevOps, with the intentional inclusion of security in several steps of the DevOps cycle. The security methods used in the DevOps cycle include

>> **Planning:** Requirements include applicable security requirements dictated by policies and controls, which themselves are aligned with applicable laws, regulations, and standards.

>> **Coding:** Developers are trained in safe coding, and their integrated development environments (IDEs) include tools used to detect and alert developers to security defects in the source code they are building or updating.

>> **Testing:** Testing the application and underlying infrastructure includes static application security testing (SAST) and dynamic application security testing (DAST), which are integrated into the build environment and operate automatically each time a new change is introduced into the environment.

>> **Operation:** Operations includes security-related activities such as data replication and backup.

>> **Monitoring:** Monitoring the environment includes performance monitoring and security monitoring performed by a security operations center (SOC).

DevSecOps embodies the concept of *Shift Security Left*, which means including security earlier in the development cycle, to the left on the arrow of time that represents the linear steps of the development cycle. Shift security left is depicted in Figure 10-2.

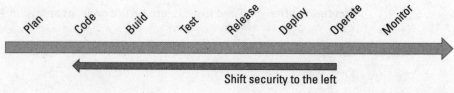

FIGURE 10-2:
The concept of Shift Security Left.

Shift security to the left

© John Wiley & Sons, Inc.

Maturity models

Organizations that need to understand and improve the quality of their software and systems development processes and practices can benchmark their SDLC by measuring its maturity. Models are available for measuring software and systems development maturity, including the following:

- » **Capability Maturity Model Integration (CMMI):** By far the most popular model for measuring software development maturity, the CMMI is required by many U.S. government agencies and contractors. The model defines five levels of maturity:

  - *Initial:* Processes are chaotic and unpredictable, poorly controlled, and reactive.

  - *Managed:* Processes are characterized for projects but are still reactive.

  - *Defined:* Processes are defined (written down) and more proactive.

  - *Quantitatively managed:* Processes are defined and measured.

  - *Optimized:* Processes are measured and improved.

  Information about the CMMI is available at https://www.isaca.org.

- » **Software Assurance Maturity Model (SAMM):** This model is an open framework geared to organizations that want to ensure that development projects include security features. More information about SAMM is available at https://owaspsamm.org.

- » **Building Security in Maturity Model (BSIMM):** This model is used to measure the extent to which security is included in software development processes. This model has four domains:

  - Governance

  - Intelligence

  - Secure software development life cycle touchpoints

  - Deployment

  Information is available at https://www.bsimm.com.

- » **Agile Maturity Model (AMM):** This software process improvement framework is for organizations that use Agile software development processes. More information about AMM is available at www.researchgate.net/publication/45227382_Agile_Maturity_Model_(AMM)_A_Software_Process_Improvement_framework_for_Agile_Software_Development_Practices.

Organizations can perform self-assessments or employ outside experts to measure their development maturity. Some organizations opt to use outside experts as a way to instill confidence in customers.

Operation and maintenance

Software and systems that have been released to operations become part of IT operations and its processes. Several operational aspects come into play, including the following:

>> **Access management:** If the application or system uses its own user access management, the person or team that fulfills access requests will do so for the application.

>> **Event management:** The application or system will be writing entries to one or more audit logs or audit logging systems. Personnel will review these logs, or (better) these logs will be tied to a security information and event management (SIEM) system to notify personnel of actionable events.

>> **Vulnerability management:** Periodically, personnel test the application or system to see whether it contains security defects that could lead to a security breach. The types of tests that may be employed include security scans, vulnerability assessments, and penetration tests. For software applications, tests could also include static and dynamic code reviews.

>> **Performance and capacity management:** The application or system may be writing performance-related entries in a logging system, or external tools may measure the response time of key system functions. This phase helps ensure that the system is healthy, usable, and not consuming excessive or inappropriate resources.

>> **Audits:** To the extent that an application or system is in scope for security or privacy audits, operational aspects of an application or system are examined by internal or external auditors to ensure that the application or system is managed properly and operating correctly. This topic is expanded later in this chapter.

From the time a software application or system is placed into production, development continues, but typically at a slower pace. During this phase, additional development tasks may be needed, such as

>> Minor feature updates

>> Bug fixes

>> Security patching and updating

>> Custom modifications

Finally, at the end of a system's service life, the system is decommissioned, which typically involves one of three outcomes:

>> **Migration to a replacement system:** Data in the old system may be migrated to a replacement system to preserve business records so that transaction history during the era of the old system may be viewed in its replacement.

>> **Coexistence with a replacement system:** The old system may be modified to operate in rea- only mode, permitting users to view data and records in the old system. Organizations that take this path keep an old system for a few months to a year or longer. This option usually is chosen when the cost of migrating data to the new system exceeds the cost of keeping the old system running.

>> **Shutdown:** In some instances, an organization discontinues use of the system. The business records may be archived for long-term storage if requirements or regulations dictate doing so.

TIP

The operations and maintenance activities described in this section may be part of an organization's DevOps processes. We discuss this topic later in this chapter.

Change management

Change management is the formal business process that ensures all changes made to a system receive formal review and approval from all stakeholders before implementation. Change management gives everyone a chance to voice their opinions and concerns about any proposed change so that the change goes as smoothly as possible, with no surprises or interruptions in service.

Change management is discussed in greater detail in Chapter 9.

REMEMBER

The process of approving modifications to a production environment is called *change management.*

WARNING

Don't confuse the concept of change management with configuration management (discussed later in this chapter). The two concepts are distinctly different.

Integrated product team

In the context of software development, an *integrated product team* (IPT) is a multidisciplinary group of people whose mission is to develop and operate an information system. An IPT attempts to remove barriers between developers, operations, and users, who are often isolated in organizations.

A development organization that implements IPT will have not just developers, but also operations and end users on a system development team. This alignment intends to develop and reinforce synergies between these (and other) groups, to ensure better outcomes in the form of systems that better meet users' needs and that can be operated and monitored.

Identify and Apply Security Controls in Software Development Ecosystems

Development environments are the collection of systems and tools used to develop and test software and systems before their release to production. Particular care is required in securing development environments to ensure that security vulnerabilities and back doors are not introduced into the software created there. These safeguards also protect source code from theft by adversaries.

CROSS REFERENCE

This section covers Objective 8.2 of the Software Development Security domain in the CISSP Exam Outline (May 1, 2021).

KEEP DEVELOPERS OUT OF PRODUCTION ENVIRONMENTS

Software developers should not have access to production environments in an organization. This practice is required by regulations and standards, including PCI DSS, NIST SP800-53, and ISO/IEC 27002.

Different personnel should be installing updated software in production environments. Developers can put installable software on a staging system for trained operations personnel to install and verify proper operation.

Developers may on occasion require read-only access to production environments so that they can troubleshoot problems. Even this read-only access should be disabled, however, except during actual support cases.

Programming languages

A riddle in the programming and software engineering profession goes like this:

Question: What's the best programming language to use for a specific project on <insert topic here>?

Answer: The language known to the programmer.

Alternative answer: The language that is presently available for use.

The meaning of the riddle is this: Programs are written by developers using familiar and available languages, with two limitations:

>> The chosen language is not always the best fit for the chosen purpose.

>> The developer's expertise in the chosen language will vary.

The result in these situations: The resulting programs may have defects that could make them vulnerable to attack. And depending upon the programming language chosen for a project, the availability of tools to identify these defects may be widely available, may have limited availability, or are not available at all.

You may have started this section thinking that the selection of a programming language has little consequence on the project's outcome. Still, we hope that by now you realize that the opposite is true. The selection of a programming language puts the project on a long-term trajectory that will help determine the long-term success of the project, based on the following factors:

>> Expertise of the developer in writing secure code

>> Availability of tools to identify defects in source code

Libraries

In the context of software development, *libraries* are collections of source code or object code that are used in software development projects. Libraries may be purchased from commercial organizations, obtained as open-source, or developed in-house.

Increasingly, organizations employ libraries for new development projects, which results in most of an organization's software having been developed by other parties. Software in general is undergoing continuous improvement in terms of functionality, leading to enterprise programs containing a much larger source code base, most of which was developed elsewhere.

Organizations' use of source code libraries can act as a force multiplier, resulting in a smaller team of developers creating far more powerful applications. This situation is not without costs and risks, however. Organizations that use external code libraries must develop processes to continually ensure that these libraries are free of exploitable defects.

The use of libraries brings a compliance-related concern: licensing and attribution. The terms and conditions of many libraries require developers to include attribution for the use of a software library. Figure 10-3 shows a portion of the component attributions for Microsoft Word for Mac.

Tool sets

Many tools are involved in the software development process, including IDEs, repositories, compilers, code scanners, build systems, testing tools, and release systems. From a security perspective, there are two principal objectives:

>> **Protection of the tool and its environment:** Tools and the environments containing them must be protected from unauthorized access and unauthorized changes.

>> **Security of the software being developed:** Organizations must take every required measure to ensure that the confidentiality and integrity of software being developed are never compromised.

FIGURE 10-3:
An example of
software library
attributions for a
software
application.

Microsoft Word for Mac

Microsoft Excel for Mac

Microsoft PowerPoint for Mac

Microsoft Outlook for Mac

© 2021 Microsoft. All rights reserved.

Portions of this software are copyright © 2008 The FreeType Project (www.freetype.org). All rights reserved.

Includes RSA Data Security, Inc. MD5 Message Digest Algorithm. Portions derived from the RSA Data Security, Inc. MD5 Message Digest Algorithm. Copyright © 1990 RSA Data Security, Inc. All rights reserved.

Certain portions copyright © 1998-2009 Marti Maria, at notice. All Rights Reserved.

Import/Export Converters© 1988-1998 DataViz, Inc. at www.dataviz.com. All Rights Reserved.

WASTE Text Engine © 1993 – 1999 Macro Piovanelli.

Zip Code Conversion © 1997 – 2004 Advance Software Corp. All Rights Reserved.

Applicable to Word for Mac only – Compare Versions © 1993-2000 Advanced Software, Inc. All Rights Reserved.

Certain templates developed for Microsoft Corporation by Impressa Systems, Santa Rosa, California.

Import/Export Converters© 1988-1998 DataViz, Inc. at www.dataviz.com. All Rights Reserved.

These two objectives can be accomplished through

>> **Policies:** Security and operational policies should define the security of tooling environments, who has access to these tools, what controls should be in place, and what audits and reviews should occur.

>> **Standards:** Details on the configuration of tools and supporting environments will help ensure that these environments are secure and operated properly.

>> **Monitoring:** As in any critical business environment, security events, configuration changes, access changes, successful and unsuccessful logins, and other events should be logged on the organization's SIEM system so that security operations personnel can be notified of situations that could be signs of misbehavior or intrusion.

>> **Audits and reviews:** Periodic examinations of tools, their settings, access controls, and the business processes they support will help identify gaps requiring attention and remediation.

WARNING

Software vendors are popular targets for cybercriminal organizations that want to infiltrate large numbers of organizations. Attacks on companies such as RSA, Solar Winds, Accellion, and Atlassian are familiar examples. Rigorous and continuous diligence is called for.

Integrated development environment

An *integrated development environment* (IDE) is a software tool that developers use to write, test, debug, and run software. Example IDEs include Eclipse, Microsoft Visual Studio, and GNU Emacs. Many IDEs also perform local version control, allowing a developer to revert to an earlier version of source code and to check code in and out from a central code repository.

Many IDEs can be integrated with security tools, such as Veracode, that can scan source code to alert developers to security and other defects. Catching errors while a developer is coding helps reinforce safe coding practices.

To prevent supply chain attacks, organizations need to make sure that developers' IDEs are protected from unauthorized tampering, including introducing malicious code into a software program that could be later used to compromise organizations that use the program.

Runtime

As critical as it is to protect software development and testing environments, one cannot overlook a program's runtime environment. Several factors make security an ongoing issue requiring continuous diligence in a runtime environment. We'll discuss a few of those concerns here:

>> **Input vulnerabilities:** Let's start with the obvious. Throughout this chapter, we've been discussing the measures needed to make sure that programs do not have exploitable vulnerabilities, which could permit an attacker to cause a program to malfunction or take control of a program or the system on which the program is running. All those measures taken during design, development, and testing are critical. Beyond that, using a *web application firewall* (WAF) to protect web-based applications is an effective first line of defense. Penetration testing of new or changed applications can be effective as well, but a pen test is a "point in time" assessment instead of the continuous protection provided by the WAF.

>> **Mobile code:** Some applications bring in code from libraries in your environment or other environments in real time. At times, it can be difficult to know the contents of the code and whether it can be trusted. Using only digitally signed code ensures its authenticity but not its security. (In a different context, even hackers use HTTPS on their phishing sites.)

>> **Code obfuscation and symbol table:** No matter where your program is running (whether on a so-called protected server or an end-user system), there's the risk that an attacker will grab your program's binaries and attempt to reverse-engineer the program to learn its secrets, such as how it obtains

encryption keys. Various code obfuscation techniques can make it more difficult for an attacker to succeed. You might also consider compiling the program so that it does not contain a symbol table. This measure makes it more difficult for an attacker to reverse-engineer the program and possibly learn its secrets.

>> **Untrusted endpoints:** The majority of today's applications are mobile apps, applications running on laptops and desktop computers, and web-based applications. In all cases, untrusted endpoints are involved. Developers need to keep this fact in mind and consider performing threat modeling to better understand how their applications could be attacked and misused. A common approach is to imagine that an end user's machine has been compromised by an attacker who is attempting to observe or even alter the user's running of the application in an effort to steal secrets.

Continuous integration/ continuous delivery

Many organizations have put *continuous integration/continuous delivery (CI/CD)* environments in place as a part of their DevOps environment. As opposed to producing software changes in large, waterfall-type projects or sprints, CI/CD environments are purpose-built to enable a larger number of smaller changes to be made in an environment. CI/CD is essential for organizations that require agility and responsiveness in their software environment.

CI/CD relies heavily on automation, particularly for code review, code inspection, security testing, and functional testing. Fewer human eyes are looking at changes in the CI/CD pipeline.

Organizations that implement CI/CD for its speed may overlook the need for careful, continuous scrutiny of the CI/CD pipeline. Otherwise, defects can easily be introduced (deliberately or not) without being noticed, and intruders may have an easier time compromising systems as well. Often, security is sacrificed for speed in a move to CI/CD, but things don't have to be this way.

Security orchestration, automation, and response

As the number of actionable alerts and the velocity of attacks increase, security operations teams are straining to keep up and protect their organizations from compromise. This situation has led to the introduction of security orchestration, automation, and response (SOAR) platforms, most often integrated with an organization's SIEM system, where security alerts are collected and analyzed.

Here is an example of a SOAR tool taking action:

> A SIEM receives an alert from an IPS that suggests that a bot or a hacker is trying to brute-force login to a server in the DMZ. The SOAR platform will get this alert and look up the location of the IP address. When the SOAR platform determines that the attacking system is in a foreign country, the SOAR platform directs the firewall to block all traffic from that IP address for 48 hours.

Without a SOAR platform, a security analyst would have had to use manual tools to determine the location of the IP address and then ask a firewall engineer to block the IP address. These manual steps might have taken anywhere from 10 minutes to more than an hour, but with a SOAR platform, the attack was blocked automatically in seconds.

SOAR platforms can be set up to protect applications by responding more quickly than humans can. Given the increase in attack velocity, SOAR can make the difference between an attempted break-in and a successful break-in.

SOAR is discussed in more detail in Chapter 9.

Software configuration management

Software configuration management (SCM) is the practice of tracking and controlling changes to software programs, including source code. SCM embraces the principles of configuration management regarding the management and use of a repository for tracking and storing changes, and security controls that limit who can view or make software changes.

SCM is governed by change management — the process used to control changes made to an environment. In the context of software development, SCM is the recordkeeping system for the changes made to software. It should be driven by a defect management process, in which management decides which defects will be addressed at any given time.

A broader discussion of configuration management appears in Chapter 9.

REMEMBER

The process of managing the changes being made to systems is called *change management*. The process of recording modifications to a production environment is called *configuration management*. The process of recording modifications in a software development environment is called *software configuration management*.

Code repositories

During and after development, program source code resides in a central source code repository, sometimes known as a *repo*. Source code must be protected from both unauthorized access and unauthorized changes. Controls to enforce this protection include

- **System hardening:** Intruders must be kept out of the OS itself. This process includes all the usual system hardening techniques and principles for servers, as discussed in Chapter 5.

- **System isolation:** The system should be reachable by authorized personnel — no one else. It should not be reachable from the Internet or capable of accessing the Internet for any reason. The system should function only as a source code repository, not for other purposes.

- **Restricted developer access:** Only authorized developers and other personnel should have access to source code.

- **Restricted administrator access:** Only authorized personnel (ideally, *not* developers!) should have administrative access to the source code repository software, the underlying operating system, and components such as database management systems.

- **No direct source code access:** No one should be able to access source code directly. Instead, everyone should access it via the check-out process in its management software.

- **Limited, controlled check-out:** Developers should be able to check out modules only when they are specifically authorized to be. This process can be automated through integration with a software defect tracking system.

- **Restricted access to critical code:** Few developers should have access to the most critical code, including code used for security functions such as authentication, session management, and encryption.

- **No bulk access:** Developers should not under any circumstances be able to check out all modules. (This restriction exists primarily to prevent theft of intellectual property.)

- **Retention of all versions:** The source code repository should maintain copies of all previous source code versions so that modules can be rolled back as needed.

- **Check-in approval:** All check-ins should require approval from another person to prevent a developer from unilaterally introducing defects or back doors into a program.

- **Activity reviews:** The activity logs for a source code repository should be reviewed periodically to ensure that there are no unauthorized check-outs or check-ins, and that all check-ins represent only authorized changes to source code.

Application security testing

The sheer size and complexity of software applications, together with the fact that some defects invariably slip through, result in relentless attacks on applications. Manual and automatic testing are needed to root out all identifiable defects, with the goal of zero defects on software being released into production. Less rigor will practically guarantee that defects will be discovered and exploited.

Various types of software testing are discussed in this section.

Code reviews

A *code review* (also known as a *peer review)* is performed by a developer who examines the code changes made by another developer. The purposes of a code review include

>> **Identification of defects:** A peer may recognize security defects and functional defects that the developer inadvertently placed in the program.

>> **Identification of improper logic:** A peer can double-check program logic and flow to confirm that the developer coded the change correctly.

>> **Identification of violations of coding standards:** A peer can check to see whether the new or changed source code complies with the organization's coding standards.

>> **Transparency:** When a developer knows that one or more peers will be examining their source code, they are less likely to sneak a back door or other malicious feature into a program.

Static application security testing

Static application security testing (SAST) represents a class of tools used to examine software source code. SAST tools identify various defects, including security defects that could enable intrusion and compromise of a running system. SAST tools are often built into IDEs as well as software build environments.

SAST tooling is useless if its output is ignored. Organizations that use SAST must develop policies and procedures to govern when and how builds and releases are deferred until defects can be fixed. Otherwise, SAST tools create noise and are disregarded.

Although SAST tooling can be run as a stand-alone function, SAST is frequently integrated into a DevOps or DevSecOps environment, where static code testing on new and changed code is performed automatically. Defects, then, are raised as software defects, the most serious of which may delay release until they are remediated.

Static application security testing is sometimes known as *white-box testing*, meaning that all available information, including source code, is available for testing.

Dynamic application security testing

Dynamic application security testing (DAST) represents a class of tools used to test running software programs. DAST tools execute a software program being tested and provide keyboard inputs as though the tool were a human user. The DAST tool repeatedly provides various types of inputs to discover exploitable weaknesses in the program being tested. DAST tools are used primarily to test web-based applications and mobile applications.

CODING STANDARDS

Better organizations develop and publish software coding standards, which specify various rules concerning the contents of source code. Topics of coding standards may include

- **Open-source:** Coding standards should address the organization's stance on the use of open-source code.

- **Safe and unsafe functions:** The use of safe and unsafe functions should be addressed in coding standards so that developers know what functions to avoid. Unsafe functions often lack boundary and type checking; their use may make a program vulnerable to attack.

- **Encryption:** Coding standards should specify the encryption algorithms, implementation, and libraries to be used.

- **Input validation:** Coding standards should address the techniques or functions to sanitize inputs to prevent buffer overflow and other attacks.

- **Session management:** Coding standards should cite the techniques to be used for establishing and enforcing session management.

- **Security:** Coding standards may specify methods for referencing or storing login credentials needed for a running program to authenticate to another program.

- **Indentation:** No coding standard is complete without addressing the critical issue of indents. Should be they be two characters? Four? Eight? How about an odd number or a prime number, just to be interesting?

Like SAST testing, DAST testing can be run on a stand-alone basis or as a part of software build automation. As with DAST, defects found in SAST testing can hold up release if they are severe enough. Each organization sets the standard for the severity level of defects that require remediation before release, as opposed to defects that can be fixed in a future release.

DAST is sometimes known as *black-box testing*, meaning that very little or no information is available to the tester other than the program (or URL) itself.

Assess the Effectiveness of Software Security

U.S. President Ronald Reagan was well known for the phrase "Trust, but verify." We take this saying a little further: "Don't trust until verified." This credo applies to many aspects of information security, including software.

CROSS REFERENCE

This section covers Objective 8.3 of the Software Development Security domain in the CISSP Exam Outline (May 1, 2021).

Initial and periodic security testing of software is an essential part of developing (or acquiring) and managing software throughout its life span. The reason for periodic testing is that researchers (both white-hat and black-hat) are always finding new ways of exploiting software programs that were once considered to be secure.

Other facets of security testing are explored in lurid detail in Chapter 8.

Auditing and logging of changes

Logging changes is an essential aspect of system and software behavior. The presence of logs facilitates troubleshooting, verification, and reconstruction of events.

Two types of changes are important here:

>> **Changes performed *by* the software:** Mainly, changes made to data. As such, a log entry will include "before" and "after" values, as well as other essentials, including user, date, time, and transaction ID, and configuration changes that alter software behavior.

>> **Changes made *to* the software:** Generally, changes made to the actual software code. In most organizations, this process involves change management and configuration management processes. While investigating system problems, however, you shouldn't discount the possibility of unauthorized changes. The absence of change management records is not evidence of the absence of changes.

Log data for these categories may be stored locally or in a central repository, such as a configuration management database (CMDB) or a security information and event management system (SIEM). Appropriate personnel should be notified promptly when actionable events take place, as discussed more fully in Chapter 9.

Risk analysis and mitigation

Risk analysis of software programs and systems is essential for identifying, analyzing, and treating risks. The types of risks that will likely be included are

>> **Known vulnerabilities:** What vulnerabilities can be identified, how they can be exploited, and whether the software has any means of being aware of attempted exploitation and defending itself.

>> **Unknown vulnerabilities:** Vulnerabilities that have yet to be discovered. If you're unsure of what we mean, just imagine any of several widely available programs that seem to be plagued by new vulnerabilities month after month. Software with that kind of track record certainly has undisclosed vulnerabilities. We won't shame those products by listing them here.

>> **Transaction integrity:** Whether the software works properly and produces the correct results in all cases, including unintentional and deliberate misuse and abuse. Manual or automated auditing of software programs can identify transaction calculation and processing problems, but humans often spot them too.

Tools that are used to assess the vulnerability of software include

>> **Security scanners:** These tools, including DAST tools scan an entire website or web application. They examine form variables, hidden variables, cookies, and other web-page features to identify vulnerabilities.

>> **Website security tools:** These tools are used to examine web pages manually for vulnerabilities that scanners often can't find.

>> **Source code scanning tools:** These tools examine program source code and identify vulnerabilities that security scanners often cannot see.

Information about software vulnerability testing tools can be found at `https://owasp.org/www-community/Vulnerability_Scanning_Tools`.

Another approach to discovering vulnerabilities and design defects uses a technique known as threat modeling. *Threat modeling* involves a systematic, detailed analysis of a program's interfaces, including user interfaces, APIs, and interaction with the underlying database management and operating systems. The analysis involves studying these elements to understand how they could be used, misused, and abused by insiders and adversaries. Information about threat modeling tools can be found at `https://owasp.org/www-project-threat-model`.

The STRIDE threat classification model is also handy for threat modeling. STRIDE stands for the following:

>> Spoofing of user identity

>> Tampering

>> Repudiation

>> Information disclosure

>> Denial of service

>> Elevation of privilege

Mitigating software vulnerabilities generally means updating source code (if you have it!) or applying security patches. Patches can't always be obtained and applied right away, however, which means implementing temporary work-arounds or relying on security in other layers, such as a web application firewall (WAF).

Mitigation of transaction integrity issues may require manual adjustments to affected data or work-arounds in associated programs.

Assess Security Impact of Acquired Software

Every organization acquires some (or all) of its software from other entities. Any acquired software related to the storage or processing of sensitive data needs to be understood from a security perspective so that an organization is aware of the risks associated with its use.

CROSS REFERENCE

This section covers Objective 8.4 of the Software Development Security domain in the CISSP Exam Outline (May 1, 2021).

Some use cases bear further discussion:

>> **Commercial off-the-shelf:** Confirming the security of commercial tools is usually more difficult than confirming the security of open-source software because the source code usually is not available to examine. Depending on the type of software, automated scanning tools may help, but testing is often a manual effort. Some vendors voluntarily permit security consulting firms to examine their software for vulnerabilities and permit customers to view test results (sometimes in summary form). Responsible vendors voluntarily undergo audits such as SOC 1 and SOC 2 or pursue ISO 27001 certification to give customers more confidence in their security controls, including those related to software development.

>> **Open-source:** Many security professionals fondly recall those blissful days when we all trusted open-source software, believing that the examination of source code by many caring, talented people would surely root out security defects. Security vulnerabilities in OpenSSL, jQuery, and MongoDB, and other software, however, have burst that bubble. Now it is obvious that we need to examine open-source software with as much scrutiny as any other software. Organizations need to maintain an accurate, up-to-date inventory of all open-source code in use and develop a means of staying informed on security-related issues with open-source code. For each application, organizations develop a *software bill of materials (SBOM)*, a detailed inventory of application source code, including the origins of each part. This practice aids in the identification of security defects in source code that, if unmitigated, could result in a security incident or breach.

>> **Third party:** The term *third party* is most often associated with the practice known as *third-party risk management* (TPRM), a process used to assess vendors and service providers. TPRM is discussed in detail in Chapter 3.

» **Managed services:** Many organizations are migrating to cloud-based services that have several types of offerings:

- *Software as a Service (SaaS):* A software vendor hosts its software on its computers (or in an Infrastructure as a Service [IaaS] environment) and enables its customers to access and run the software over a network connection (usually, the Internet). Example SaaS vendors include SAP Concur and Webex. SaaS vendors generally undergo periodic penetration tests and external audits such as SOC 1 or SOC 2, and they make those test and audit reports available to corporate customers on request. Some SaaS vendors permit customers to perform their own penetration testing.

- *Platform as a Service (PaaS):* A software vendor hosts software platforms on its own computers (or in an IaaS environment), and permits customers and other vendors to host or integrate other programs and applications as part of the overall platform. Example PaaS offerings include Salesforce and Microsoft 365 (formerly Office 365). Like SaaS vendors, PaaS vendors often undergo penetration tests and external audits, and make the reports available to corporate customers.

- *IaaS:* A vendor hosts an environment where customers can set up and run virtual machines running operating systems, virtual network devices, and virtual storage systems. Customers develop their own designs for their environments and set them up in nearly the same way as though they were implementing these environments on their own computing, network, and storage hardware. In IaaS environments, customers are responsible for the security of all virtual systems, including operating systems and network devices; all patching, scanning, identity management, network architecture, and other activities are customers' responsibility. IaaS vendors often provide SOC 1, SOC 2, or ISO 27001 certifications to customers who want to better understand the vendors' security.

Define and Apply Secure Coding Guidelines and Standards

Organizations that develop software, whether for their own use or as products for use by other organizations, need to develop policies and standards regarding the development of source code to reduce the number of vulnerabilities that could lead to errors, incidents, and security breaches. Even organizations that use tools to find vulnerabilities in source code (and at run time) would benefit from such practices for two reasons:

>> The time to fix application vulnerabilities is reduced.

>> Some application vulnerabilities may not be discovered by tools or code reviews but could still be exploited by an adversary, leading to an error, incident, or breach.

CROSS REFERENCE

This section covers Objective 8.5 of the Software Development Security domain in the CISSP Exam Outline (May 1, 2021).

Security weaknesses and vulnerabilities at the source-code level

Software development organizations must have standards, processes, and tools in place to ensure that all developed software is free of defects, including security vulnerabilities that could lead to system compromise and data tampering or theft. The types of defects that need to be identified include

>> **Buffer overflow:** In this attack, a program's input field is deliberately over-flowed in an attempt to corrupt the running software program and permit the attacker to force the program to run arbitrary instructions. A buffer overflow attack gives an attacker partial or complete control of the target system, thereby enabling them to access, tamper with, destroy, or steal sensitive data.

>> **Injection attacks:** An attacker may be able to manipulate the application through a SQL injection or script injection attack, with various results, including access to sensitive data.

>> **Escalation of privileges:** An attacker may trick the target application or system into raising the attacker's level of privilege, allowing them to access sensitive data or take control of the target system.

>> **Improper authentication (authentication bypass):** Authentication that is not airtight may be exploited by an attacker who may compromise or bypass authentication. Doing authentication correctly means writing resilient code and avoiding features that would give an attacker an advantage (such as telling the user that the user ID is correct but the password is not).

>> **Improper session management:** Session management that is not pro-grammed and configured correctly may be exploited by intruders, which could lead to session hijacking through a session replay attack.

>> **Improper use of encryption:** Strong encryption algorithms can be ineffective if they are not implemented properly, which would make it easy for an attacker to attack the cryptosystem and access sensitive data. Remedying includes not only proper use of encryption algorithms, but also proper encryption key management.

>> **Gaming:** This general term refers to faulty application or system design that may permit a user or intruder to use the application or system in ways not intended by its owners or designers. Criminals might use an image-sharing service, for example, to pass messages via steganography.

>> **Memory leaks:** This type of defect occurs when a program fails to release unneeded memory, resulting in growing memory requirements in a running program growing until available resources are exhausted.

>> **Trap door:** This is a feature in a program that performs an undocumented function, typically a security bypass.

>> **Race conditions:** This type of defect involves two or more programs, processes, or threads, each of which accesses and manipulates a resource as though it had exclusive access to the resource. This defect can cause an unexpected result with one or more programs, processes, or threads.

The Open Web Application Security Project (OWASP) addresses these and other weaknesses in detail at https//owasp.org. OWASP publishes a top-ten web application vulnerabilities list every three years, updated to reflect changing trends.

Security of application programming interfaces

Application programming interfaces (APIs) are components of software programs used for data input and data output. An API has an accompanying specification (documented or not) that defines functionalities, input and/or output fields, data types, and other details. Typically, an API is used for nonhuman interaction between programs. Although you would consider a web interface to be a human-readable interface, an API is considered to be a machine-readable interface.

APIs exist in many places: operating systems, subsystems (such as web servers and database management systems), utilities, and application programs. APIs are also implemented in computer hardware for components, such as memory and peripheral devices, disk drives, network interfaces, keyboards, and display devices.

In software development, a developer can create an API from scratch (not recommended) or acquire an API by obtaining source code modules or libraries with APIs built in, such as REST. An API can be part of an application used to transfer data between applications, in bulk or transaction by transaction.

APIs need to be secure so that they do not become the means through which an intruder can covertly obtain sensitive data or cause the target system to malfunction or crash. Four primary means of ensuring an API security include

- **Secure design:** Each API needs to be implemented so that it carefully examines and sanitizes all input data (to defeat any attempts at injection or buffer overflow attacks, as well as program malfunctions). In APIs that require authentication, the API should be implemented so that authentication bypass attacks cannot succeed. Output data must also be examined so that the API does not output any noncompliant or malicious data.

- **Security testing:** Each API needs to be thoroughly tested to ensure that it functions correctly and resists attacks. Automated tools such as SAST and DAST (discussed earlier in this chapter) are commonly used to identify defects in APIs.

- **Monitoring:** APIs should log common activity as well as errors. All such log entries should be sent to the organization's SIEM, which correlates events and generates alarms when personnel action is required.

- **External protection:** In the case of a Web Services API, an API gateway and/or a WAF may protect an API from attack. Such an option may not be available, however, if the API uses other protocols. Packet filtering firewalls do not protect APIs from logical attacks because firewalls do not examine the contents of packets — only their source and destination IP addresses and ports.

Secure coding practices

The purpose of secure coding practices is the reduction of exploitable vulnerabilities in tools, utilities, and applications. The practice of secure coding isn't just about secure coding; it involves many other considerations and activities. Following are some of the factors related to secure coding:

- **Tools:** From the selection and configuration of IDEs to the use of SAST and DAST, tools can be used to detect the presence of source code defects, including security vulnerabilities. The earlier such defects are found, the less effort it takes to correct them and stay on schedule.

- **Processes:** As discussed earlier in this chapter, software development processes need to be designed and managed with security in mind. Processes define the sequence of events. In the context of software and systems development, security-related steps such as security requirements definition, peer reviews, and the use of vulnerability scanning tools ensure that all the right steps are taken to make sure that source code is reasonably free of defects.

- **Training:** Software developers and engineers are more likely to write good, safe code when they know how to. Training in secure development is essential. Very few universities include secure development in their undergraduate computer science programs. Secure coding generally is not part of university training, so organizations must fill the gap.

» **Incentives:** Money talks. Providing incentives of some form will help software developers pay more attention to whether they're producing code with security vulnerabilities. People like the carrot more than the stick, so perhaps rewards for the fewest defects per calendar quarter or year is a good start.

» **Selection of source code languages:** The selection of source code languages and policies on the use of open-source code come into play. By design, some coding languages are more secure (we might say safe) than others. The C language, for example, as powerful as it is, has few protective features, so software developers must be more skilled in and knowledgeable about writing safe, secure code.

Developed in the 1970s, the C language was created during an era when there was more trust. But either Brian Kernighan or Dennis Ritchie, who are the co-creators of C, allegedly said, "We [Unix and C] won't prevent you from doing something stupid, as that restriction might also prevent you from doing something good." We have been unable to confirm whether one of them said this or not. The quote may have come from a book, a lecture, or a pub after a few pints of ale. The point is, some languages are safer than others. We're sorry for the rabbit hole. Well, mostly sorry.

OPEN WEB APPLICATION SECURITY PROJECT

OWASP has published a short list of security standards that organizations have adopted, most notably the Payment Card Industry Data Security Standard (PCI DSS). The top ten software risks cited by OWASP are broken access control, cryptographic failures, injection, insecure design, security misconfiguration, vulnerable and outdated components, identification and authentication failures, software and data integrity failures, security logging and monitoring failures, and server-side request forgery (SSRF).

Earlier versions of the OWASP top ten software vulnerabilities included missing function level access control, cross-site request forgery, malicious file execution, information leakage, improper error handling, and insecure communications, which are also important security considerations for any software development project.

Removal of these risks makes a software application more robust, reliable, and secure. You can find out more about OWASP — and even join or form a local chapter — by visiting https://owasp.org.

Software-defined security

Software-defined security is a security model in which security mechanisms are defined and controlled by software. Put another way, in the software-defined security model, security hardware devices such as firewalls, spam filters, IPSes, and web content filters are implemented as software-based virtual machines. Software-defined security is closely related to *network function virtualization* (NFV), in which network devices such as routers, firewalls, and IPSes are implemented as virtual network devices instead of physical appliances.

Organizations can implement software-defined security or network function virtualization in both the private cloud and the public cloud. There is no reason to assume that software-defined security can be implemented only in IaaS environments such as Amazon Web Services or Microsoft Azure.

Software-defined security is considered to be a force multiplier that enables organizations to adapt quickly to evolving business architecture and threats by changing security architecture as quickly as engineers can click, drag and drop, and type. But software-defined security as a force multiplier can just as easily result in catastrophic errors that needlessly expose an organization to active threats when an engineer makes an error in the software-defined security UI.

3

The Part of Tens

Chapter **11**

Ten Ways to Prepare for the Exam

S o much information, so little time! In this chapter, we recommend ten (mostly) long-term planning tips for helping you prepare for that special day. (No, not *that* special day; read *Wedding Planning For Dummies*, by Marcy Blum and Laura F. Kaiser, for that one.) We're talking about the CISSP exam here.

Know Your Learning Style

As you anticipate your study and preparation for the CISSP exam, it's important for you to understand your personal learning style. You might prefer a long-term study plan as opposed to a one-week boot-camp training course, for example, or you may learn better by participating in a study group or by studying and reading alone in a quiet room. Your studying might be more fruitful if you do it in short, frequent sessions (say, 30 minutes a couple of times a day), or in less frequent and longer marathons (such as four hours a few nights a week).

To make the most of the tips in this chapter, you need to know in advance what works best for you so you can customize your study plan and pass the CISSP exam with flying colors!

Get a Networking Certification First

The Communication and Network Security domain is the most complex and comprehensive domain tested on the CISSP exam. Although its purpose is to test your security knowledge, you must have a strong understanding of communications and networking. For this reason, we strongly advise that you earn a networking certification.

If you already have a networking certification, you should find most of the information in the Communication and Network Security domain to be very basic. In this case, a quick review that focuses on security concepts (particularly the methods of attack) should be sufficient for this domain. We dedicate Chapter 6 of this book to the Communication and Network Security domain.

TIP

If you haven't taken a computer-based examination before, getting a networking certification first will also familiarize you with the testing-center location and environment, as well as the general format of computer-based exams. You can take a generic practice computer-based exam at https://home.pearsonvue.com to get used to how these exams work.

Register Now

Go online and register for the CISSP exam at https://home.pearsonvue.com/isc2 — now! Committing yourself to a test date is the best cure for procrastination, especially because the test costs $749 (U.S.). Setting your date can help you plan and focus your study efforts.

Make a 60-Day Study Plan

After you register for the CISSP exam, commit yourself to a 60-day study plan. Your work experience and professional reading should span a much greater period, of course, but for your final preparations for the exam, plan on a 60-day period of intense study and review.

Exactly how intensely you study depends on your experience and learning ability, but plan on a minimum of 2 hours a day for 60 days. If you're a slow learner or reader, or find yourself weak in many areas, plan on 4 to 6 hours a day and more on the weekends. Regardless, try to stick to the 60-day plan. If you feel that you need 360 hours of study, you might be tempted to spread this time out over a 6-month period for 2 hours a day. But committing to 6 months of intense study is much harder (on you, as well as your family and friends) than committing to 2 months. In the end, you'll likely find yourself studying only as much as you would have in a 60-day period.

Get Organized and Read

A wealth of security information is available for the CISSP candidate. Studying everything is impractical, however. Instead, get organized, determine your strengths and weaknesses, and then read!

Begin by downloading *The Ultimate Guide to the CISSP* from the (ISC)² website (`www. isc2.org/Certifications/CISSP`) to get an idea of the subjects on which you'll be tested and to find helpful links to official CISSP study resources. Then read this book, use the online practice at `www.dummies.com`. (See the introduction for more information.) *CISSP For Dummies* is written to provide the CISSP candidate an excellent overview of all the broad topics covered in the exam. Next, focus on the areas that you've identify as being your weakest. Read or review the pertinent chapters in this book. If necessary, obtain additional references on specific topics. Finally, in the last week before your exam, go through all your selected study materials at least once. Review or read *CISSP For Dummies* and your study notes one more time, and complete as many practice questions as you can.

TIP

You can download the free (ISC)² *Official CISSP Flash Cards* from the (ISC)² website at `www.isc2.org/Training/Self-Study-Resources`. You can also purchase the (ISC)² Official CISSP Study App from the Apple App Store or Google Play.

Join a Study Group

You can find strength in numbers. Joining a study group or creating your own can help you stay focused and provide a wealth of information from the broad perspectives and experiences of other IT and security professionals. You can find a study group, discussion forums, and many other helpful resources at `https://community.isc2.org/t5/CISSP-Study-Group/gh-p/CISSP_Study_Group_Hub`.

Take Practice Exams

No practice exams are available that exactly duplicate the CISSP exam. And forget about brain dumps (actual test questions and answers that others have unscrupulously posted on the Internet); in addition to possibly being wrong, brain dumps violate the CISSP exam's nondisclosure agreement. But many resources are available for practice questions. You may find that some practice questions are too hard, others are too easy, and some are just plain irrelevant. Nevertheless, the repetition of practice questions can help reinforce important information that you need to know to answer questions on the CISSP exam. For this reason, we recommend taking as many practice exams as possible and using the results to focus on your weak areas. There is an (ISC)² official practice questions book and several resources available at www.isc2.org/Training/Self-Study-Resources.

Take a CISSP Training Seminar

You can take an official (ISC)² CISSP Training Seminar. The online, instructor-led, classroom-based, and private on-site training seminars are intense five-day sessions that will definitely have you eating, drinking, and sleeping CISSP. The online training seminar gives you the same benefits as the classroom-based or private on-site training seminars with a more flexible schedule, including options for weekday, weekend, evening, and self-paced courses. Schedules and additional information are available at www.isc2.org/training.

Adopt an Exam-Taking Strategy

It'll be difficult to assess whether you're going too fast or too slow as you work through the exam questions, because the test is adaptive. You'll have a minimum of 100 questions and a maximum of 150 questions. If you're going too slow on the exam — perhaps 2½ hours have gone by, and you've answered only 50 questions — rushing through the remaining questions could make matters worse. If you start making careless mistakes and getting more wrong answers, it's likely that you'll get more questions rather than fewer, so the test and the clock will both be working against you. On the other hand, if you rush through the exam — perhaps you're 30 minutes into the exam, and you've already answered 80 questions — you can't go back to check or change your answers.

With an adaptive exam, you need to develop a more in-depth strategy than simply managing the clock. Think about what you'll do when you don't know the answer to a question. How will you eliminate answer choices to make a better guess? What will you do if you start feeling overwhelmed by panic or anxiety? Have a strategy to deal with these and other possible scenarios during the exam.

Take a Breather

The day before the exam, relax, and plan for a comfortable night's rest. If you've been cramming for the exam, set your study materials aside. At that point, you either know the material or you don't!

Chapter **12**

Ten Test-Day Tips

Well, your big day has finally arrived. After months of study and mind-numbing stress, you cram all night before the exam, skip breakfast because you're running late, and then forget everything you know because you have a splitting headache for the next three hours (six hours if you're taking the non-English, form-based exam) while sitting for your exam! That isn't exactly a recipe for success — but the following ten test-day tips can definitely get you on the right track.

Get a Good Night's Rest

The night before the exam isn't the time to do any last-minute cramming. Studies have proved that a good night's rest is essential to doing well on an exam. Have a nice dinner, whatever that means for you, as long as it nourishes you and gives you energy for the next day. (We recommend going for some carbohydrates and avoiding anything spicy.) Then get to bed early. Save the all-night party for the day after the exam.

Take it from us: The CISSP and similar exams can bring you to the very brink of mental exhaustion.

Dress Comfortably

You should dress in attire that's comfortable. Remember, this exam is a *three-hour* exam. It's also a good idea to dress in layers; the exam room could be warmer or cooler than you're used to.

Consider wearing loafers or other shoes that you can easily slip off. (But please be considerate of others: Wear clean socks!)

Eat a Good Meal

Try to get *something* down before sitting for the CISSP exam. Three hours can feel like an eternity on an empty stomach.

Arrive Early

Absolutely, *under no circumstances*, don't arrive late for this exam. Make sure that you know where the testing center is located, what the traffic is like at that time of the day, and where you can park. You may even want to do a dry run before the test day to be sure you know what delays you might encounter, particularly if you're not familiar with the area where the exam is being administered.

Bring Approved Identification

(ISC)² requires two forms of identification (ID) to take your CISSP exam. You'll be asked to provide a primary and a secondary ID when you check in at your test center. (ISC)² also requires you to submit to a palm vein scan. Go to www.isc2. org/Exams/Exam-Day for full details.

REMEMBER

The testing center will verify your identity when you arrive for your exam. You need to bring your driver's license, government-issued ID, or passport — the only forms of ID that are accepted — and the name on your ID must *exactly* match the name you used to register for your exam.

Bring Snacks and Drinks

Check with your testing center (https://home.pearsonvue.com/isc2) regarding its rules about consuming snacks and drinks in the testing area. If refreshments are permitted, bring a small bag that holds enough food and drink to get you through the exam. A *big* bottle of water is essential. Also consider bringing a soda, some snacks, a sandwich, or energy bars — whatever you like to snack on that replenishes and renews you without making you too thirsty.

Bring Prescription and Over-the-Counter Medications

Again, check with your testing center, and notify the test administrator if you must take any prescription medication during the exam. Nothing can ruin your chances of succeeding on the CISSP exam like a medical emergency! Also, if you're taking any over-the-counter meds — such as acetaminophen, nasal spray, or antacids to eliminate any annoying inconveniences such as a headache, heartburn, or a gastrointestinal malady — be sure to take them before you start the exam. A box of tissues might also be appropriate if you have a cold (or if you feel like crying when you see the exam!).

Leave Your Mobile Devices Behind

Testing day is the one day when your office and family members will have to do without you. Turn off your mobile phone and anything else that beeps or buzzes; even better, leave it locked and hidden in your car or at home. Most test centers have lockers that you can use to store personal belongings, but you should confirm that this is the case before your exam. You don't want to rush through your exam because you're worried that your mobile phone has left you for a new owner.

Take Frequent Breaks

Three hours is a long time. Be sure to get up and walk around during the exam, if you're permitted to do so. If not, at least stretch your legs, curl your toes, crack your knuckles, rest your eyes (but don't fall asleep!), and roll your neck — whatever you need to do (within reason) to keep the blood flowing throughout your body. We recommend taking short, frequent breaks throughout the exam and then getting back to the task at hand. You might even incorporate breaks into your test-taking strategy. You might answer 30 or 40 questions and then take a short break. At the very least, close your eyes and take a big breath.

Also, if you find your mind wandering or if you have trouble focusing, take a break. Burnout and fatigue can lead to careless mistakes or indifference. If you feel these symptoms coming on, take a break.

Be careful not to overdo your breaks, however. Stick to frequent but short breaks, and you'll be fine.

Guess — As a Last Resort

Guessing is a desperate approach to test-taking, but it can be effective when all else fails. An unanswered question is definitely wrong, so don't leave any questions unanswered. If you must guess, try to eliminate as many obviously wrong answers as possible. If you can eliminate two possible choices that are definitely wrong, you have a 50/50 chance of getting the answer right.

When all else fails, go with your gut feeling! Research has shown time and again that your first guess is often correct. So unless you find that you misread a question — maybe you missed a key word like *not* or *all* in the question — avoid the temptation to change an answer without a compelling reason.

Glossary

3DES (Triple DES): An enhancement to the original DES algorithm that uses multiple keys to encrypt plaintext. Officially known as the Triple Data Encryption Algorithm (TDEA or Triple DEA). *See also* Data Encryption Standard (DES).

3G: The first widely used standard for digital mobile communications used in cellular networks.

4G: *See* Long-Term Evolution (LTE).

5G: The fifth generation of mobile communications protocols used in cellular networks, using higher bandwidths than 4G.

AAA: Shorthand for *authentication, authorization, and accountability controls.*

abstraction: A process that involves viewing an application from its highest-level functions, which makes lower-level functions abstract.

acceptance testing: The verification of proper functionality of a software program or system. *See also* user acceptance testing (UAT).

access card. *See* key card.

access control: The capability to permit or deny the use of an *object* (a passive entity, such as a system or file) by a *subject* (an active entity, such as a person or process).

access control list (ACL): Lists the specific rights and permissions assigned to a subject for a given object.

access management: The life cycle process concerned with the management of user access to information and systems.

Access Matrix Model: Provides object access rights (read/write/execute or R/W/X) to subjects in a DAC system. An access matrix consists of ACLs and capability lists. *See also* access control list (ACL) *and* discretionary access control (DAC).

accountability: The capability of a system to associate users and processes with their actions.

accreditation: Official, written approval for the operation of a specific system in a specific environment, as documented in a certification report.

accumulation of privileges: See aggregation (2).

acquisition: (1) The process of purchasing another organization. (2) The process of purchasing information systems hardware or software. (3) The process of obtaining data from an external source.

active assailant: Any situation in which a person is threatening to harm others at a workplace or other location where people are gathered.

active-active: A clustered configuration in which all the nodes in a system or network are load-balanced, synchronized, and active. If one node fails, the other nodes continue providing services seamlessly.

active-passive: A clustered configuration in which only one node in a system or network is active. If the primary node fails, a passive node becomes active and continues providing services, usually after a short delay.

Address Resolution Protocol (ARP): The network protocol used to query and discover the MAC address of a device on a LAN.

address space: A range of discrete addresses allocated to a network host, device, disk sector, or memory cell.

administrative controls: The policies and procedures that an organization implements as part of its overall information security strategy.

administrative laws: Legal requirements passed by government institutions that define standards of performance and conduct for major industries (such as banking, energy, and health care), organizations, and officials.

Advanced Encryption Standard (AES): A block cipher based on the Rijndael cipher, which replaced DES. *See also* Data Encryption Standard (DES).

Advanced Evolved High Speed Packet Access (HSPA+): Two mobile protocols that extended the performance of 3G networks. *See also* 3G.

after-action review (AAR): A post-incident review of incident response to identify potential improvements in detection or response.

agreement. *See* contract.

aggregation: (1) A database security issue that describes the act of obtaining information classified at a high sensitivity level by combining other items of low-sensitivity information. (2) The unintended accumulation of access privileges by people who transfer from role to role in an organization over time.

Agile: A software development methodology known for its iterative approach to the development of a system.

Agile Maturity Model (AMM): A framework for measuring the maturity of Agile software development processes and practices. *See also* Agile, maturity model.

air gap: The process of placing components on separate networks, or of removing network connectivity from specific components, to prevent communication between specific components.

analytic attack: An attack on a cryptosystem that uses algebraic manipulation in an attempt to reduce the complexity of the algorithm.

Annualized Loss Expectancy (ALE): A standard, quantifiable measure of the impact that a realized threat will have on an organization's assets. ALE is determined by the formula Single Loss Expectancy (SLE) × Annualized Rate of Occurrence (ARO) = ALE. *See also* Single Loss Expectancy (SLE) *and* Annualized Rate of Occurrence (ARO).

Annualized Rate of Occurrence (ARO): The estimated annual frequency of occurrence for a specific threat or event.

anonymization: An irreversible deidentification procedure in which specific identifiers that relate personal information to a specific person are removed. *See also* deidentification *and* pseudonymization.

antivirus software: Software that's designed to detect and prevent computer viruses and other malware from entering and harming a system.

applet: A component in a distributed environment (various components are located on separate systems) that's downloaded into and executed by another program, such as a web browser.

application firewall: A firewall that inspects OSI Layer 7 content to block malicious content from reaching or leaving an application server. *See also* web application firewall (WAF).

Application Layer (OSI model): Layer 7 of the OSI model. *See also* Open Systems Interconnection (OSI) model.

Application Layer (TCP/IP model): Layer 4 of the TCP/IP model. *See also* TCP/IP model.

application-level firewall: *See* application firewall.

application penetration test: A penetration test of a software application. *See also* penetration test.

application programming interface (API): A specification for input data and output data for a nonhuman interface in an information system.

application scan: An automated test used to identify weaknesses in a software application.

application software: Computer software that a person uses to accomplish a specific task.

application whitelisting: A mechanism used to control which applications are permitted to execute on a system. *See also* whitelisting.

archive: In a public key infrastructure, an archive is responsible for long-term storage of archived information from the Certificate Authority. *See also* Certificate Authority (CA) *and* public key infrastructure (PKI).

artificial intelligence (AI): The ability of a computer to interact with and learn from its environment and to automatically perform actions without being explicitly programmed.

asset: A resource, process, product, system, or program that has some value to an organization and therefore must be protected. Assets can be hard goods, such as computers and equipment, but can also be information, programs, and intellectual property.

asset classification: Policy that defines sensitivity levels, hardening standards, and handling procedures for assets at each level.

asset inventory: The process of tracking assets in an organization.

asset valuation: The process of assigning a financial or relative value to an organization's information assets.

asymmetric key system (or asymmetric algorithm; public key): A cryptographic system that uses two separate keys: one key to encrypt information and a different key to decrypt information. These key pairs are known as *public* and *private keys.*

Asynchronous Transfer Mode (ATM): A very high-speed, low-latency, packet-switched communications protocol.

Attached Resource Computer NETwork (ARCNET): An early physical LAN cabling standard that is no longer in common use.

attack surface reduction: The effort to reduce the number of systems, devices, and components that are potentially exploitable.

attack tree: A diagram that depicts types of attacks and their progression.

attribute-based access control (ABAC): An access control model in which a subject is granted access to an object based on subject attributes, object attributes, and environmental considerations.

audit: The independent verification of any activity or process.

audit trail: The auxiliary records that document transactions and other events.

authenticated scan: A vulnerability scan that attempts to log in to a device, system, or application during its search for exploitable vulnerabilities.

authentication: The process of verifying a subject's claimed identity in an access control system.

authentication bypass: Any attack on a system that attempts to gain access to the system without providing authentication credentials.

Authentication Header (AH): In Internet Protocol Security, a protocol that provides integrity, authentication, and nonrepudiation. *See also* Encapsulating Security Payload (ESP) and Internet Protocol Security (IPsec).

authority to operate (ATO): Formal approval to use a new or changed system in a production environment.

authorization (or establishment): The process of defining and granting the rights and permissions granted to a subject (what you can do).

automatic controls: Controls that are not performed manually.

automatic external defibrillator (AED): A portable defibrillator that can be used by untrained personnel to diagnose and treat arrhythmia, otherwise known as a heart attack.

autonomous system number (ASN): An identifier used to assign publicly accessible network address space to organizations.

availability: The process of ensuring that systems and data are accessible to authorized users when they need it.

backdoor: Malware that enables a person to bypass normal authentication to gain access to a compromised system. *See also* malware.

background check: The process of verifying a person's professional, financial, and legal history, usually in connection with employment.

backup: The process of making copies of critical information in the event of a later event that results in the loss of that information.

baseline: A process that identifies a consistent basis for an organization's security architecture, taking into account system-specific parameters, such as different operating systems.

Bell-LaPadula model: A formal confidentiality model that defines two basic properties: the simple security property (ss property) and star property (* property). *See also* simple security property (ss property) *and* star property (* property).

best evidence: Original, unaltered evidence, which is preferred by the court over secondary evidence. *See also* best evidence rule *and* evidence.

best evidence rule: As defined in the Federal Rules of Evidence; states that "to prove the content of a writing, recording, or photograph, the original writing, recording, or photograph is (ordinarily) required." *See also* evidence.

Biba model: A formal integrity model that defines two basic properties: the simple integrity property and star integrity property (*-integrity property). *See also* simple integrity property *and* star integrity property (*-integrity property).

biometrics: Any of various means used, as part of an authentication mechanism, to verify the identity of a person. Types of biometrics used include fingerprints, palm prints, signatures, retinal scans, voice scans, and keystroke patterns.

birthday attack: A type of attack that attempts to exploit the probability of two messages using the same hash function and producing the same message digest. *See also* hash function.

black-box testing: A security test wherein the tester has no previous knowledge of the system being tested. *See also* dynamic application scanning tool (DAST).

blacklisting: A mechanism that explicitly blocks access based on the presence of an item in a list. *See also* whitelisting.

blackout: A complete loss of electric power.

block cipher: An encryption algorithm that divides plaintext into fixed-size blocks of characters or bits and then uses the same key on each fixed-size block to produce corresponding ciphertext.

Bluetooth: A wireless technology standard for data exchange over short distances between fixed and mobile devices.

bollard: A post used to divert traffic from a building, area, or road.

bot: A target computer that is infected by malware and is part of a botnet. *See also* botnet *and* malware.

breach: An action resulting in unauthorized disclosure of confidential information or damage to a system.

breach attack simulation: A type of penetration test in which defenses and incident response are tested.

bridge: A network device that forwards packets to other networks.

bring your own device (BYOD): A mobile device policy that permits employees to use their personal mobile devices in the workplace for work-related and personal business.

broadcast: A type of network protocol whereby packets are sent from a source to every node on a network.

broken windows theory: A theory that suggests that broken windows, trash, and other visible signs of physical damage and neglect invite criminal elements and result in further criminal activity.

brownout: Prolonged drop in voltage from an electric power source, such as a public utility.

brute-force attack: A type of attack in which the attacker attempts every possible combination of letters, numbers, and characters to crack a password, passphrase, encryption key, or personal identification number.

buffer (or stack) overflow attack: A type of attack in which the attacker enters an out-of-range parameter or intentionally exceeds the buffer capacity of a system or application to effect a denial of service (DoS) attack or exploit a vulnerability.

Building Security In Maturity Model (BSIMM): A maturity model for benchmarking software development processes.

bus: A network topology in which all devices are connected to a single cable.

business continuity plan (BCP): A set of procedures to be followed in the event of a business interruption to ensure the continuation of critical business processes.

business impact analysis (BIA): A risk analysis that, as part of a business continuity plan, describes the impact on business operations that the loss of various IT systems would impose.

California Consumer Privacy Act (CCPA): A state law that defines privacy rights and consumer protections for California residents.

California Privacy Rights Act (CPRA): A state law that amends the CCPA.

Caller ID: The protocol used to transmit the calling party's telephone number to the called party's telephone equipment during the establishment of a telephone call.

Caller ID spoofing: The use of a device or service to alter the Caller ID of an outgoing call, used by callers to impersonate others for the purpose of perpetrating fraud. *See also* Caller ID.

Capability Maturity Model Integration (CMMI): A maturity model for software development and other IT practices, including information security.

card key: *See* key card.

Center for Internet Security Critical Security Controls (CIS CSC): A cybersecurity controls framework.

certification: A formal methodology that uses established evaluation criteria to conduct comprehensive testing and documentation of information system security safeguards, both technical and nontechnical, in a given environment.

Certificate Authority (CA): In a public key infrastructure, the CA issues certificates, maintains and publishes status information and Certificate Revocation Lists, and maintains archives. *See also* public key infrastructure (PKI).

chain of custody (or chain of evidence): Procedures that provide accountability and protection for evidence throughout that evidence's entire life cycle.

Challenge-Handshake Authentication Protocol (CHAP): A remote access control protocol that uses a three-way handshake to authenticate both a peer and a server. *See also* three-way handshake.

change management: The formal business process that ensures that all changes made in a system are properly requested, reviewed, approved, tested, and implemented.

Children's Online Privacy Protection Act (COPPA): A U.S. law protecting information about children under the age of 13.

choose your own device (CYOD): A mobile device policy that permits employees to select their preferred mobile device from a list of devices that have been approved by the organization.

chosen plaintext attack: An attack technique in which the cryptanalyst selects the plaintext to be encrypted and then analyzes the resulting ciphertext.

C-I-A: Confidentiality, integrity, and availability.

cipher: A cryptographic transformation.

ciphertext: A plaintext message that has been transformed (encrypted) into a scrambled message that's unintelligible.

ciphertext-only attack: A method of cryptanalysis in which the attacker has access only to ciphertext.

circuit-switched network: Any of several telecommunications network designs that provide a dedicated physical circuit path between endpoints.

circumstantial evidence: Relevant facts that can't be directly or conclusively connected to other events but about which a reasonable inference can be made. *See also* evidence.

civil (or tort) law: Legal codes that address wrongful acts committed against a person or business, either willfully or negligently, resulting in damage, loss, injury, or death. Unlike criminal law, U.S. civil law cases are determined based on a preponderance of evidence, and punishments are limited to fines.

Clark-Wilson model: A formal integrity model that addresses all three goals of integrity (preventing unauthorized users from making any changes, preventing authorized users from making unauthorized changes, and maintaining internal and external consistency) and identifies special requirements for inputting data.

classification: The process of assigning a security label to a document that defines how the document should be handled.

closed system: A system that uses proprietary hardware and/or software that may not be compatible with other systems or components. *See also* open system.

cloud: Internet-based network, computing, and application infrastructure available on demand.

cloud access security broker (CASB): Systems used to enforce policy regarding the use of cloud-based resources.

cluster: A system or network configuration containing multiple redundant nodes for resiliency. *See also* active-active *and* active passive.

clustering (or key clustering): Generating identical ciphertext messages from a plaintext message by using the same encryption algorithm but different encryption keys.

coaxial cable: A network medium consisting of a single solid-wire core that is surrounded by an insulation layer and a metal foil wrap.

COBIT: Formerly Control Objective for Information and Related Technologies. An IT controls and process framework developed by ISACA (formerly Information Systems Audit and Control Association).

code of ethics: A formal statement that defines ethical behavior in a given organization or profession.

code review: The examination of source code to identify defects.

coercion: Compelling a person to provide evidence involuntarily through intimidation, trickery, or bribery.

cold site: An alternative computer facility that has electricity, heating, air conditioning, and ventilation but no computer equipment onsite. *See also* hot site, reciprocal site, *and* warm site.

collision: (1) A network event in which two nodes simultaneously transmit frames. (2) An event in which two different messages produce the same message digest.

collision domain: A portion of a network that would receive broadcast packets sent from one of its nodes.

common vulnerability scoring system (CVSS): An industry-standard method for determining the severity of a vulnerability identified by a vulnerability scan, penetration test, or other means.

Common Criteria: An international effort to standardize and improve existing European and North American information systems security evaluation criteria.

common law: A legal system, originating in medieval England, based on custom and judicial precedent.

community cloud: As defined by the National Institute of Standards and Technology, a cloud infrastructure "provisioned for exclusive use by a specific community of consumers from organizations that have shared concerns." *See also* cloud.

compensating controls: Controls that are implemented as an alternative to other preventive, detective, corrective, deterrent, or recovery controls.

compensatory damages: Actual damages to the victim, including attorney/legal fees, lost profits, and investigative costs.

compliance: Conformance to rules, including laws, regulations, standards, policies, and legal agreements.

compliance risk: Any risk identified that is a consequence of failing to comply with a policy, law, regulation, or other legal obligation.

Computer Incident Response Team (CIRT) or Computer Emergency Response Team (CERT): A team that comprises people who are properly trained in incident response and investigation.

concealment cipher: A technique of hiding a message in plain sight. The key is knowing where the message lies.

concentrator: *See* hub.

conclusive evidence: Incontrovertible and irrefutable . . . you know, a smoking gun. *See also* evidence.

confidentiality: The concept that information and functions should be accessed only by authorized subjects.

confidentiality agreement: *See* nondisclosure agreement (NDA).

configuration management: The process of recording all changes to information systems.

configuration management database (CMDB): A repository that is used to store all configuration changes made to an information system.

container: An isolated instance in a running operating system in which a software application is executed.

containerization: A method of virtualization in which several isolated operating zones are created in a running operating system so that application programs and data can execute independently within their respective containers.

content-distribution network (CDN): A system of distributed servers that delivers cached web pages and other static content to a user from the nearest geographic location to the user. Also known as a *content delivery network.*

continuing professional education (CPE): Training classes and other activities that further a person's skills and knowledge in a profession.

Continuity of Operations Planning (COOP): Disaster recovery planning and business continuity planning blended into a single coordinated activity.

continuous improvement: Intentional practices that result in the gradual improvement of people, processes, and technology.

continuous integration and continuous deployment (CI/CD): A development and operations environment supported by automation such that changes to application source code and infrastructure configuration are built, integrated, and deployed automatically.

continuous monitoring: Real-time or near-real-time examination of a process or system. *See also* monitoring.

contract: A legally binding document, signed by two or more parties, that describes rights and duties.

control: A safeguard or countermeasure that helps prevent or mitigate a security risk.

control assessment: An examination of a control to determine its effectiveness.

control framework: An organized collection of controls.

control self-assessment (CSA): An activity wherein a control owner is prompted to assert the effectiveness of a control, usually through answering questions and submitting evidence.

controller: *See* data controller.

copyright: A form of legal protection granted to the author(s) of "original works of authorship," both published and unpublished.

corporate owned personally enabled (COPE): A practice of issuing computing devices to employees when personal use, in addition to business use, is permitted.

corrective controls: Controls that remedy violations and incidents or improve existing preventive and detective controls.

corroborative evidence: Evidence that supports or substantiates other evidence presented in a legal case. *See also* evidence.

corroborative inquiry: An audit technique in which auditors ask other personnel about a particular control to see whether their responses align with those of control owners.

countermeasure: A device, control, or action required to reduce the impact or probability of a security incident.

covert channel: An unintended communications path, which may be a covert storage channel or a covert timing channel.

crime prevention through environmental design (CPTED): A philosophy for the inclusion of security and physical design of defensible spaces.

criminal law: Defines crimes committed against society, even when the actual victim is a business or person. Criminal laws are enacted to protect the general public. Unlike civil cases, U.S. criminal cases are decided when a party is guilty beyond a reasonable doubt. Punishments may include fines, incarceration, and even execution.

criticality assessment (CA): The part of a business impact analysis that ranks the criticality of business processes and IT systems. *See also* business impact analysis (BIA).

Crossover Error Rate (CER): In biometric access control systems, the point at which the FRR equals the FAR, stated as a percentage. *See also* False Accept Rate (FAR; or Type II Error) *and* False Reject Rate (FRR; or Type I Error).

cross-site request forgery (CSRF): An attack in which an attacker attempts to trick a victim into clicking a link to perform an action that the victim would not otherwise perform.

cross-site scripting (XSS): An attack in which an attacker attempts to inject client-side script into web pages viewed by other intended victims.

cryptanalysis: The science of deciphering ciphertext without using the cryptographic key.

cryptocurrency: A form of digital currency, such as Bitcoin, that uses encryption to control the creation of currency and verify the transfer of funds independent of a central bank or authority.

cryptography: The techniques and algorithms used for encrypting and decrypting information, such as a private message, to protect its confidentiality, integrity, and/or authenticity.

cryptologist: A practitioner of cryptology.

cryptology: The science that encompasses both cryptography and cryptanalysis.

cryptomining: Computer processing to validate cryptocurrency transactions, resulting in a financial reward for the owner of the computer performing the processing.

cryptoperiod: The length of time for which a specific encryption key is authorized for use.

cryptosystem: The hardware or software components that transform plaintext into ciphertext (encrypts) and back into plaintext (decrypts).

cryptovariable (or key): A secret value applied to a cryptographic algorithm. The strength and effectiveness of the cryptosystem is largely dependent on the secrecy and strength of the cryptovariable.

culpable negligence: A legal term that may describe an organization's failure to follow a standard of due care in the protection of its assets and thereby expose the organization to a legal claim. *See also* due care.

custodian: A person who has day-to-day responsibility for protecting information assets.

cutover test: *See* full interruption test.

cybercrime: Any criminal activity in which computer systems or networks are targeted or used as tools.

Cybersecurity Maturity Model Certification (CMMC): An assessment program for evaluating the security of service providers providing services to U.S. government agencies.

data carrier equipment (DCE): A device used to establish, maintain, and terminate communications between a data source and its destination in a network. *See also* data terminal equipment (DTE).

data classification: Policy that defines sensitivity levels and proper handling procedures for data at each level and in various handling scenarios.

data collection: The process of receiving data from a subject.

data controller: An organization that directs the storage and processing of information, as defined by the General Data Protection Regulation and other privacy laws.

data destruction: Any means used to remove data from a storage medium.

data discovery: Tools that scan stored data on systems to determine the presence of specific types of data.

data encapsulation: In networking, the wrapping of protocol information from the OSI model layer immediately above in the data section of the layer immediately below. *See also* Open Systems Interconnection (OSI) model.

data encryption key (DEK): An encryption key used to encrypt and decrypt data. *See also* key encryption key (KEK).

Data Encryption Standard (DES): A commonly used symmetric key algorithm that uses a 56-bit key and operates on 64-bit blocks. *See also* Advanced Encryption Standard (AES).

Data Link Layer: Layer 2 of the OSI network model. *See also* Open Systems Interconnection (OSI) model.

data loss prevention (DLP): An application or device used to detect or prevent the unauthorized storage or transmission of sensitive data.

data maintenance: Any activity where data is being reviewed, updated, corrected, or discarded.

Data Over Cable Service Interface Specification (DOCSIS): A communications protocol for transmitting high-speed data over an existing TV cable system.

data processor: As defined by the General Data Protection Regulation, an organization or entity that processes information at the direction of a data controller. *See also* General Data Protection Regulation (GDPR).

data protection officer (DPO): A person responsible for the development and management of a data privacy program, as directed by the European General Data Privacy Regulation and other privacy laws. *See also* General Data Protection Regulation (GDPR).

data recovery: The process of retrieving data from backup media in the event of an error or malfunction.

data remanence: Residual data that remains on storage media or in memory after the data has been deleted.

data retention: The activities supporting an organization's effort to retain specific sets and types of data for minimum and/or maximum periods.

data subject: An identifiable natural person.

data terminal equipment (DTE): A device that communicates with a DCE in a network. *See also* data carrier equipment (DCE).

data warehouse: A special-purpose database used for decision support or research purposes.

database management system (DBMS): Restricts access by different subjects to various objects in a database.

datagram: The protocol data unit for the User Datagram Protocol. *See also* protocol data unit (PDU), User Datagram Protocol (UDP).

deciphering: *See* decryption.

decryption: The process of transforming ciphertext into plaintext.

deep packet inspection (DPI): An advanced method of examining and managing network traffic.

defense in depth: The principle of protecting assets by using layers of dissimilar mechanisms.

Defense Information Technology Security Certification and Accreditation Process (DITSCAP): A program that formalizes the certification and accreditation process for U.S. Department of Defense information systems.

deidentification: Any procedure through which specific identifiers about a data subject are removed or replaced. *See also* anonymization, masking, *and* pseudonymization.

deluge: A type of water-based fire suppression in which large amounts of water are sprayed into an area.

Deming cycle: The conceptual life cycle model that consists of Plan, Do, Check, Act.

demonstrative evidence: Evidence that is used to aid the court's understanding of a legal case. *See also evidence.*

denial of service (DoS): An attack on a system or network with the intention of making the system or network unavailable for use.

design review: An examination of the design of a system to ensure it complies with policies, standards, and secure practices.

destructware: Malware that functions similar to ransomware, except that the attacker has no intention of extracting a ransom payment and, therefore, no decryption key is available to recover the encrypted data.

detective controls: Controls that are intended to identify violations and incidents.

deterrent controls: Controls that are intended to discourage violations.

DevOps: The culture and practice of improved collaboration between software developers and IT operations.

DevSecOps: The integration of security practices within DevOps. *See also* DevOps.

Diameter: The successor protocol to RADIUS for remote authentication. *See also* Remote Authentication Dial-In User Service (RADIUS).

dictionary attack: A focused type of brute-force attack in which a predefined word list is used. *See also* brute-force attack.

Diffie-Hellman: A key-exchange algorithm based on discrete logarithms.

digital certificate: A certificate that binds an identity with a public encryption key.

digital forensics: The science of conducting a computer incident investigation to determine what has happened.

digital rights management (DRM): A tool or technique used to enforce the use, modification, and distribution of software or data.

digital signature: A cryptographic method used to verity the authenticity and integrity of a message.

Digital Signature Standard (DSS): Published in Federal Information Processing Standard (FIPS) 186-1, DSS specifies two acceptable algorithms in its standard: The RSA Digital Signature Algorithm and the Digital Signature Algorithm (DSA). *See also* NIST *and* Rivest, Shamir, Adleman (RSA).

digital subscriber line (xDSL): A high-bandwidth communications protocol delivered over analog telecommunications voice lines.

digital watermarking: A technique used to verify the authenticity of an image or data. A watermark may be conspicuous or hidden.

direct evidence: Oral testimony or a written statement based on information gathered through the witness's five senses that proves or disproves a specific fact or issue. *See also* evidence.

disaster: Any natural or human-made event that may cause the interruption of business operations.

disaster recovery plan (DRP): A set of procedures to be followed in the event of a business interruption to ensure the recovery of critical assets and information systems.

discovery sampling: A sampling technique in which an auditor selects additional samples to find a single exception.

discretionary access control (DAC): An access policy determined by the owner of a file or other resource. *See also* mandatory access control (MAC) system.

distributed denial of service (DDoS): An attack in which the attacker initiates simultaneous denial-of-service attacks from many systems.

Distributed Network Protocol (DNP3): A set of communications protocols used between components in process automation systems (such as public utilities).

distribution frame: (1) A room in which telephone and data cabling is terminated. (2) The componentry used for terminating telephone and data cabling. *See also* main distribution frame (MDF), intermediate data frame (IDF).

DNS cache poisoning: A type of attack, also known as DNS spoofing, that exploits vulnerabilities in DNS to divert Internet traffic away from legitimate destination servers to fake servers. *See also* Domain Name System (DNS).

DNS hijacking: An attack technique used to redirect DNS queries away from legitimate DNS servers. *See also* Domain Name System (DNS).

documentary evidence: Evidence that is used in legal proceedings, including originals and copies of business records, computer-generated and computer-stored records, manuals, policies, standards, procedures, and log files. *See also* evidence.

domain: A collection of users, computers, and resources that have a common security policy and single administration.

domain homograph attack: A type of spoofing attack in which the attacker uses similar-looking keyboard characters to deceive computer users about the actual remote system they are communicating with, such as by replacing a Latin *O* with a Cyrillic *O* in a website address.

Domain Name System (DNS): A hierarchical, decentralized directory service database that converts domain names to IP addresses for computers, services, and other computing resources connected to a network or the Internet.

domain name system security extensions (DNSSEC): Specifications for securing certain kinds of information provided by DNS as used on IP networks.

drift: The gradual change in a system's configuration from an established baseline or standard.

drop: *See* voltage drop.

drug screen: A test for the presence of drugs and controlled substances, usually as a part of pre-employment screening. *See also* background check.

dry pipe: A fire suppression system in which sprinkler pipes are not filled with water until fire suppression is necessary. *See also* wet pipe.

due care: The steps that an organization takes to implement security best practices.

due diligence: The prudent management and execution of due care.

dumpster diving: The process of examining garbage with the intention of finding valuable goods or information.

duress alarm: A hidden alarm trigger that personnel can use to summon help in an emergency.

dwell time: The elapsed time between the onset of a security incident and the organization's realization that an incident has occurred (or is occurring).

dynamic application scanning tool (DAST): A tool used to identify vulnerabilities in a software application that works by executing the application and attempts various means to compromise the application.

dynamic password: A password that changes at some regular interval or event.

east–west traffic: Network communications between systems within a network.

eavesdropping: Listening to network traffic to obtain content or learn more about communications.

edge computing: A method used to optimize cloud computing by processing data at the edge of the network, near the source of the data.

egress monitoring: Any practice of monitoring outbound traffic to discover potential intrusion or data leakage.

electric generator: A machine used to generate electricity locally in the event of interruption of electric utility power.

electromagnetic interference (EMI): Electrical noise generated by the different charges among the three electrical wires (hot, neutral, and ground) and can be *common-mode noise* (caused by hot and ground) or *traverse-mode noise* (caused by hot and neutral).

electronic protected health-care information (ePHI): Any patient related health information as defined by HIPAA. *See also* Health Insurance Portability and Accountability Act (HIPAA).

electrostatic discharge (ESD): A sudden flow of electricity between two objects.

emanations: Unintentional emissions of electromagnetic or acoustic energy from a system.

emergency power off (EPO): A switch that can be used to remove electric power from nearby equipment in case of fire or electric shock.

employment agreement: A legal agreement between an employer and employee that stipulates the terms and conditions of employment.

employment candidate screening: *See* background check.

employment termination: The cessation of employment for one or more employees in an organization.

encapsulation: The process of layering protocol information at different levels of a protocol stack.

Encapsulating Security Payload (ESP): A protocol that provides confidentiality (encryption) and limited authentication. *See also* Authentication Header (AH) *and* Internet Protocol Security (IPsec).

encryption: The process of transforming plaintext into ciphertext.

end of life (EOL): A date after which hardware or software product is considered to be unviable.

end of support (EOS): A date after which a product manufacturer no longer supports a hardware or software product.

end-to-end encryption: A process by which packets are encrypted at the original encryption source and decrypted only at the final decryption destination.

endpoint: A general term referring to a desktop computer, laptop or notebook computer, or mobile device.

Enhanced Mobile Broadband (eMBB): A standard used in 5G mobile communications.

enticement: Luring someone toward certain evidence after that person has already committed a crime.

entitlement: Access rights assigned to employees based on job title, department, or other established criteria.

entrapment: Encouraging someone to commit a crime that the person may have had no intention of committing.

ephemeral account: *See* just-in-time (JIT) access.

escalation of privilege: An attack technique in which the attacker uses some means to bypass security controls to attain a higher privilege level on the target system.

Escrowed Encryption Standard (EES): Divides a secret key into two parts and places those two parts into escrow with two separate, trusted organizations. Published by NIST in FIPS PUB 185 (1994). *See also* NIST.

espionage: The practice of spying or using spies to obtain proprietary or confidential information.

Ethernet: A common bus-topology network transport protocol.

ethics: Professional principles and duties that guide decisions and behavior. *See also* code of ethics.

e-vaulting: The practice of backing up data to a cloud-based data storage provider.

event management: The life cycle process concerned with the receipt, logging, and alerting of security and operational events in a system or environment.

evidence: Information obtained in support of an investigation or incident.

evidence life cycle: The various phases of evidence, from initial discovery to its final disposition. The evidence life cycle has the following five stages: collection and identification; analysis; storage, preservation, and transportation; presentation in court; and return to victim (owner).

Exclusive Or (XOR): A binary operation applied to two input bits. If the two bits are equal, the result is zero. If the two bits are not equal, the result is one.

exigent circumstances: If probable cause exists and the destruction of evidence is imminent or human lives are at stake, property or people may be searched and/or evidence may be seized by law enforcement personnel without a search warrant.

exploit: (1) Software or code that takes advantage of a vulnerability in an operating system (OS) or application and causes unintended behavior in the OS or application, such as privilege escalation, remote control, or denial of service. (2) Action taken by a subject, system, or program that uses a vulnerability to gain illicit access to an object.

exposure factor (EF): A measure, expressed as a percentage, of the negative effect or impact that a realized threat or event would have on a specific asset.

Extensible Authentication Protocol (EAP): A remote access control protocol that implements various authentication mechanisms, including MD5, S/Key, generic token cards, and digital certificates. Often used in wireless networks.

facilities classification policy: Policy that defines sensitivity levels and protection controls for work locations at each classification level.

Fagan inspection: A structured process that is used to find defects in design documents, specifications, and source code.

fail closed: A control failure that results in all accesses being blocked.

fail open: A control failure that results in all accesses being permitted.

fail securely: A concept similar to fail closed and fail open that dictates that the failure of a system or control should result in the system or control being in a secure state.

failover: A failure mode in which the system automatically transfers processing to a hot backup component, such as a clustered server, if a hardware or software failure is detected.

fail-safe: A failure mode in which program execution is terminated and the system is protected from compromise if a hardware or software failure is detected.

fail-soft (or resilient): A failure mode in which certain noncritical processing is terminated and the computer or network continues to function in a degraded mode, if a hardware or software failure is detected.

False Accept Rate (FAR; or Type II Error): In biometric access control systems, the percentage of unauthorized users who are incorrectly granted access. *See also* Crossover Error Rate (CER) *and* False Reject Rate (FRR; or Type I Error).

False Reject Rate (FRR; or Type I Error): In biometric access control systems, the percentage of authorized users who are incorrectly denied access. *See also* Crossover Error Rate (CER) *and* False Accept Rate (FAR; or Type II Error).

fault: Momentary loss of electric power.

fault injection: Any of several techniques used to test a system to see how it will behave under stress.

fault-tolerant: A system that continues to operate after the failure of a computer or network component.

Federal Information Processing Standard (FIPS): Standards and guidelines published by the U.S. National Institute of Standards and Technology (NIST) for federal computer systems. *See also* NIST.

Federal Privacy Act of 1974: A U.S. law requiring the protection of personal information by U.S. government agencies.

Federal Risk and Authorization Management Program (FedRAMP): The required process for U.S. federal government agencies when procuring cloud-based services.

federated identity management (FIM): A system whereby multiple organizations share a common identity management system.

federation of identity (FIdM): The standards, technologies, and tools used to facilitate the portability of identity across separately managed organizations.

fence: *See* security fence.

Fiber Distributed Data Interface (FDDI): A star topology, token-passing, network transport protocol.

fiber optic cable: A network medium consisting of glass or plastic strands that carry light signals.

Fibre Channel over Ethernet (FCoE): A communications protocol that encapsulates Fibre Channel frames over 10 Gigabit Ethernet (or faster) networks.

fiduciary: A person in a legal or moral position of trust and sound management, such as a company board member.

firewall: A device or program that controls traffic flow between networks.

first aid: Techniques used to treat injuries to personnel prior to receiving medical care.

forensics (or computer forensics): The science of conducting a computer crime investigation to determine what's happened and who's responsible for what's happened. One major component of computer forensics involves collecting legally admissible evidence for use in a computer-crime case.

fourth-party risk: A concern within third-party risk management in which third-party service organizations employ their own third parties, thereby increasing risk.

frame: The protocol data unit of the Ethernet protocol. *See also* protocol data unit (PDU), Ethernet.

frame relay (FR): A packet-switched network protocol used to transport WAN communications.

fraud: Any deceptive or misrepresented activity that results in illicit personal gain.

full interruption test: A test of a disaster recovery or business continuity plan in which contingency procedures and systems are used to conduct live business transactions.

functional requirements: The required visible characteristics of a program or system.

fuzzing: A software testing technique in which many different combinations of input strings are fed to a program in an attempt to elicit unexpected behavior.

gaming: Using a system for a purpose other than its intended purpose.

gateway: A system, connected to a network that performs any real-time translation or interface function, such as a system that converts Microsoft Exchange email to Lotus Notes email.

General Data Protection Regulation (GDPR): A law that strengthens data protection for European Union (EU) citizens and addresses the export of personal data outside the EU.

geographic diversity: A characteristic of electric utility and telecommunication facilities in which two or more connections to a facility are available.

geolocation: Any technique used to determine the location of a device.

Global Positioning System (GPS): A U.S. government-owned global system of satellites that provide geolocation and time information to GPS receivers anywhere on or near Earth that has an unobstructed line of sight to four or more GPS satellites.

goals: Specific milestones that an organization hopes to accomplish.

golden-ticket attack: An attack on a Kerberos system in which an attacker is able to forge valid ticket granting tickets and use them to access network resources.

governance: Policies and processes that ensure that executive management is fully informed and in control of some aspect of an organization.

Gramm-Leach-Bliley Act (GLBA): A U.S. law that defines privacy requirements for customers of financial services institutions.

guard dog: A trained canine that is accompanied by a security guard as part of an active work area protection plan.

guest: (1) An instantiation of an operating system within a virtual environment. *See also* virtualization. (2) A visitor to a commercial work facility.

guidelines: Similar to standards but considered to be recommendations rather than requirements.

hacker: Formerly a term describing a computer hobbyist; now commonly used to refer to a person with criminal intent who breaks into computers and networks.

hacktivist: A person who attacks organizations' systems based on ideological motivations.

hardening: Changing the architecture and/or configuration of a system to make it more resistant to attack.

hardening standard: A written document describing security configuration settings for applicable systems.

hardware: The physical components in a computer system.

hash function: A mathematical function that creates a unique representation of a larger set of data (such as a digest). Hash functions are often used in cryptographic algorithms and to produce checksums and message digests. *See also* message digest.

Health Information Technology for Economic and Clinical Health (HITECH) Act: A U.S. federal act that expanded the use of health-care information systems and of privacy requirements protecting healthcare information.

Health Insurance Portability and Accountability Act (HIPAA): A U.S. federal act that addresses security and privacy requirements for medical systems and information.

hearsay evidence: Evidence that isn't based on the witness's personal, firsthand knowledge but was obtained through other sources.

hearsay rule: Under the Federal Rules of Evidence, hearsay evidence is normally not admissible in court. Computer evidence is an exception to the hearsay rule.

heat detector: A device that is used to detect heat from a fire.

heating, ventilation, and air conditioning (HVAC): Environmental controls that ensure that temperature and humidity remain within acceptable levels.

heterogeneous environment: A systems environment that consists of a variety of types of systems. *See also* homogeneous environment.

hextel: Thirty-two hexadecimal numbers grouped into eight blocks of four decimal digits.

high availability (HA): A system's architecture and design that ensures a higher degree of availability than that of an individual system.

high-performance computing (HPC): The use of supercomputers for solving problems requiring large quantities of computation.

High-Speed Serial Interface (HSSI): A point-to-point WAN connection protocol.

homogeneous environment: A systems environment that consists largely of one type of system. *See also* heterogeneous environment.

honeynet: A large deployment of honeypots, also referred to as a *honeyfarm*. *See also* honeypot.

honeypot: A decoy system deployed by a security administrator to discover the attack methods of potential hackers.

host-based intrusion detection system (HIDS): An intrusion detection system designed to detect intrusions through examination of activities on a host system. *See also* intrusion detection system.

hot site: A fully configured alternative computer facility that has electrical power, HVAC, and functioning file/print servers and workstations. *See also* cold site, reciprocal site, *and* warm site.

hub: A network device used to connect several LAN devices. Also known as a *concentrator*.

human–machine interface (HMI): Features of a system or device designed to interact with a person, in which information is entered via switches, buttons, keys, microphones, or cameras, and/or information imparted to a user via a display, sound, or touch.

hybrid cloud: As defined by the National Institute of Standards and Technology, a cloud infrastructure composed of "two or more distinct cloud infrastructures (private, community, or public)."

hybrid risk analysis: Risk analysis that combines quantitative and qualitative risk analysis techniques.

Hypertext Transfer Protocol (HTTP): An application protocol used to transfer data between web servers and web browsers.

Hypertext Transfer Protocol Secure (HTTPS): The HTTP protocol encrypted with SSL or TLS. *See also* Hypertext Transfer Protocol.

hypervisor: In a virtualized environment, the supervisory program that controls allocation of resources and access to communications and peripheral devices. *See also* virtualization.

identification: The means by which a user claims a specific, unproven identity to a system. *See also* authentication.

identity and access management (IAM): The processes and procedures that support the life cycle of people's identities and access privileges in an organization.

identity as a service: A centralized, usually external service provider that provides tools for user identification.

identity management (IdM): The processes and procedures that support the life cycle of people's identities in an organization.

improper authentication: *See* authentication bypass.

inactivity timeout: A mechanism that locks, suspends, or logs off a user after a predetermined period of inactivity.

indicators of compromise (IOCs): An artifact observed on a network or in an operating system that is likely to be associated with a breach attempt.

industrial control system (ICS): Systems and devices used to monitor and/or control industrial machinery.

inference: The ability of users to figure out information about data at a sensitivity level for which they're not authorized.

information custodian (or custodian): The person who has day-to-day responsibility for managing and protecting information assets.

information flow model: A lattice-based model in which each object is assigned a security class and value, and their direction of flow is controlled by a security policy.

information owner (or owner): The person who decides who's allowed access to a file and what privileges are granted.

information security continuous monitoring (ISCM): The ongoing awareness of information security, vulnerabilities, and threats in support of organizational risk management decisions.

information security management system (ISMS): A set of processes and activities used to manage an information security program in an organization. ISMS is defined in ISO/IEC 27001.

Information Technology Security Evaluation Criteria (ITSEC): Formal evaluation criteria that address confidentiality, integrity, and availability for an entire system.

infrastructure as code: The concept that the infrastructure underlying a software application, including operating systems and database management systems, is part of the systems development and release process.

Infrastructure as a Service (IaaS): A cloud-based environment in which customers implement various types of virtual machines, including server operating systems and network devices.

injection attack: An attack against a system involving the use of malicious input.

inquiry: An audit technique in which an auditor interviews personnel to learn how a process or system is used.

inrush: Initial electric power surge experienced when electrical equipment is turned on.

inspection: An audit technique in which an auditor examines an information system, business process documentations, or business records.

Institute of Electrical and Electronics Engineers (IEEE): A technical professional organization that develops technical standards and promotes the advancement of technology.

integrated development environment (IDE): A software program used by developers to compose, debug, test, and run software programs.

integrated product team (IPT): A multidisciplinary team with the mission of designing, developing, and managing an information system.

Integrated Services Digital Network (ISDN): A low-bandwidth communications protocol that operates over analog telecommunications voice lines.

integration test: A test of software components to ensure that they work together properly.

integrity: A concept that safeguards the accuracy and completeness of information and processing methods, and ensures that modifications to data aren't made by unauthorized users or processes; unauthorized modifications to data aren't made by authorized users or processes; and data is internally and externally consistent, meaning that a given input produces an expected output.

intellectual property: Includes patents, trademarks, copyrights, and trade secrets.

interface testing: Tests that are performed on application programming interfaces and other human and nonhuman interfaces.

Intermediate distribution frame (IDF): (1) A room in which telephone and data cabling for one floor or portion of building is terminated. (2) The componentry used for terminating telephone and data cabling. *See also* distribution frame, main data frame (MDF).

International Electrotechnical Commission (IEC): A standards organization that defines and publishes international standards for electrical, electronic, and related technologies.

International Organization for Standardization (ISO): An international body for creating standards. ISO is derived from the Greek word *isos,* meaning "equal."

International Telecommunications Union (ITU): A United Nations agency responsible for coordinating worldwide telecommunications operations and services.

Internet: The worldwide, publicly accessible network that connects the networks of organizations.

Internet Assigned Numbers Authority (IANA): The organization that assigns AS numbers to organizations. *See also* autonomous system number (ASN).

Internet Control Message Protocol (ICMP): An Internet protocol used to transmit diagnostic messages.

Internet Control Message Protocol (ICMP) flood: An attack in which a large number of ICMP packets are sent to a target network in an attempt to incapacitate the network.

Internet Engineering Task Force (IETF): An international, membership-based, not-for-profit organization that develops and promotes voluntary Internet standards.

Internet Key Exchange (IKE): A set of protocols used to establish a security association between systems using the IPsec protocol. *See also* security association, Internet Protocol Security.

Internet Layer: Layer 2 of the TCP/IP model. *See also* TCP/IP model.

Internet of Things (IoT): The network of physical, connected objects embedded in electronics, operating systems, software, sensors, and network connectivity.

Internet Protocol (IP): The Open Systems Interconnection (OSI) Layer 3 protocol that's the basis of the modern Internet.

Internet Protocol version 4 (IPv4): The original and still widely used Layer 4 protocol that is the basis of the Internet and most organizations' internal networks.

Internet Protocol version 6 (IPv6): The replacement of IPv4 that provides a larger address space and additional functionality to provide security, multimedia support, plug and play, and backward compatibility with IPv4.

Internet Protocol Security (IPsec): An Internet Engineering Task Force open-standard virtual private network protocol for secure communications over local area networks, wide area networks, and public Internet Protocol–based networks.

Internet Relay Chat (IRC): An Application Layer protocol that facilitates communication in text form using a client-server network.

Internet Security Association and Key Management Protocol (ISAKMP): An Internet key exchange protocol.

Internet Small Computer Systems Interface (iSCSI): A communications protocol that enables the SCSI protocol to be sent over LANs, WANs, or the Internet. *See also* Small Computer Systems Interface (SCSI).

Internetwork Packet Exchange (IPX): A network packet-oriented protocol that's the basis for Novell Netware networks. IPX is analogous to IP.

interprocess communication (IPC): Any of several mechanisms through which separate processes can communicate.

intrusion detection system (IDS): A hardware or software application that detects and reports on suspected network or host intrusions.

intrusion prevention system (IPS): A hardware or software application that both detects and blocks suspected network or host intrusions.

investigation: A study and analysis of an event, including the identification of evidence, to determine the facts related to the event.

IT Infrastructure Library (ITIL): An industry standard of IT service management processes.

iteration: *See* sprint.

JavaScript: A high-level, dynamic, lightweight interpreted programming language used to make web pages interactive and provide online programs.

job description: A formal description of a position's roles and responsibilities.

job rotation: The practice of moving employees from one position to another for cross-training and security reasons.

judgmental sampling: A sampling technique in which individual items are chosen by the auditor.

just-in-time (JIT) access: A procedure in which temporary, granular access to an application or resource is granted when needed to perform a specific task or function.

Kerberos: A ticket-based authentication protocol, in which tickets are used to identify users, developed at the Massachusetts Institute of Technology.

key card: A type of building or room access control in which personnel wave a key card (also known as an access card) in front of a key card reader to unlock a door.

key change: The practice of replacing an encryption key in a cryptosystem.

key clustering: An occurrence where the encryption of a single plaintext message using two different encryption keys results in the same ciphertext.

key control: Safeguards and procedures for protecting an encryption key.

key disposal: The practice of securely disposing an encryption key so that it cannot be recovered.

key distribution: The practice of moving an encryption key from the point of generation to the point of use and storage.

key encryption key (KEK): An encryption key used to encrypt and decrypt data encryption keys. *See also* data encryption key (DEK).

key escrow: The practice of storing an encryption key with a third party in the event that the original encryption key is lost.

key generation: The practice of creating a new encryption key.

key installation: The act of placing an encryption key in a cryptosystem.

key logging: The practice of recording keystrokes, usually for illicit purposes, such as acquiring user IDs, passwords, and other confidential information.

key management: Practices and procedures used to manage encryption keys.

key performance indicator (KPI): A measurable value that evaluates how successful an organization is in achieving a specific objective or activity.

key risk indicator (KRI): A metric used to indicate the level of risk associated with a particular activity or course of action.

key vault: A centralized vault that securely stores shared account credentials.

keyspace: The range of all possible values for an encryption key.

known-plaintext attack: An attack technique in which the cryptanalyst has a given plaintext message and the resulting ciphertext.

latency: The time required for an operation to complete.

lattice-based access controls: A method for implementing mandatory access controls in which a mathematical structure defines greatest lower-bound and least upper-bound values for a pair of elements, such as subject and object.

Layer 2 Forwarding Protocol (L2F): A virtual private network protocol similar to Point-to-Point Tunneling Protocol.

Layer 2 Tunneling Protocol (L2TP): A virtual private network protocol similar to Point-to-Point Tunneling Protocol and Layer 2 Forwarding Protocol.

least privilege: A principle requiring that a subject is granted only the minimum privileges necessary to perform an assigned task.

Li-Fi: A wireless communication technology that uses light to transmit data.

Lightweight Directory Access Protocol (LDAP): An Internet Protocol and data storage model that supports authentication and directory functions.

link encryption: Packet encryption and decryption at every node along the network path; requires each node to have separate key pairs for its upstream and downstream neighbors.

Link Layer: Layer 1 of the TCP/IP model. *See also* TCP/IP model.

live forensics: Techniques used to gather forensic information from a running system.

load balancer: A device that routes incoming messages to a pool of one or more destinations.

log review: The examination of a system or event log.

logic bomb: A program, or portion thereof, designed to perform some malicious function when a predetermined circumstance occurs. *See also* malware.

Long-Term Evolution (LTE): A mobile telecommunications protocol for IP communications over cellular networks.

machine learning (ML): A method of data analysis that enables computers to analyze a data set and automatically perform actions based on the results without being explicitly programmed.

main distribution frame (MDF): (1) A room in which telephone and data cabling for an entire building is terminated. (2) The componentry used for terminating telephone and data cabling. *See also* distribution frame, intermediate data frame (IDF).

maintenance hook: A backdoor that allows a software developer or vendor to bypass access control mechanisms to perform maintenance. These backdoors are often well known and pose a significant security threat if not properly secured.

malware: Malicious software that typically damages, takes control of, or collects information from a computer. This classification of software broadly includes viruses, worms, ransomware Trojan horses, logic bombs, spyware, and (to a lesser extent) adware.

managed security service (MSS): Security-related services provided by a service provider, typically involving monitoring or management of information systems.

management review: Activities whereby management reviews a program or process.

mandatory access control (MAC) system: A type of access control system in which the access policy is determined by the system rather than by the owner. *See also* discretionary access control (DAC).

man-in-the-middle attack: A type of attack in which an attacker intercepts messages between two parties and forwards a modified version of the original message.

mandatory vacation: A practice by some organizations that requires each worker to take at least one vacation (usually, an entire week) at least once per year to provide the organization opportunities to detect fraud.

mantrap: A physical access control method consisting of a double set of locked doors or turnstiles to prevent tailgating. *See also* bollard *and* sally port.

manual controls: Controls that are not performed automatically and, therefore, require human action.

marking: Affixing a human-readable classification label on a document, device, or data storage object. *See also* tagging.

mashup: A web application employing content from multiple sources and displayed through a single user interface.

masking: A technique used to conceal the contents of data.

Massive Machine-Type Communications (mMTC): A standard used in 5G mobile communications.

maturity model: A technique used to assess the maturity of an organization and the capability of its processes.

maximum tolerable downtime (MTD): An extension of a criticality assessment that specifies the maximum period of time that a given business process can be inoperative before experiencing unacceptable consequences. *See also* criticality assessment.

maximum tolerable outage (MTO): The maximum period of time that a given business process can be operating in emergency or alternative processing mode.

maximum tolerable period of disruption (MTPD): *See* maximum tolerable downtime (MTD).

mean time between failures (MTBF): The amount of time, usually measured in hours, that a component is expected to continuously operate before experiencing a failure.

media controls: Controls that are used to manage information classification and physical media.

meet-in-the-middle attack: A type of attack in which an attacker encrypts known plaintext with each possible key on one end, decrypts the corresponding ciphertext with each possible key, and then compares the results in the middle.

memory leak: A software defect that results in a program's continuing to allocate memory.

mesh: A network design in which all nodes are connected to all other nodes.

message digest: A condensed representation of a message that is produced by using a one-way hash function. *See also* hash function.

metropolitan area network (MAN): A network that extends across a large area, such as a city.

microsegmentation: Techniques used to isolate groups of systems or individual systems using network access controls such as firewalls.

microservices: Software-based services running on various systems in a distributed environment.

mission statement: A statement that defines an organization's (or organizational unit's) reason for existence.

mobile app: An application that runs on a mobile device and has the capability to interact with the user, communicate over the Internet, and store data locally.

mobile code: A software architecture in which code is moved from a repository to a system for execution.

mobile device: A general term encompassing devices such as smartphones, tablets, phablets, and wearables that run operating systems such as iOS, Android, and Windows 10.

mobile device management (MDM): Software used to manage the administration of mobile devices such as smartphones, phablets, and tablets.

monitoring: Activities that verify processes, procedures, and systems.

monoalphabetic substitution: A cryptographic system that uses a single alphabet to encrypt and decrypt an entire message.

MoSCoW (Must have, Should have, Could have, Won't have) method: A prioritization method used to classify the importance of requirements in a business project.

multicast: A type of network protocol whereby packets are sent from a source to multiple destinations.

multifactor authentication: Any authentication mechanism that requires two or more of the following factors: something you know, something you have, or something you are.

multiprotocol label switching (MPLS): An extremely fast method of forwarding packets through a network by using labels inserted between Layer 2 and Layer 3 headers in the packet.

Multipurpose Internet Mail Extensions (MIME): An IETF standard that defines the format for messages that are exchanged between email systems over the Internet. *See also* IETF.

National Computer Security Center (NCSC): A U.S. government organization within the National Security Agency that is responsible for evaluating computing equipment and applications that are used to process classified data.

National Information Assurance Certification and Accreditation Process (NIACAP): Formalizes the certification and accreditation process for U.S. government national security information systems.

National Institute of Standards and Technology (NIST): A federal agency within the U.S. Department of Commerce that is responsible for promoting innovation and competitiveness through standards, measurement science, and technology.

near-field communications (NFC): A wireless communications protocol that operates over distances of up to 4 centimeters.

need to know: Status defines the essential information a person needs to perform their assigned job function.

NetBIOS: A TCP/IP protocol that allows applications to communicate over a network.

Network Access Layer: Layer 1 of the TCP/IP model. *See also* TCP/IP model.

network address translation (NAT): The process of converting internal, privately used addresses in a network to external, public addresses.

network-based intrusion detection system (NIDS): An intrusion detection system designed to detect intrusions through examination of network traffic. *See also* intrusion detection system.

network file system (NFS): A TCP/IP protocol used to provide access to file systems on remote computers.

network function virtualization (NFV): The practice of implementing network devices as virtual machines instead of hardware-based systems.

Network Layer: Layer 3 of the OSI model. *See also* Open Systems Interconnection (OSI) model.

network penetration test: A penetration test that targets systems and network devices on a network. *See also* penetration test.

network sprawl: A phenomenon wherein virtual network elements are created, generally without approval or with limited planning and control, in an environment such as the cloud.

next-generation firewall (NGFW): A network security platform that fully integrates traditional firewall and network intrusion prevention capabilities with other advanced security functions that provide deep packet inspection for complete visibility, accurate application, content, and user identification, and granular policy-based control. *See also* deep packet inspection (DPI), intrusion prevention system (IPS).

noncompete agreement: A legal agreement in which an employee agrees not to accept employment in a competing organization.

nondisclosure agreement (NDA): A legal agreement in which one or more parties agrees to refrain from disseminating confidential information related to other parties.

nonfunctional requirements: The characteristics of a program or system that are not apparent to an end user.

noninterference model: Ensures that the actions of different objects and subjects aren't seen by, and don't interfere with, other objects and subjects on the same system.

nonrepudiation: The inability for a user to deny an action; their identity is positively associated with that action.

north–south traffic: Network communications between systems within a network and systems outside the network.

Oakley Key Exchange Protocol: A key agreement protocol implemented by Cisco in ISAKMP to facilitate Diffie-Hellman Key Exchange.

obfuscation: A technique in which data is scattered, rearranged, or hidden to make it more difficult to identify and exploit.

object: A passive entity, such as a system or file.

object reuse: The process of protecting the confidentiality of objects that are reassigned after initial use. *See also* Trusted Computer System Evaluation Criteria (TCSEC).

objectives: Specific milestones that an organization wants to perform to meet its goals. *See also* goals.

observation: An audit technique in which an auditor passively observes activities performed by personnel or information systems.

on-premises: Information systems, applications and data that is physically located in an organization's own information processing center.

one-time pad: A cryptographic keystream that can be used only once.

one-time password: A password that's valid for only one login session.

one-way function: A problem that's easy to compute in one direction but not in the reverse direction.

open message format: A message encrypted in an asymmetric key system by using the sender's private key. The sender's public key, which is available to anyone, is used to decrypt the message. This format guarantees the message's authenticity. *See also* secure and signed message format, secure message format.

open source: A software licensing methodology wherein source code is freely available.

open system: A vendor-independent system that complies with an accepted standard, which promotes interoperability among systems and components made by different vendors. *See also* closed system.

Open Systems Interconnection (OSI) model: The seven-layer reference model for networks. The layers are Physical, Data Link, Network, Transport, Session, Presentation, and Application.

Open Web Application Security Project (OWASP): A not-for-profit organization dedicated to web application security.

OpenID Connect (OIDC): A standards-based authentication protocol built on the OAuth framework.

operating system (OS): Software that controls computer hardware and resources and facilitates the operation of application software. *See also* application software.

operational level agreement (OLA): An agreement specifying operational support parameters between support groups in an organization. *See also* service-level agreement (SLA).

operational technology (OT): Network and computing infrastructure supporting industrial control systems or supervisory control and data acquisition environments.

opt-in: A choice made by a data subject to desires that an organization include the data subject in its uses of personal information. Commonly, this means inclusion in marketing campaigns.

opt-out: A choice made by a data subject who desires that an organization discontinue specific uses of the data subject's personal information. Commonly, this means removal from marketing campaigns.

optical disk: Media such as CD-ROM and DVD-ROM used to read and write information.

Orange Book: *See* Trusted Computer System Evaluation Criteria (TCSEC).

outsourcing: The use of an external organization (third party) to perform some aspect of business operations.

over the top (OTT): A term describing cloud-based media services such as videoconferencing, texting, and "television" content, bypassing telecommunications and cable operators.

owner: A person in an organization who's responsible for management of an asset, including classification, handling, and access policy.

packet: The protocol data unit of the Internet Protocol. See also protocol data unit (PDU), Internet Protocol (IP).

packet-filtering firewall: A type of firewall that examines the source and destination addresses of an incoming packet, and then either permits or denies the packet based on an ACL. *See also* access control list (ACL).

packet sniffing: A type of attack in which an attacker uses a sniffer to passively capture network packets and analyze their contents.

packet-switched network: Any of several telecommunications network technologies in which packets transport data between sender and receiver.

parallel test: A test of a business continuity or disaster recovery plan in which contingency procedures are performed in parallel with normal procedures.

parity bit: A technique used to detect errors in a bit pattern.

pass the hash: An authentication-bypass attack on a system in which the attacker authenticates with stolen NTLM or LanMan hashes instead of plaintext passwords.

passphrase: A string of characters consisting of multiple words that a subject provides to an authentication mechanism to authenticate to a system. *See also* password.

password: A string of characters (a word or phrase) that a subject provides to an authentication mechanism to authenticate to a system.

Password Authentication Protocol (PAP): A remote access control protocol that uses a two-way handshake to authenticate a peer to a server when a link is initially established.

password cracking: An attack in which an attacker has been able to obtain password hashes and attempts to crack the hashes to obtain plaintext passwords.

patch: A corrective fix for a program or system to correct a defect.

patch management: The use of procedures and tools to apply patches to target systems.

patent: As defined by the U.S. Patent and Trademark Office, "the grant of a property right to the inventor."

Payment Card Industry Data Security Standard (PCI DSS): A standard set of requirements developed for the protection of personal data related to credit, debit, and cash card transactions.

peer review: Any instance in which a worker checks the work performed by another. *See also* code review.

pen tester: A person who performs a penetration test.

penetration test: A test involving automated and manual techniques that is used to identify potential software vulnerabilities. Also known as *pen testing*.

performance management: The life cycle process concerned with the measurement and management of information processing resources.

permutation cipher: *See* transposition cipher.

personally identifiable information (PII): Information (such as name, address, Social Security number, birthdate, place of employment, and so on) that can be used on its own or with other information to identify, contact, or locate a person.

personal identification number (PIN): A numeric-only passcode, usually used when only a numeric keypad (versus an alphanumeric keyboard) is available. *See also* password.

pharming: A phishing attack that targets a specific organization. *See also* phishing.

phishing: A social-engineering cyberattack technique widely used in identity-theft crimes. An email, purportedly from a known legitimate business (typically, financial institutions, online auctions, retail stores, and so on), requests the recipient to verify personal information online at a forged or hijacked website. *See also* pharming *and* spear phishing.

phone tap. *See* wiretap.

physical controls: Controls that ensure the safety and security of the physical environment.

physical evidence: *See* real evidence.

Physical Layer: Layer 1 of the OSI model. *See also* Open Systems Interconnection (OSI) model.

physical penetration test: An evaluation of physical security controls in the form of an attack simulation.

plain old telephone system (POTS): A slang term for analog telephone service.

plaintext: A message in its original readable format or a ciphertext message that's been properly decrypted (unscrambled) to produce the original readable plaintext message.

Plan, Do, Check, Act: *See* Deming cycle.

Platform as a Service (PaaS): A cloud-based environment in which customers can implement applications within a software ecosystem.

plenum: The space between a false ceiling and the actual ceiling in a building.

Point-to-Point Protocol (PPP): A protocol used in Remote Access Service servers to encapsulate Internet Protocol packets and establish dial-in connections over serial and Integrated Services Digital Network links.

Point-to-Point Tunneling Protocol (PPTP): A virtual private network protocol designed for individual client–server connections.

policy: A formal high-level statement of an organization's objectives, responsibilities, ethics and beliefs, and general requirements and controls.

port scan: A test used to determine which Transmission Control Protocol/Internet Protocol and User Datagram Protocol service ports on a system are active. *See also* Transmission Control Protocol (TCP), Internet Protocol (IP), User Datagram Protocol (UDP).

power surge: *See* surge.

pre-action: A type of water-based fire suppression system that is a hybrid of dry pipe and wet pipe. *See also* dry pipe, wet pipe.

Presentation Layer: Layer 6 of the OSI model. *See also* Open Systems Interconnection (OSI) model.

preventive controls: Controls that are intended to prevent unwanted events.

printer steganography: A technique in which printers include a hidden (barely visible) machine identification code on every page of printed matter, which permits the identification of an individual printer.

privacy: In information security, the protection and proper handling of personal information.

private branch exchange (PBX): A system for managing telephones and telephone communications in a business environment.

private cloud: As defined by the National Institute of Standards and Technology, a cloud infrastructure "provisioned for exclusive use by a single organization comprising multiple consumers." *See also* cloud.

private key cryptography: A cryptographic method that requires parties to exchange a secret key to communicate.

private network address: Addresses on TCP/IP networks that are not routable on the Internet and are used for private, internal networks.

privilege creep: *See* aggregation (2).

privilege escalation: *See* escalation of privilege.

privileged access management (PAM): (1) Business processes and procedures concerning the provisioning of privileged credentials to administrative personnel. (2) Tools used to manage privileged credentials and privileged access to systems and devices.

privileged entity controls: The mechanisms that provide and monitor privileged access to hardware, software, and data.

procedures: Detailed instructions about how to implement specific policies and meet the criteria defined in standards.

process isolation: An operating system feature whereby different user processes are unable to view or modify information related to other processes.

processor: *See* data processor.

promiscuous mode: A setting on a network adapter that passes all network traffic to the associated device for processing, not just traffic that is specifically addressed to that device. *See also* sniffing.

Protected Extensible Authentication Protocol (PEAP): An open standard used to transmit authentication information in a protected manner.

protected health information (PHI): Any information about health status, provisioning of health care, or payment for health care collected by a covered entity (such as a health-care provider or insurance company) that can be linked to a specific person.

protection domain: Prevents other programs or processes from accessing and modifying the contents of an address space that has already been assigned to an active program or process.

protection rings: A security architecture concept that implements multiple domains that have increasing levels of trust near the center.

protocol data unit (PDU): The unit of data used at a particular layer of a communications protocol.

provisioning: (1) The act of creating a user account on a system or network. (2) The act of applying configuration changes to a system or network device.

proximate causation: An action taken or not taken as part of a sequence of events that result in negative consequences.

proxy server: A system that transfers data packets from one network to another.

prudent-man rule: A rule under the Federal Sentencing Guidelines that requires senior corporate officers to perform their duties in good faith, in the best interests of the enterprise, and with the care and diligence that ordinary, prudent people in a similar position would exercise in similar circumstances.

pseudonymization: An irreversible deidentification procedure whereby a specific identifier is replaced by other values to make it less identifiable to the original data subject. See *also* anonymization *and* de-identification.

public cloud: As defined by the National Institute of Standards and Technology, a cloud infrastructure "provisioned for open use by the general public." *See also* cloud.

public key cryptography: A cryptographic method that permits parties to communicate without exchanging a secret key in advance.

public key infrastructure (PKI): A system that enables secure e-commerce through the integration of digital signatures, digital certificates, processes, procedures, and other services necessary to ensure confidentiality, integrity, authentication, nonrepudiation, and access control.

punitive damages: Determined by a jury and intended to punish the offender.

qualitative risk analysis: A risk analysis that expresses risks and costs in qualitative terms versus quantitative terms (such as high, medium, and low). *See also* risk analysis.

quality of service (QoS): The ability to prioritize various types of voice and data traffic based on operational needs such as response time, packet loss, and jitter.

quantitative risk analysis: A risk analysis that includes estimated costs. *See* also risk analysis.

quantum computing: An emerging computing processor design that uses the properties of quantum states to perform computation.

quarantine: A general term referring to the process of isolating a resource for security reasons.

race condition: A situation in which two programs, processes, or threads are accessing or manipulating a resource as though they are doing so exclusively, thereby leading to an unexpected outcome.

radio frequency (RF) emanations: Unintentional emissions of electromagnetic energy from a system.

radio frequency interference (RFI): Electrical noise caused by electrical components, such as fluorescent lighting and electric cables.

rainbow table: A database of hashes and their corresponding passwords.

ransomware: Malware that encrypts files on a target system and demands a ransom payment, usually cryptocurrency, to retrieve the key to decrypt the files. A permutation of ransomware also threatens to publish the plaintext data. *See also* malware *and* cryptocurrency.

read-through: *See* tabletop.

real (or physical) evidence: Tangible objects from the actual crime, such as the tools or weapons used and any stolen or damaged property. *See also* evidence.

reciprocal site: An alternative computer facility with systems that can be used in an emergency. *See also* cold site, hot site, warm site.

recovery controls: Controls that restore a process or system to its pre-event state.

recovery point objective (RPO): The maximum period of time in which data may be lost if a disaster occurs.

recovery time objective (RTO): The period of time in which a business process must be recovered (during a disaster) to ensure the survival of the organization.

Red Book: *See* Trusted Computer System Evaluation Criteria (TCSEC).

Reduced-Instruction-Set-Computing (RISC): A microprocessor instruction set architecture that sues a smaller, simpler instruction set than Complex-Instruction-Set-Computing, which makes RISC more efficient. *See also* Complex-Instruction-Set-Computing (CISC).

reduction analysis: A step in threat modeling designed to reduce duplication of effort.

redundancy: Multiple systems, nodes, or network paths that provide the same functionality for resiliency and availability in the event of failure.

redundant array of independent disks (RAID): A collection of one or more hard drives in a system for purposes of improved performance or reliability.

reference monitor: An abstract machine (a theoretical model for a computer system or software program) that mediates all access to an object by a subject.

referential integrity: A property of a database management system in which all data relationships (such as indexes, primary keys, and foreign keys) are sound.

Registration Authority (RA): In a public key infrastructure, the RA is responsible for verifying certificate contents for the Certificate Authority. *See also* Certificate Authority (CA), public key infrastructure (PKI).

remediation: Corrective action taken to resolve an issue identified in an assessment.

remote access: The capability for a user in a remote location to establish a logical connection to a private internal network.

Remote Access Service (RAS): A remote access protocol typically used over dial-up facilities.

Remote Authentication Dial-In User Service (RADIUS): An open-source, User Datagram Protocol–based client–server protocol used to authenticate remote users.

remote backup: A backup operation in which the target backup media is located in a remote location. *See also* e-vaulting.

Remote Desktop Protocol (RDP): A proprietary Microsoft protocol used to connect to another computer over a network connection.

Remote Procedure Call (RPC): A TCP/IP network protocol used to direct the execution of application code on another computer.

repeater: A device that boosts or retransmits a signal to extend the range of a wired or wireless network.

reperformance: An audit technique in which an auditor performs tasks or transactions on their own to see whether the results are correct.

replay attack: An attack on a system in which an attacker is able to replay captured data such as login credentials in an attempt to break in o the system.

replication: The process of copying data transactions from one system to another.

repo: *See* repository (2).

repository: (1) In a public key infrastructure, a system that accepts certificates and Certificate Revocation Lists from a Certificate Authority and distributes them to authorized parties. *See also* Certificate Authority (CA), public key infrastructure (PKI). (2) In a software development environment, a system used to store software source code.

requirements: A list of one or more required characteristics of a system, generally used as a guide to later design and development phases.

resilience: The ability of a process or system to continue operation despite various harmful effects.

restricted algorithm: A cryptographic algorithm that must be kept secret to provide security of a cryptosystem.

retention: The period of time (whether a minimum or maximum) that an organization will retain documents and business records.

retention schedule: A policy that defines retention requirements for various types of business records.

Reverse Address Resolution Protocol (RARP): A protocol used by diskless workstations to query and discover their own IP addresses by using machine addresses (known as a media access control, or MAC, address).

Rijndael: The encryption algorithm used by the Advanced Encryption Standard. *See also* Advanced Encryption Standard (AES).

ring: A network topology in which all devices are connected to a closed loop.

risk acceptance: Accepting a risk or residual risk as is, without mitigating or transferring it.

risk analysis: A method used to identify and assess threats and vulnerabilities in a business, process, system, or activity as part of a risk assessment. *See also* risk assessment.

risk appetite: The highest level of risk that an organization will accept.

risk assessment: A study of risks associated with a business process, information system, work facility, or other object of study.

risk assignment (or transference): *See* risk transfer.

risk avoidance: Eliminating risk through discontinuation of the activity related to the risk.

risk-based access control: A process wherein an information system presents authentication challenges that are commensurate with the user's security profile (such as geolocation and device type).

risk identification: The process of examining assets, threats, and vulnerabilities that result in knowledge of a risk.

risk management: The process life cycle that includes risk assessment and risk treatment.

risk mitigation: Reducing risk to a level that's acceptable to an organization.

risk reduction: Mitigating risk by implementing the necessary security controls, policies, and procedures to protect an asset.

risk tolerance: The variation from risk appetite that an organization is willing to accept.

risk transfer: Transferring the potential loss associated with a risk to a third party, such as an insurance company.

risk treatment: The formal decision-making process for the management of identified risks.

Rivest, Shamir, Adleman (RSA): A key transport algorithm based on the difficulty of factoring a number that's the product of two large prime numbers.

role-based access control (RBAC): A method for implementing discretionary access controls in which access decisions are based on group membership, according to organizational or functional roles.

rootkit: Malware that provides privileged (root-level) access to a computer. *See also* malware.

rotation of duties: *See* job rotation.

router: A network device that forwards packets between separate networks.

routing protocol: A network protocol used by routers to communicate information about internetwork connections, facilitating proper routing of packets toward their destinations.

RSA: *See* Rivest, Shamir, Adleman.

rubber hose attack: An attack on a cryptosystem in which the attacker uses coercion to compel the owner of a cryptosystem to relinquish the encryption key.

rule-based access control: A method for applying mandatory access control by matching an object's sensitivity label and a subject's sensitivity label to determine whether access should be granted or denied. *See also* role-based access control.

safeguard: A control or countermeasure implemented to reduce the risk or damage associated with a specific threat.

sag: A short drop in voltage.

sally port: A secure, controlled entrance to a facility.

salvage: In disaster recovery planning, the operations performed to repair or replace damaged facilities and equipment.

sampling: Any of several techniques in which individual items are selected for examination during an audit.

sandbox: A mechanism for isolating a program or system.

Sarbanes-Oxley (SOX): A U.S. law that attempts to prevent fraudulent accounting practices and errors in U.S. public corporations and mandates data retention requirements.

scan: A technique used to identify vulnerabilities in a system or network, usually by transmitting data to it and observing its response.

screen saver: An image or pattern that appears on a display, usually as part of an inactivity timeout. *See also* inactivity timeout.

screening router: A firewall architecture that consists of a router that controls packet flow through the use of access control lists. *See also* access control list (ACL), firewall.

script injection: An attack in which the attacker injects script code in the hope that the code will be executed on a target system.

script kiddie: A person who does not have any programming or hacking skills but uses scripts, malware, exploits, and other hacking tools developed by others to attack an endpoint or network.

Scrum: A common implementation of the Agile systems-development methodology.

search warrant: A signed court order that permits law enforcement to search for and seize specific evidence.

secondary evidence: A duplicate or copy of evidence, such as a tape backup, screen capture, or photograph. *See also* evidence.

secret key cryptography: *See* symmetric key cryptography.

secure access service edge (SASE): An identity-based collection of cloud-based systems to protect endpoints.

secure and signed message format: A message encrypted in an asymmetric key system by using the recipient's public key and the sender's private key. This encryption method protects the message's confidentiality and guarantees the message's authenticity. *See also* open message format, secure message format.

secure by default: A principle of architecture and engineering that requires system settings and options be set to secure defaults.

secure by deployment: A principle of architecture and engineering that ensures the protection of the implementation process and results in a secure system.

secure by design: A principle of architecture and engineering that promotes the integration of security concepts and features into the design of a system.

Secure Hypertext Transfer Protocol (S-HTTP): An Internet protocol that provides a method for secure communications with a webserver. S-HTTP is now considered to be obsolete. *See also* Hypertext Transfer Protocol (HTTP) *and* Hypertext Transfer Protocol Secure (HTTPS).

Secure Key Exchange Mechanism (SKEME): A key exchange mechanism developed by IBM in the 1990s.

secure message format: A message encrypted in an asymmetric key system by using the recipient's public key. Only the recipient's private key can decrypt the message. This encryption method protects the message's confidentiality. *See also* open message format, secure and signed message format.

Secure Shell (SSH): A secure character-oriented protocol that's a secure alternative to telnet and RSH. *See also* telnet.

Secure Sockets Layer (SSL): A deprecated Transport Layer protocol that provides session-based encryption and authentication for secure communication between clients and servers on the Internet. *See also* Transport Layer Security (TLS).

Security Assertion Markup Language (SAML): An XML-based, open-standard data format for exchanging authentication and authorization credentials between organizations.

Security Association (SA): In IPsec, one-way connection between two communicating parties. *See also* IPsec.

security awareness: The process of providing basic security information to users in an organization to help them make prudent decisions regarding the protection of the organization's assets.

security control assessment: An examination of one or more security controls in an organization.

security engineering: A subspecialty of engineering that focuses on security design and operations.

security fence: A fence designed to prevent and deter unauthorized people from approaching a work or storage location.

security gate: A movable gate that prevents the entrance of unauthorized personnel at a work location.

security guard: A trained person who provides deterrent and protective services at a work location.

security information and event management (SIEM): A system that provides real-time collection, analysis, correlation, and presentation of security logs and alerts.

security kernel: The combination of hardware, firmware, and software elements in a Trusted Computing Base that implements the reference monitor concept. *See also* Trusted Computing Base (TCB).

security modes of operation: Designations for U.S. military and government computer systems based on the need to protect secrets stored within them. The modes are Dedicated, System High, Multi-Level, and Limited Access.

security operations center (SOC): A facility that provides information security monitoring, assessment, defense, and remediation for enterprise compute and network resources, including on-premises and cloud environments.

security orchestration, automation, and response (SOAR): A set of capabilities that enables automated action and response when specific types of events occur. A SOAR platform is most often integrated into a security information and event management system to enable automated response when specific types of security events occur. *See also* security information and event management system (SIEM).

Security Parameter Index (SPI): In IPsec, a 32-bit string used by the receiving station to differentiate between the Security Associations terminating on that station. The SPI is located within the Authentication Header or Encapsulating Security Payload. *See also* IPSec, Security Association (SA), Authentication Header (AH), Encapsulating Security Payload (ESP).

security perimeter: The boundary that separates the Trusted Computing Base from the rest of the system. *See also* Trusted Computing Base (TCB).

security policy: Formal policy that specifies expected behavior regarding the protection and use of information.

security posture: The level of risk in an organization based on its security practices.

segregation of duties: *See* separation of duties.

Sensitive but Unclassified (SBU): A U.S. government data classification level for information that's not classified but requires protection, such as private or personal information.

sensitivity label: In a mandatory access control–based system, a subject's sensitivity label specifies that subject's level of trust, whereas an object's sensitivity label specifies the level of trust required for access to that object. *See also* mandatory access control (MAC) system.

separation of duties (SoD): A concept that ensures that no single person has complete authority for and control of a critical system or process.

Serial Line Internet Protocol (SLIP): An early Point-to-Point Protocol used to transport Internet Protocol over dial-up modems. PPP is more commonly used for this purpose.

serverless computing: A cloud-services model in which software container instances are presented to customers for the execution of software.

service-level agreement (SLA): Formal minimum performance standards for systems, applications, networks, or services.

service set identifier (SSID): The name used to uniquely identify a Wi-Fi network.

session: An individual user's dialogue, or series of interactions, with an information system.

session hijacking: Similar to a man-in-the-middle attack, except that the attacker impersonates the intended recipient instead of modifying messages in transit. *See also* man-in-the-middle attack.

Session Initiation Protocol (SIP): A TCP/IP protocol used to transport voice and video communications over a network.

Session Layer: Layer 5 of the OSI model. *See also* Open Systems Interconnection (OSI) model.

session management: A mechanism in an application that tracks and enforces separate user sessions.

shared responsibility matrix: A chart that specifies the parties that are responsible for various aspects of information security.

Shift Security Left: The concept of introducing security earlier in a development life cycle process. *See also* DevSecOps.

Short Message Service (SMS): A protocol for sending short text messages over mobile telecommunications networks.

shoulder surfing: A social engineering technique that involves looking over someone's shoulder to obtain information such as passwords or account numbers.

side-channel attack: An attack on a system in which any of several system characteristics are observed in an attempt to obtain its secrets.

simulation: A facilitated walk-through of a disaster recovery plan, business continuity plan, or incident response plan wherein the proceedings of a disaster or intrusion are scripted to provide validation and learning for participants.

simple integrity property: A state in which a subject can't read information from an object that has a lower integrity level than the subject (no read down, or NRD). *See also* Biba model.

Simple Key Management for Internet Protocols (SKIP): A protocol used to share encryption keys.

Simple Mail Transport Protocol (SMTP): A protocol used to transport email messages between email servers.

Simple Mail Transport Protocol over TLS: A protocol used to transport email messages between email servers with encryption. *See also* Secure Sockets Layer/Transport Layer Security (SSL/TLS).

simple security property (ss property): A state in which a subject can't read information from an object that has a higher sensitivity label than the subject (no read up, or NRU). *See also* Bell-LaPadula model.

gain access to a system what you know, what you have, or what you are.

single-key cryptography: *See* symmetric key system.

Single Loss Expectancy (SLE): Asset value × exposure factor (EF). A measure of the loss incurred from a single realized threat or event, expressed in dollars. *See also* exposure factor (EF).

single sign-on (SSO): A system that allows a user to present a single set of log-on credentials, typically to an authentication server, which then transparently logs the user on to all other enterprise systems and applications for which that user is authorized.

SKIP: *See* Simple Key Management for Internet Protocols (SKIP).

small computer systems interface (SCSI): A set of standards for communications between computers and peripheral devices, usually hard drives.

smartphone: *See* mobile device.

smishing: The practice of sending phishing messages through Short Message Service. *See also* phishing.

smoke detector: A device that detects the early products of combustion.

sniffing: The practice of intercepting communications for usually covert purposes. *See also* packet sniffing.

social engineering: An attack method that employs techniques such as dumpster diving, shoulder surfing, and ruses designed to trick workers into providing information or system access.

socket: A logical endpoint on a system or device used to communicate over a network to another system or device (or even on the same device).

software: Computer instructions that enable the computer to accomplish tasks. *See also* application software, operating system (OS).

Software as a Service (SaaS): As defined by the National Institute of Standards and Technology, "the capability provided to the consumer to use the provider's applications running on a cloud infrastructure."

Software Assurance Maturity Model (SAMM): A maturity model for software development.

software bill of materials (SBOM): The inventory of source code components for a given system.

software configuration management (SCM): The practice of tracking and controlling changes in software programs, including source code.

software-defined networking (SDN): A computer networking approach that abstracts higher-level network functionality from the underlying physical infrastructure.

software-defined security: A security model in which security functions are defined and controlled by software.

software-defined wide area networking (SD-WAN): The use of software-defined networking in a wide-area network.

software development life cycle (SDLC): The business-level process used to develop and maintain software. *See also* systems development life cycle (SDLC).

software escrow agreement: A legal agreement between a software manufacturer and its customer(s) wherein the software manufacturer will maintain a copy of its original software source code with a third-party software escrow company. In the event that the software manufacturer ceases to operate as a going concern (or other events defined in the software escrow agreement), the software escrow company will release the original source code to the customers that are party to the software escrow agreement.

source code: Human-readable machine instructions that are the basis of system and application software.

source code repository: A system used to store, manage, and protect application or system software source code.

source code review: *See* code review.

spam (or Unsolicited Commercial Email [UCE]): Junk email, which currently constitutes about 85 percent of all worldwide email.

spear phishing: A phishing attack that's highly targeted, such as at a particular organization or part of an organization. *See also* phishing.

spike: A momentary rush of electric power.

SPIM: Spam that is delivered via instant messaging.

SPIT: Spam that is delivered via Internet telephony.

spoofing: A technique used to forge TCP/IP packet information or email header information. In network attacks, IP spoofing is used to gain access to systems by impersonating the IP address of a trusted host. In email spoofing, the sender address is forged to trick an email user into opening or responding to an email (which usually contains a virus or spam).

sprint: A short interval, usually two weeks, during which a development team develops features during a systems development project.

SQL injection: A type of attack wherein the attacker injects SQL commands into a computer input field in the hope that the SQL command will be passed to the database management system.

stand-alone power system (SPS): An off-the-grid electricity system for generation, storage, and regulation, which is used in facilities that are not equipped with an electricity distribution system.

standards: Specific, mandatory requirements that further define and support high-level policies.

star: A network topology in which all devices are directly connected to a central hub or concentrator.

star integrity property (*-integrity property): A state in which a subject can't write information to an object that has a higher integrity level than the subject (no write up, or NWU). *See also* Biba model.

star property (* property): A state in which a subject can't write information to an object that has a lower sensitivity label than the subject (no write down, or NWD). *See also* Bell-LaPadula model.

state attack: An attack in which the attacker attempts to steal other users' session identifiers to access a system by using the stolen session identifier.

state machine model: An abstract model in which a secure state is defined and maintained during transitions between secure states.

stateful inspection firewall: A type of firewall that captures and analyzes data packets at all levels of the Open Systems Interconnection model to determine the state and context of the data packet and whether it's to be permitted access to the network.

static application scanning tool (SAST): A tool used to identify vulnerabilities in a software application that works by examining the application's source code in search for exploitable vulnerabilities.

static password: A password that's the same for each login.

statistical attack: An attack on a cryptosystem through exploitation of a statistical weakness.

statistical sampling: A sampling technique in which individual items are chosen at random.

statutory damages: Mandatory damages determined by law and assessed for violating the law.

steganography: The art of hiding the very existence of a message, such as in a picture.

stream cipher: An encryption algorithm that operates on a continuous stream of data, typically bit by bit.

Stream Control Transmission Protocol (SCTP): A TCP/IP Layer 4 message-oriented protocol that provides for proper message sequencing and congestion control. *See also* Transmission Control Protocol (TCP), User Datagram Protocol (UDP).

strong authentication: *See* multifactor authentication.

Structured Query Language (SQL): A computer language used to manipulate data in a database management system.

subject: An active entity, such as a person or a process.

substitution cipher: Ciphers that replace bits, characters, or character blocks in plaintext with alternative bits, characters, or character blocks to produce ciphertext.

supervisor mode: A level of elevated privilege, usually intended for only system administration use. *See also* user mode.

Supervisory Control and Data Acquisition (SCADA): An industrial automation system that operates with coded signals over communication channels to provide remote control of equipment. *See also* industrial control system (ICS).

supply chain risk management (SCRM): Activities that identify and analyze risks associated with suppliers and other third parties. *See also* third-party risk management (TPRM).

surge: A prolonged rush of electric power.

surge protector: A device that protects electronic equipment from power surges and spikes.

surge suppressor: *See* surge protector.

switch: An intelligent hub that transmits data only to individual devices on a network, rather than all devices (in the way that hubs do). *See also* hub.

Switched Multimegabit Data Service (SMDS): A high-speed, packet-switched, connectionless-oriented, datagram-based technology available over public switched networks.

symmetric key system (or symmetric algorithm, secret key, single key, private key): A cryptographic system that uses a single key to both encrypt and decrypt information.

Synchronous Optical Networking (SONET): A telecommunications carrier-class protocol used to communicate digital information over optical fiber.

synthetic transaction: A mechanized transaction executed on a system or application to determine its ability to perform transactions properly.

system access control: A control that prevents a subject from accessing a system unless the subject can present valid credentials.

system hardening. *See* hardening.

system high mode: A state in which a system operates at the highest level of information classification.

system test (software development): A test of all the modules of an application or program. *See also* unit test.

systems development life cycle (SDLC): The business-level process used to develop and maintain information systems. *See also* software development life cycle (SDLC).

tabletop (review): A group review of a disaster recovery plan, business continuity plan, or incident response plan.

tactics, techniques, and procedures (TTPs): An approach to cyber threat intelligence that analyzes the patterns and methods of a threat actor or group of threat actors to develop more effective security responses.

tagging: Applying a machine-readable classification label on a document, device, or data storage object. *See also* marking.

Take-Grant model: A security model that specifies the rights that a subject can transfer to or from another subject or object.

tape library: A hardware system consisting of magnetic tape read and write equipment, as well as robotics to position individual tape volumes in read/write drives.

TCP/IP model: A four-layer networking model, originally developed by the U.S. Department of Defense.

technical (or logical) controls: Hardware and software technology used to implement a control.

technical debt: The extent of an organization's use of unsupported hardware and software and the effort required to perform upgrades to systems and applications.

telnet: A deprecated network protocol used to establish a command line interface on another system over a network. *See also* Secure Shell (SSH).

Terminal Access Controller Access Control System (TACACS): A User Datagram Protocol–based access control protocol that provides authentication, authorization, and accounting.

termination: *See* employment termination.

test coverage analysis: A measurement of the percentage of objects that have been included in a test.

texting: *See* Short Message Service (SMS).

Third Generation Partnership Project (3GPP): The consortium of standards organizations that develop mobile telecommunications protocols and standards.

third party: An organization to which some portion of business operations are outsourced. *See also* outsourcing, third-party risk management (TPRM).

third-party risk management (TPRM): Activities and analysis to identify third parties and the risks associated with their use. *See also* supply chain risk management (SCRM).

threat: Any natural or human-made circumstance or event that can have an adverse or undesirable impact, whether minor or major, on an organizational asset.

threat analysis: The study of an identified threat and its potential impact on an asset.

threat hunting: The proactive search for indicators of compromise in a network or system.

threat intelligence: Any human- or machine-readable information about known intrusion techniques.

threat modeling: A systematic process used to identify likely threats, vulnerabilities, and countermeasures for a specific application and its potential abuses during the design phase of the application (or software) development life cycle.

three-way handshake: The method used to establish and tear down network connections in the Transmission Control Protocol.

token: A hardware device used in two-factor authentication.

Token Ring: A star-topology network transport protocol.

trade secret: Proprietary or business-related information that a company or person uses and has exclusive rights to.

trademark: As defined by the U.S. Patent and Trademark Office, a trademark is "any word, name, symbol, or device, or any combination, used, or intended to be used, in commerce to identify and distinguish the goods of one manufacturer or seller from goods manufactured or sold by others."

transborder data flow: The transfer of electronic data across national borders.

transient: A momentary electrical line noise disturbance.

transitive trust: The phenomenon where a user inherits access privileges established in a domain environment.

Transmission Control Protocol (TCP): A connection-oriented network protocol that provides reliable delivery of packets over a network.

Transport Layer (OSI model): Layer 4 of the OSI model. *See also* Open Systems Interconnection (OSI) model.

Transport Layer (TCP/IP model): Layer 3 of the TCP/IP model. *See also* TCP/IP model.

Transport Layer Security (TLS): An OSI Layer 4 (Transport) protocol that provides session-based encryption and authentication for secure communication between clients and servers on the Internet. *See also* Open Systems Interconnection (OSI) model.

transposition cipher: Ciphers that rearrange bits, characters, or character blocks in plaintext to produce ciphertext.

trap door: A feature within a program that performs an undocumented function (usually a security bypass, such as an elevation of privilege).

triple DES (3DES): A variation of the Data Encryption Standard algorithm.

Trojan horse: A program that purports to perform a given function but actually performs some other (usually malicious) function. *See also* malware.

trust but verify: A concept in which a policy or control is examined for effectiveness.

trusted computer system: A system that employs all necessary hardware and software assurance measures and meets the specified requirements for reliability and security.

Trusted Computer System Evaluation Criteria (TCSEC): Commonly known as the *Orange Book*, a formal systems evaluation criteria developed for the U.S. Department of Defense by the National Computer Security Center as part of the Rainbow Series.

Trusted Computing Base (TCB): The total combination of protection mechanisms within a computer system — including hardware, firmware, and software — that are responsible for enforcing a security policy.

Trusted Network Interpretation (TNI): Commonly known as the *Red Book* (of the Rainbow Series), addresses confidentiality and integrity in trusted computer/communications network systems. *See also* Trusted Computer System Evaluation Criteria (TCSEC).

trusted path: A direct communications path between the user and the Trusted Computing Base that doesn't require interaction with untrusted applications or operating system layers.

Trusted Platform Module (TPM): A hardware module in a computer that performs cryptographic functions.

trusted recovery: Safeguards to prevent the disclosure of information during the recovery of a system after a failure.

twinaxial cable: A network medium consisting of two solid wire cores that are surrounded by an insulation layer and a metal foil wrap.

twisted-pair cable: A network medium consisting of four to eight twisted pairs of insulated conductors.

two-factor authentication: An authentication method that requires two ways of proving identity. *See also* multifactor authentication.

Ultra Reliable and Low Latency Communication (URLLC): A standard used in 5G mobile communications.

unauthenticated scan: A vulnerability scan that does not log in to a device, system, or application during its search for exploitable vulnerabilities.

unicast: A type of network protocol whereby packets are sent from a source to a single destination node.

Unified Communications as a Service (UCaaS): The use of cloud-based PBX and VoIP systems.

unified threat management (UTM): A security appliance that integrates various security features such as firewall, antimalware, and intrusion prevention capabilities into a single platform.

uninterruptible power supply (UPS): A device that provides continuous electrical power, usually by storing excess capacity in one or more batteries.

unit test: A test performed on an individual source code module.

USA PATRIOT Act (Uniting [and] Strengthening America [by] Providing Appropriate Tools Required [to] Intercept [and] Obstruct Terrorism Act of 2001): A U.S. law that expands the authority of law enforcement agencies for the purpose of combating terrorism.

user: A person who has access to information and/or information systems.

user acceptance testing (UAT): Testing of systems and applications by end users so that they can verify correct functionality; also, the environments in which such testing takes place.

user and entity behavior analytics (UEBA): A process used to detect malicious activity and potential breaches or intrusions by creating a baseline of normal user and entity activity and analyzing anomalies.

user behavior analytics (UBA): *See* user and entity behavior analytics (UEBA).

User Datagram Protocol (UDP): A network protocol that doesn't guarantee packet delivery or the order of packet delivery over a network.

user entitlement: The data access privileges that are granted to an individual user.

user mode: A level of privilege, usually intended for ordinary users. *See also* supervisor mode.

Vernam cipher: *See* one-time pad.

version control: The tracking of a system or data set for the purpose of recording changes.

virtual desktop infrastructure (VDI): A desktop operating system running within a virtual machine on a physical host server.

virtual extensible local area network (VxLAN): An extension of VLAN capabilities for use in large networks.

virtual local area network (VLAN): A logical network that resides within a physical network.

virtual machine: An instantiation of an operating system running within a hypervisor.

virtual private network (VPN): A private network used to communicate privately over public networks. VPNs typically use encryption and encapsulation to protect and simplify connectivity.

virtual tape library (VTL): A disk-based storage system that is used like a magnetic tape library system in backup and restore operations. *See also* tape library.

virtualization: The practice of running one or more separate, isolated operating system "guests" within a computer system.

virtualization (or VM) sprawl: The rapid creation of virtual machines without proper security and operations controls.

virus: A set of computer instructions whose purpose is to embed itself within another computer program to replicate itself. *See also* malware.

Voice over Internet Protocol (VoIP): Telephony protocols that are designed to transport voice communications over TCP/IP networks.

Voice over Long-Term Evolution (VoLTE): A protocol used for voice calls over Long-Term Evolution telecommunications networks using smartphones.

Voice over Misconfigured Internet Telephone (VOMIT): A tool used to intercept voice calls on VoIP networks.

Voice over Wi-Fi (VoWiFi): A protocol used for voice calls over Wi-Fi networks.

voltage drop: A decrease in electric voltage, typically from a public utility.

vulnerability: The absence or weakness of a safeguard in an asset, which makes a threat potentially more harmful or costly, more likely to occur, or likely to occur more frequently.

vulnerability assessment: The use of tools and techniques to identify vulnerabilities in an application, information system, facility, business process, or other object of study.

vulnerability management: The life cycle process used to identify and remediate vulnerabilities in information systems.

vulnerability scan: The use of an automated tool or technique to identify vulnerabilities in a target system or network.

vulnerability scanning tool: A software program designed to scan a device, system, or application to identify exploitable vulnerabilities.

walk-through: (1) A facilitated review of a process or procedure. (2) An audit activity. *See* inquiry.

wardialing: A brute-force attack that uses a program to automatically dial a large block of phone numbers (such as an area code), searching for vulnerable modems or fax machines.

wardriving: A brute-force attack that involves driving around looking for vulnerable wireless networks.

warm site: An alternative computer facility that's readily available and equipped with electrical power, heating, air conditioning, ventilation, and computers but not fully configured. *See also* cold site, hot site, reciprocal site.

waterfall: The software development process in which each phase is performed independently and in sequence.

watering-hole attack: An attack on browsers in which malware is installed on a web server and downloaded to users' browsers.

web application firewall (WAF): A device used to protect a web server from web application attacks such as script injection and buffer overflow.

web content filter: A system or application that permits and blocks Internet access to websites based on a defined policy.

wet pipe: A fire suppression system in which sprinkler pipes are always filled with water. *See also* dry pipe.

whaling: The practice of sending phishing messages to targeted executives in an organization.

white-box testing: A security test in which the tester has complete knowledge of the system being tested. *See also* static application security testing (SAST).

whitelisting: A mechanism that explicitly permits access based on the presence of an item in a list.

Wideband Code Division Multiple Access (W-CDMA): A mobile wireless communications standard that is a part of the 3G group of standards. *See also* 3G.

Wi-Fi: A technology used for wireless local area networking with devices based on the IEEE 802.11 standards. *See also* Institute of Electrical and Electronics Engineers (IEEE).

Wi-Fi Calling: A protocol used to transport smartphone telephone calls over Wi-Fi networks.

Wi-Fi Protected Access (WPA): A means of encrypting communications over 802.11 networks.

Wired Equivalent Privacy (WEP): A means of encrypting communications; specifically, 802.11/Wi-Fi networks. WEP is obsolete.

wireless intrusion detection system (WIDS): A network intrusion detection system that focuses on wireless networks.

Wireless LAN (WLAN): *See* Wi-Fi.

wiretap: Any technique to overhear or record a telephone conversation.

wiring closet: *See* distribution frame.

work factor: The difficulty (in terms of time, effort, and resources) of breaking a cryptosystem.

work from home (WFH): A work model in which an employee spends most work hours at their residence and performs their work duties remotely.

worker: An all-inclusive term that includes full and part-time employees, temporary employees, contractors, consultants, and others in an organization who have access to workplaces or information systems.

worm: Malware that usually has the capability to replicate itself from computer to computer without the need for human intervention. *See also* malware.

X.25: The first wide-area, packet-switching network.

XML (Extensible Markup Language): A human- and machine-readable markup language.

zero trust (ZT): A security and systems architecture approach in which systems, end-points, or people are considered to be untrusted or unverifiable.

Zigbee: A collection of high-level communication protocols for use in small, low-power personal area networks and smart home automation.

Index

A

abstraction, 209, 509

abuse/misuse case testing, 391

acceptability, 354

acceptable use policies (AUPs), 123

acceptance testing, 509

access control
  about, 177, 208, 264
  defined, 509
  models for, 196–199
  shared responsibility and, 195
  techniques for, 366–370

access control list (ACL), 369, 509

access management, 222, 223, 474, 509

Access Matrix model, 198, 509

Access Points (APs), 308–309

access provisioning lifecycle, 370–372

access rights/permissions, 369

accessibility, as a consideration for choosing locations, 260

accidents, in disaster recovery (DR) plan, 449

account management, 395

accountability, 202, 358–359, 509

accreditation, 205–208, 509

acquisitions, 146–148, 510

active assailant response, 266

active hubs, 315

active IDS, 325

active-active, 510

active-assailant, 510

active-passive, 510

ActiveX, 466

activities, logging and monitoring, 419–424

ad hoc, 310

Adaptive Chosen Ciphertext Attack (ACCA), 255

Adaptive Chosen Plaintext Attack (ACPA), 255

adaptive exam, 17

Address Resolution Protocol (ARP), 301, 510

address space, 510

Adleman, Len (Dr), 241

administrative controls, 136, 510

administrative laws, 66, 510

administrative management and control
  about, 120
  compliance, 125
  consultant controls, 124–125
  contractor controls, 124–125
  employment agreements and policies, 123
  employment candidate screening, 120–123
  employment termination processes, 123–124
  onboarding, 123–124
  privacy, 125
  transfers, 123–124
  vendor controls, 124–125

Advanced Encryption Standard (AES), 240, 510

Advanced Evolved High Speed Packet Access (HSPA+), 510

advisory policies, 95

after-action review (AAR), 510

age, as a criteria for commercial data classification, 157

agent of change, 32

aggregation, 187, 216, 428, 510

Agile, 464–466, 510

Agile Maturity Model (AMM), 473, 510

Agile Project Management For Dummies (Layton), 466

air gap, 511

Amazon Web Services (AWS), 194–195

American Accounting Association (AAA), 58

American Bar Association (ABA), 94

American Council on Education's College Credit Recommendation Service (ACE CREDIT), 15

American Institute of Certified Public Accountants (AICPA), 58

American National Standards Institute (ANSI), 10, 239

American Society for Industrial Security (ASIS), 25, 30, 35

American Standard Code for Information Interchange (ASCII), 282

analog signaling, 314

analysis
  of evidence, 414
  of test output, 400–404

analytic attack, 257, 511

Annualized Loss Expectancy (ALE), 130, 511

Annualized Rate of Occurrence (ARO), 130, 511

anomaly-based endpoint protection, 329

anomaly-based IDS, 420

anonymization, 192, 511

antimalware, 442

antivirus software, 511

Anton Piller order, 413

anycast, 301

applet, 511

application firewall, 511

Application Layer (Layer 7) (OSI Reference Model), 279–281, 511

Application Layer (TCP/IP Model), 315, 511

application penetration test, 386, 511

application programming interfaces (APIs), 341, 492–493, 511

application scan, 511

application security testing (AST), 484–486

application software, 222, 225, 511

application virtualization, 337

application whitelisting, 329, 511

application-level gateway, 320–321

applications, 342–343, 426

apprenticeship program, 150

archive, 248, 512

area identifiers, 289

artifacts, 419

artificial intelligence (AI), 442, 512

aspirating devices, 270

assessment, in disaster recovery (DR) plan, 455

asset check-in/check-out log, 265

asset classification, 512

asset inventory, 165–166, 425, 512

Asset Security domain

  about, 153

  data protection, 176–177

  determining data security controls and compliance requirements, 172–177

  determining ownership, 164–165

  establishing handling requirements, 162–163

  identifying and classifying assets, 153–161

  identifying and classifying information, 153–161

  maintaining ownership, 164–165

  managing data life cycle, 167–171

  provisioning resources, 164–167

  retention, 171–172

assets

  about, 126

  controlling physical and logical access to, 340–343

  defined, 512

  identifying and classifying, 153–161

  valuation of, 127–128, 512

Associate of (ISC)² certification, 33

assurance, 201–202

asymmetric algorithm, 512

asymmetric algorithm cryptography. *See* asymmetric key cryptography

asymmetric key cryptography

  about, 242–245

  Diffie-Hellman key exchange, 246–247

  El Gamal, 247

  Elliptic Curve (EC), 247

  Merkle-Hellman (Trapdoor) Knapsack, 247

  quantum computing, 247

  RSA algorithm, 245

asymmetric key system, 512

asynchronous communication, 308

Asynchronous Transfer Mode (ATM), 305, 512

Attached Resource Computer NETwork (ARCNET), 300, 512

attack surface reduction, 512

attack tree, 144, 512

attacks. *See specific types*

attenuation, 312

attestation, as a function of TPM, 209

attribute-based access control (ABAC), 369–370, 512

audit trail, 512

audits/auditing, 202, 405, 474, 486–487, 512

authenticated scans, 382, 512

authentication

  cryptography and, 229

  defined, 512

  improper, 491

  single factor, 345–358

authentication, authorization, and accountability (AAA) controls, 509

authentication bypass, 512

Authentication Header (AH), 296, 513

authenticity, 52

authority to operate (ATO), 513

authorization, 365–370, 513

automated external defibrillators, 266

automatic controls, 136, 513

automatic external defibrillator (AED), 513

automation, 426

autonomous system (AS), 290

autonomous system number (ASN), 513

availability, 51–52, 446, 447, 513

awareness, 148–150, 399, 456

AWS Certified Security - Specialty, 37

B

backdoor, 513

background check, 513

backup media encryption, 445

backup verification data, 397–399, 437

backups, 109–111, 161, 444–445, 513

bare metal hypervisors, 336

baseband signaling, 311

baselines, 95–96, 425, 513

Basic Input-Output System (BIOS), 283

bastion host, 323

beam devices, 270

Beaver, Kevin (author)

  Hacking For Dummies, 386

behavior-based endpoint protection, 329

behavior-based IDS, 326–327

Bell-LaPadula model, 197, 513

best evidence, 409, 513

drag and drop questions, 18

DREAD technique, 183

drift, 524

drug screen, 524

dry-pipe system, 270, 524

dual-homed gateways, 323

due care, 60–61, 63–64, 524

due diligence, 60–61, 524

Dummies (website), 4

dumpster diving, 524

duress alarm, 524

dwell time, 524

dynamic application scanning tool (DAST), 525

dynamic application security testing (DAST), 485

dynamic DLP, 176

dynamic password, 351, 525

dynamic routing protocol, 288

E

earning certifications, 33–40

east-west traffic, 525

eavesdropping, 332, 385, 525

Economic Espionage Act (1996), 73

edge computing systems, 225–226, 525

education, 148, 150–151

efficiency tolerance, 364

egress monitoring, 422, 525

EIA/TIA-232-F standard, 314

El Gamal, 247

Elad, Joel (author)

LinkedIn For Dummies, 4th Edition, 39

electric generator, 525

electrical anomalies, 272–273

electrical hazards, 272–274

electrical noise, 272–273

electricity, 263, 272–274

Electromagnetic Interference (EMI), 273, 525

Electronic Code Book (ECB), 240

electronic health records (EHRs), 90

electronic protected health information (ePHI), 86, 525

electronic signatures, 358

electronic vaulting, 399

electrostatic discharge (ESD), 272–273, 525

Elliptic Curve (EC), 247

emanations, 214, 525

embedded devices, assessing and mitigating vulnerabilities in, 224–225

emergency food/water, 266

Emergency Power Off (EPO) switch, 267, 272, 525

emergency response, 108

emergency supplies, 454

employment agreements and policies, 123, 525

employment candidate screening, 120–123

employment termination processes, 123–124, 525

Encapsulating Security Payload (ESP), 296, 525

encapsulation, 525

enciphering. *See* encryption

encryption, 210–211, 230–232, 485, 491, 525

end of support (EOS), 172, 526

end-of-life (EOL), 171–172, 526

endpoint security, 328–330, 526

end-to-end encryption, 230, 526

end-user, 56–57

Enhanced Mobile Broadband (eMBB), 526

Enterprise Risk Management - Integrated Framework, 142

enticement, 411, 526

entitlement, 186–187, 428, 526

entrapment, 411, 526

environmental considerations, 267–268

equipment

in disaster recovery (DR) plan, 40

failure of, 263

error checking/recovery, 284

escalation of privilege, 526

escorts, 342

Escrowed Encryption Standard (EES), 249, 526

espionage, 526

essential practices, 120

Ethernet, 300, 313, 526

ethical disclosure, 403–404

ethics. *See* professional ethics

European Information Technology Security Evaluation Criteria (ITSEC), 204–205

European Union (EU), 83

European Union General Data Protection Regulation (GDPR), 66–67, 92–93

evacuation plans, 448, 454

evacuation signage/drills, 266

evaluation assurance levels (EALs), 205

evaluation criteria, for selecting controls, 200–205

e-vaulting, 111, 526

even-parity bit, 239

event management, 474, 526

events, speaking at, 27

evidence, 526

evidence collection/handling

about, 408

admissibility of evidence, 411–412

rules of evidence, 409–411

types of evidence, 408–409

evidence lifecycle, 526

evidence storage, 264–265

exams

after the, 20–21

fee for, 17

overview of, 17–20

planning tips for, 499–503

practice, 15

preparing for, 12–16

question types, 18–19

registering for, 16–17

Internet Architecture Board (IAB), 48

Internet Assigned Numbers Authority (IANA), 290, 533

Internet Control Message Protocol (ICMP), 297, 533

Internet Engineering Task Force (IETF), 533

Internet Key Exchange (IKE), 297, 533

Internet Layer (TCP/IP Model), 315, 533

Internet Message Access Protocol (IMAP), 280

Internet of Things (IoT), 52, 154–155, 221, 533

Internet Protocol (IP), 290, 533

Internet Protocol Security (IPsec), 296–297, 534

Internet Protocol version 4 (IPv4), 534

Internet Protocol version 6 (IPv6), 534

Internet Relay Chat (IRC), 534

Internet Security Association and Key Management Protocol (ISAKMP), 297, 534

Internet Small Computer Systems Interface (iSCSI), 534

Internetwork Packet Exchange (IPX), 290, 534

internetworks, 287

interprocess communication (IPC), 534

intrusion detection and prevention systems (IDPSs), 325–327, 419–420, 441

intrusion detection systems (IDS), 325–327, 534

intrusion prevention systems (IPSs), 325–327, 534

investigations, 408–419, 534

ionization devices, 270

IP reputation services, 441

iris pattern, 356

ISACA, 30

ISC²

attending events, 26

being an active member of, 25–26

focus groups, 28

helping at conferences, 27

joining a chapter, 26

publications of, 27

volunteer opportunities, 26–30

voting in elections, 25

website, 501

ISC² Blog, 27

ISC² Code of Professional Ethics, 46–47

ISC² community, 28

ISC² Congress, 25, 27

ISO/IEC 17024 standard, 10

ISO/IEC 27002, *Information Technology - Security Techniques - Code of Practice for Information Security Management*, 95

ISO/IEC 27005, 142

IT Disaster Recovery Planning For Dummies (Gregory), 40, 120

IT Governance Institute (ITGI), 57

IT Infrastructure Library (ITIL), 534

J

JavaScript, 534

jitter, 286

job description, 534

job rotation, 431–433, 534

joining local security chapters, 30–31

Joint Photographic Experts Group (JPEG), 282

joint tenants, as a consideration for choosing locations, 261

judgmental sampling, 139, 534

just-in-time (JIT) access, 363, 535

juvenile laws, 69

K

Keep It Simple, Stupid (KISS) principle, 189

Kerberos, 373–376, 535

Kerberos exploitation, 257

key card access systems, 342, 535

key change, 249, 535

key clustering, 233, 516, 535

key control, 249, 535

key disposal, 249, 535

key distribution, 248, 535

key encryption key (KEK), 535

key escrow, 249, 535

key generation, 248, 535

key installation, 248, 535

key logging, 535

key management, 248–249, 535

key perfomance indicators (KPIs), 396–397, 535

key recovery, 249

key risk indicators (KRIs), 396–397, 535

key storage, 249

key vault, 535

keyed invoices, 105

keyspace, 232, 535

keystroke dynamics, 357

knowledge-based IDS, 326–327

Known Plaintext Attack (KPA), 255, 535

L

labels, 201

latency, 232, 535

lattice-based access controls, 368, 536

The Law-Abiding Citizen Fallacy, 45–46

Layer 2 Forwarding Protocol (L2F), 303, 335, 536

Layer 2 Tunneling Protocol (L2TP), 303–304, 335, 536

Layton, Mark (author)

Agile Project Management For Dummies, 466

Scrum For Dummies, 466

leading by example, 56

learning style, knowing your, 499

S

trusted facility management, 202

Trusted Facility Manual (TFM), 202

Trusted Network Interpretation (TNI), 204, 559

trusted path, 202, 560

trusted people, 69

Trusted Platform Module (TPM), 209, 560

trusted recovery, 202, 560

trusted subject, 197

Tunnel mode (IPsec), 296

twinaxial cable, 311–312, 560

twisted-pair cable, 312, 560

two-factor authentication, 560

Twofish Algorithm, 241

Type 1 error, 353

type accreditation, 207

Type II error, 353

typing dynamics, 357

U

U.K. Data Protection Act (DPA), 91–92

Ultra Reliable and Low Latency Communication (URLLC), 560

unauthenticated scans, 382, 560

Unclassified government data classification, 159

unconstrained data item (UDI), 198

unicast, 301, 560

Unified Communications as a Service (UCaaS), 560

unified threat management devices (UTMs), 312–322, 560

Uninterruptible Power Supply (UPS), 112, 272, 560

unit test, 560

United Nations Commission on International Trade Law (UNCITRAL), 83

unsecured protected health information (PHI), 90

unshielded twisted-pair cable (UTP), 312, 313

untrusted endpoints, 481

U.S. CAN-SPAM Act (2003), 78

U.S. Child Pornography Prevention Act (CPPA) (1996), 76

U.S. Communications Assistance fo Law Enforcement Act (1994), 75

U.S. Computer Fraud and Abuse Act (1986), 64, 72–74

U.S. Computer Security Act (1987), 75

U.S. Defense Information Security Agency, 193

U.S. Department of Defense (DoD), 158, 197, 200

U.S. Economic Espionage Act (EEA) (1996), 75

U.S. Electronic Communications Privacy Act (ECPA) (1986), 74

U.S. Federal Emergency Management Agency (FEMA), 98

U.S. Federal Information Systems Modernization Act (FISMA, 2014), 78

U.S. Federal Privacy Act (1974), 88

U.S. Federal Sentencing Guidelines (1991), 75

U.S. Gramm-Leach-Bliley Financial Services Modernization Act (GLBA) (1999), 89–90

U.S. Health Information Technology for Economic and Clinical Health Act (HITECH) (2009), 90

U.S. Health Insurance Portability and Accountability Act (HIPAA) (1996), 50, 88–89

U.S. Homeland Security Act (2002), 78

U.S. Identity Theft and Assumption Deterrence Act (2003), 79

U.S. Intelligence Reform and Terrorism Prevention Act (2004), 79

U.S. National Institute of Standards and Technologies (NIST), 175

U.S. Patent and Trademark Office (PTO), 83

U.S. Sarbanes-Oxley Act (SOX) (2002), 78

USA PATRIOT Act (2001), 73–74, 76–77, 560

use case testing, 391

Use stage, 173

useful life, as a criteria for commercial data classification, 157

user acceptance testing (UAT), 560

user accounts, asset management and, 156

user and entity behavior analysis (UEBA), 424, 561

user behavior analytics (UBA), 561

User Datagram Protocol (UDP), 286, 561

user endpoints, asset management and, 154

user entitlement, 561

user identity, 360

user mode, 561

user stories, 465

users, 56–578, 168, 560

utilities

about, 112

as a consideration for choosing locations, 261

considerations for, 266–267

in disaster recovery (DR) plan, 40, 448

loss of, 264

V

v. (versus), 64

V.24 ITU-T standard, 314

V.35 ITU-T standard, 314

About the Authors

Lawrence C. Miller, CISSP, has worked in information security and technology management for more than 20 years. He received his MBA from Indiana University and has earned numerous technical certifications throughout his career. He has previously worked in Vice President and Director level positions at several small to mid-sized companies in various industries. He served as a chief petty officer in the U.S. Navy in various roles and is a veteran of Operations Desert Shield/Storm. He is the author of more than 130 other *For Dummies* Custom Edition books.

Peter H. Gregory, CISSP, CISM, CISA, CRISC, CIPM, CDPSE, CCSK, DRCE, is the author of more than 40 books on security and technology, including *Solaris Security* (Prentice Hall), *Getting An Information Security Job For Dummies* (John Wiley & Sons, Inc.), *IT Disaster Recovery Planning For Dummies* (John Wiley & Sons, Inc.), and *CISA Certified Information Systems Auditor All-In-One Study Guide* (McGraw-Hill).

Peter is a career technologist and a security executive at a regional telecommunications provider. Prior to this, he held strategic security positions at Optiv Security (www.optiv.com) and Concur Technologies (www.concur.com). Peter is an advisory board member for the University of Washington and the University of South Florida for continuing education programs in cybersecurity. He is a graduate of the FBI Citizens' Academy.

Peter resides in Central Washington State and can be found at www.peterhgregory.com.

Dedication

From Lawrence C. Miller: To my wife, Michelle, and my sons, Eric and Ken.

From Peter H. Gregory: To Clément Dupuis.

Authors' Acknowledgments

Lawrence C. Miller would like to thank Amy Fandrei, Elizabeth Kuball, Peter Gregory, Peter T. Davis, and all the great people at Wiley.

Peter H. Gregory would like to thank Lindsay Lefevere, Executive Editor, Michelle Hacker, Managing Editor, and Tim Gallan, Development Editor, for their help throughout this project. There are many more people at Wiley and other organizations without whom this book could not reach readers. I don't know who you are, but I know you are out there, and I am grateful for your dedication and hard work. The expertise of subject matter experts Peter T. Davis, Thomas Egan, and Fabio Cerullo helped make the book much better. Thank you, Larry; it's great as always to work with you on security books. We've been working together on CISSP For Dummies since the first edition in 2002 — that's 20 years! Thank you, (ISC)², for your seal of approval — again, for this seventh edition.

I would like to offer special gratitude to Clément Dupuis, the founder and long-time owner of CCCure, who has helped CISSP candidates for decades and recently developed a valuable gap analysis for the most recent update of the CISSP CBK. Clement passed away in 2021 as we wrote this edition. Rest in peace, my friend.

My contribution to this book would not have been possible without support from my wife, business manager, and best friend, Rebekah Gregory. Thanks also to Carole Jelen, my literary agent, for guidance on this and other projects over the past 16 years.

Publisher's Acknowledgments

Executive Editor: Lindsay Lefevere

Development Editor: Tim Gallan

Technical Editor: Thomas Egan

Copy Editor: Keir Simpson

Production Editor: Tamilmani Varadharaj

Cover Image: © MF3d/Getty Images